Methodological Advancements in Social Impacts of Tourism Research

This book offers a fresh perspective of on some of the cutting-edge methodological approaches being used among scholars conducting work on social impacts of tourism. These works are international in focus, spanning across Europe (e.g., Austria, Croatia, Portugal, Serbia, Slovenia, Spain, and Turkey) and Asia (e.g., Hong Kong, Iran, Japan, Malaysia, Singapore, Thailand, and Turkey). The authors employ qualitative, quantitative, and mixed methods designs and some of the notable focus areas include comparative studies of residents' perspectives of tourism (i.e., involving impacts of various forms of tourism and community children's perceptions of tourism); statistical techniques such as multigroup invariance analysis and fuzzy set qualitative comparative analysis; and the employment of novel measurement tools such as the implicit association test involving residents' implicit stereotypes of visitors from multiple countries and the utilization of the human-scale development to identify residents' needs and satisfiers. This edited volume will most assuredly advance the methodological focus of research on social impacts of tourism moving forward.

This book will be of great interest to all upper-level students and researchers in tourism, planning and related fields. The chapters in this volume were originally published as a special issue of *Journal of Sustainable Tourism*.

Manuel Alector Ribeiro is Associate Professor in Tourism Management at the University of Surrey, UK; Senior Research Fellow in the School of Tourism and Hospitality Management at the University of Johannesburg, South Africa; and Research Member in the Research Centre for Tourism, Sustainability and Well-being (Cinturs) at the University of Algarve, Portugal.

Kyle Maurice Woosnam is Professor of Parks, Recreation and Tourism Management at the Warnell School of Forestry & Natural Resources, University of Georgia, USA and Senior Research Fellow in the School of Tourism and Hospitality Management at the University of Johannesburg, South Africa.

Methodological Advancements in Social Impacts of Tourism Research

Edited by
Manuel Alector Ribeiro and Kyle Maurice Woosnam

Routledge
Taylor & Francis Group
LONDON AND NEW YORK

First published 2024
by Routledge
4 Park Square, Milton Park, Abingdon, Oxon OX14 4RN

and by Routledge
605 Third Avenue, New York, NY 10158

Routledge is an imprint of the Taylor & Francis Group, an informa business

British Library Cataloguing in Publication Data
A catalogue record for this book is available from the British Library

ISBN13: 978-1-032-53690-3 (hbk)
ISBN13: 978-1-032-53692-7 (pbk)
ISBN13: 978-1-003-41316-5 (ebk)

DOI: 10.4324/9781003413165

Typeset in Myriad Pro
by Newgen Publishing UK

Publisher's Note
The publisher accepts responsibility for any inconsistencies that may have arisen during the conversion of this book from journal articles to book chapters, namely the inclusion of journal terminology.

Contents

Citation Information

The following chapters were originally published in the *Journal of Sustainable Tourism*, volume 31, issue 2 (2023). When citing this material, please use the original page numbering for each article, as follows:

Chapter 1

Do residents differentiate between the impacts of tourism, cruise tourism, and Airbnb tourism?
Evan J. Jordan, José Cabral Vieira, Carlos M. Santos and Ting-Yen (Tim) Huang
Journal of Sustainable Tourism, volume 31, issue 2 (2023), pp. 265–283

Chapter 2

Towards advancing theory and methods on tourism development from residents' perspectives: Developing a framework on the pathway to impact
Hossein G. T. Olya
Journal of Sustainable Tourism, volume 31, issue 2 (2023), pp. 329–349

Chapter 3

The effect of minority residents' attitudes and emotional solidarity on ethnic neighborhood tourism: a multigroup invariance analysis
Naho Ueda Maruyama, Manuel Alector Ribeiro and Kyle Maurice Woosnam
Journal of Sustainable Tourism, volume 31, issue 2 (2023), pp. 383–405

Chapter 4

Assessing explicit and implicit stereotypes in tourism: self-reports and implicit association test
Wai Tsz Serene Tse and Vincent Wing Sun Tung
Journal of Sustainable Tourism, volume 31, issue 2 (2023), pp. 460–482

Chapter 5

Interferences generated on the well-being of local communities by the activity of online platforms for tourist accommodation
José Maria Martín-Martín, Juan F. Prados-Castillo, Juan de Dios Jiménez Aguilera and Eva Porras González
Journal of Sustainable Tourism, volume 31, issue 2 (2023), pp. 483–503

Chapter 6

Towards sustainable tourism development in a mature destination: measuring multi-group invariance between residents and visitors' attitudes with high use of accommodation-sharing platforms
Andrés Chamarro, José Cobo-Benita and María Dolores Herrero Amo
Journal of Sustainable Tourism, volume 31, issue 2 (2023), pp. 520–537

Chapter 7

Social impacts of mega-events: a systematic narrative review and research agenda
Judith Mair, P. Monica Chien, Sarah Jane Kelly and Stephanie Derrington
Journal of Sustainable Tourism, volume 31, issue 2 (2023), pp. 538–560

Chapter 8

Exploring the neglected voices of children in sustainable tourism development: A comparative study in six European tourist destinations
Marko Koščak, Mladen Knežević, Daniel Binder, Antonio Pelaez-Verdet, Cem Işik, Vladimir Mićić, Katarina Borisavljević and Tina Šegota
Journal of Sustainable Tourism, volume 31, issue 2 (2023), pp. 561–580

Notes on Contributors

Daniel Binder, Institute of Health and Tourism Management, FH JOANNEUM University of Applied Sciences, Austria.

Katarina Borisavljević, Faculty of Economics University of Kragujevac, Serbia.

Andrés Chamarro, Department of Entrepreneurship, ESCP Business School, France.

P. Monica Chien, University of Queensland Business School, University of Queensland, Australia.

José Cobo-Benita, Department of Operations Management, ESCP Business School, France.

Stephanie Derrington, University of Queensland, Australia.

Eva Porras González, Rey Juan Carlos University, Spain.

María Dolores Herrero Amo, Department of Operations Management, ESCP Business School, France.

Ting-Yen (Tim) Huang, Department of Health and Wellness Design at the Indiana University School of Public Health – Bloomington, USA.

Cem Işik, Faculty of Tourism, Anadolu University, Turkey.

Juan de Dios Jiménez Aguilera, University of Granada, Spain.

Evan J. Jordan, Department of Health and Wellness Design, Indiana University School of Public Health – Bloomington, USA.

Sarah Jane Kelly, University of Queensland Business School, University of Queensland, Australia.

Mladen Knežević, University of Zagreb; University J. J. Strossmayer, Osijek, Croatia.

Marko Koščak, Faculty of Tourism, University of Maribor, Slovenia.

Judith Mair, Tourism Discipline Group, UQ Business School, University of Queensland, Australia.

José Maria Martín-Martín, Department of Business, International University of La Rioja; Department of Spanish and International Economics, University of Granada, Spain.

Naho Ueda Maruyama, Takasaki City University of Economics, Japan.

Vladimir Mićić, Faculty of Economics, University of Kragujevac, Serbia.

Hossein G. T. Olya, Sheffield University Management School, University of Sheffield, UK.

Antonio Pelaez-Verdet, Faculty of Tourism, University of Malaga, Spain.

Juan F. Prados-Castillo, International University of Valencia; University of Granada, Spain.

Manuel Alector Ribeiro, School of Hospitality and Tourism Management, University of Surrey, UK; School of Tourism and Hospitality Management, University of Johannesburg, South Africa; and Research Centre for Tourism, Sustainability and Well-being (Cinturs), University of Algarve, Portugal.

Carlos M. Santos, School of Business and Economics at the University of the Azores, Portugal.

Tina Šegota, Faculty of Business, University of Greenwich, UK.

Wai Tsz Serene Tse, School of Hotel and Tourism Management, The Hong Kong Polytechnic University, China.

Vincent Wing Sun Tung, School of Hotel and Tourism Management, The Hong Kong Polytechnic University, China.

José Cabral Vieira, School of Business and Economics, University of the Azores, Portugal.

Kyle Maurice Woosnam, Warnell School of Forestry & Natural Resources, University of Georgia, USA and School of Tourism and Hospitality Management, University of Johannesburg, South Africa.

Introduction: Methodological advancements in social impacts of tourism research

Manuel Alector Ribeiro and Kyle Maurice Woosnam

Introduction

As the field of social impacts of tourism research evolves, there is a growing need for methodological advancements to better comprehend and address the complex and diverse issues related to tourism development. It is crucial to highlight the innovative approaches and techniques employed to analyze and understand these intricate phenomena when examining methodological advancements in this realm.

The evolution of methods in social impacts of tourism research has been characterized by a shift toward more comprehensive and nuanced approaches. Early studies primarily utilized quantitative techniques, such as surveys and questionnaires, to gather data on residents' perceptions and attitudes toward tourism development (Andereck et al., 2005). However, as the field matured, researchers acknowledged the necessity for more in-depth analyses of the social impacts of tourism, leading to the adoption of qualitative methods, including interviews, ethnographic observations, and case studies (Deery, Jago, & Fredline, 2012; Hadinejad et al., 2019), longitudinal studies (Johnson, Snepenger, & Akis, 1994; Liang, Luo, & Bao, 2021), or recommendations to measure the actual behaviors of residents associated with tourism development initiatives, using an experimental research design (Olya, 2023).

Given these needs, there has been a growing demand to expand the scope of research settings and implement more advanced methodologies and techniques to examine intricate, multidimensional issues in the context of tourism in general, as well as within social impact of tourism research. A recent analysis of methodological trends in tourism journals, particularly in the four highly ranked journals—*Journal of Sustainable Tourism*, *Annals of Tourism Research*, *Tourism Management*, and *Journal of Travel Research*—over a 50-year period (1973–2023) demonstrates a significant increase in the complexity of methods employed in social impact research within tourism (Hadinejad et al., 2019; Nunkoo, Smith, & Ramkissoon, 2013; Woosnam & Ribeiro, 2023). Currently, numerous tourism journals with high Web-of-Science impact scores adhere to stringent methodological criteria, imposing expectations on authors like those found in leading journals in other management and business fields, such as marketing, organizational studies, and management (e.g., Aguinis, Bergh, & Molina-Azorin, 2022; Hahn & Ang, 2017; Meyer, van Witteloostuijn, & Beugelsdijk, 2017).

The hallmark of most developing fields branching out from their foundational disciplines lies in the continuous progression of methodological approaches and theoretical applications. Fundamentally, these advancements function as the driving force that propels ongoing efforts to contribute the most substantial research possible. As we enter the fifth decade of investigating the social impacts of tourism, we find ourselves at a crucial juncture. The key question to consider is not whether we need to propel our field forward, but rather, how we can achieve this. Reflecting on our past and current state will guide us in mapping out a future direction.

Tourism impacts research has evolved over time to encompass a range of social, cultural, environmental, political, and economic factors, but this was not always the case. Initially, researchers focused almost exclusively on economic impacts, while social, cultural, and environmental considerations were given less attention (Deery, Jago, & Fredline, 2012). However, methodological advancements have emerged in response to various factors, including the recognition that financial gains from tourism are not the sole objective for destinations and their residents (Nunkoo & Gursoy, 2019). This has led to a broader understanding of the complex interconnections between different types of impacts (Mason, 2016). Moreover, fewer studies now focus on destinations where tourism is a recent phenomenon, shifting the emphasis away from economic gains toward understanding the sociocultural, environmental, and political impacts of tourism. This is particularly relevant in the context of the COVID-19 pandemic and global issues of justice, war, and climate change. Furthermore, the growing appreciation for sustainability (encompassing the triple-bottom-line) in tourism research, industry, and society at large (Ruhanen, Moyle, & Moyle, 2019) has led to methodological advancements in studying tourism impacts. This is evidenced by the increasing calls in the tourism literature for the use of more robust and sophisticated methodological and statistical techniques for both data collection and data analysis within the field (Olya, 2023). As we navigate through the recovery from the COVID-19 pandemic and face unprecedented global challenges related to climate change, natural disasters, war/armed conflicts, terrorism, and political and racial divides, there is an urgent need to advance our methodological approaches in studying the social impacts of tourism. This will enable us to better understand how tourism is affecting the quality of life and well-being of local communities. Developing innovative research methodologies that can capture the complexities and dynamics of the interrelated impacts of tourism will enable a more comprehensive understanding and contribute to more sustainable tourism management and practices.

The development of social impacts of tourism research, as Woosnam and Ribeiro (2023) indicates, can be divided into five distinct stages, as shown in Figure I.1. While Deery, Jago, and Fredline (2012) identify some of these stages, we have modified and expanded their approach by reorganizing and reclassifying the stages and including a newly emerging stage. The foundation for social impacts research was established by defining crucial concepts, developing typologies and conceptual models, and giving significant attention to the attitudes of destination residents. During the first stage of social impacts of tourism research, scholars such as Doxey (1975), Rothman (1978), Butler (1980), and Mathieson and Wall (1982) played an instrumental role in laying the groundwork for understanding the heterogeneous perspectives of destination residents toward tourism. These works are considered seminal in the field, as they focused on identifying and classifying the various social impacts of tourism. In this stage, the most used method was the descriptive statistical analysis of surveys, which provided a quantitative understanding of the impacts of tourism on local communities (Milman & Pizam, 1988; Liu, Sheldon, & Var, 1987). Descriptive statistics allowed researchers to analyze and summarize the data collected through surveys and to identify key themes and concepts that informed subsequent stages of social impacts of tourism research. The second stage marked the beginning of more case

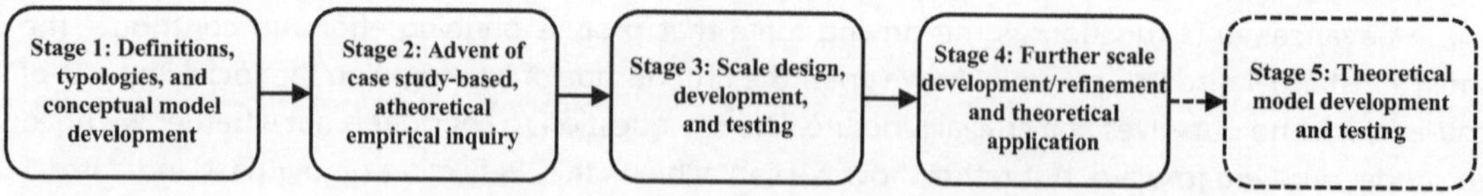

Figure 1.1. Social impacts of tourism research development stages (as adapted from Deery, Jago, & Fredline, 2012)

study-based empirical work that delineated divergent resident perspectives of tourism development. During this time, many studies were atheoretical in nature and employed methods such as factor-cluster analysis or considered a range of predictors of residents' attitudes. Notable examples of these studies include the works of Belisle and Hoy (1980), Evans-Pritchard (1989), Liu and Var (1986), and Perdue, Long, and Allen (1990).

The third stage of research on the social impacts of tourism was characterized by the development, design, and testing of measures to assess residents' attitudes toward tourism. During this stage, there was a more concerted effort in theory development related to residents' attitudes research, and the methods used played a crucial role. Notably, during this stage, early scales such as the Tourism Impact Attitude Scale (TIAS) by Lankford and Howard (1994) and the general scale by Ap and Crompton (1998) were developed using univariate and bivariate statistical methods, including factor analysis and ANOVA. However, when the goal is to conduct a more intricate analysis that delves into the connections between and among numerous dependent and independent factors, univariate and bivariate statistical approaches fall short in examining relationships among various constructs simultaneously, leading to some unexplained interactions (Crowley & Fan, 1997).

The fourth stage of social impacts of tourism research involved the continued development and refinement of scales, as well as greater theoretical applications. However, to address the shortcomings of univariate and bivariate statistical approaches and to address complex research problems, tourism researchers and scholars must employ multivariate statistical analysis. Loehlin and Beaujean (2017) observe that:

> "Scientists ... rarely have the luxury of the simple bivariate experiment, in which a single independent variable is manipulated, and the consequences are observed for a single dependent variable. Even those scientists who think they do are often mistaken ... A full experimental analysis would again turn out to be multivariate, with several alternative experimental manipulations on the one side, and numerous alternative response measures on the other" (p. 1).

This period was marked by a shift toward considering the social impacts of tourism within the context of sustainability. With the proliferation of computers, more sophisticated statistical software became available, facilitating the implementation of robust statistical techniques such as structural equation modeling (SEM) (Nunkoo & Ramkissoon, 2012). As Shah and Goldstein (2006, p. 149) suggest, SEM is a statistical approach that "specifies, estimates, and evaluates models of linear relationships among a set of observed variables in terms of a generally smaller number of unobserved variables." However, according to Loehlin and Beaujean (2017, p. 1), "until the advent of powerful digital computers and associated software, the use of these methods tended to be restricted to a dedicated few." Nevertheless, in the last few decades, the proliferation of computers and the explosive growth in the use of statistical software packages such as LISREL, SPSS, AMOS, Mplus, Lavaan, SmartPLS, and R have contributed to the increasing use of SEM in social science, particularly in tourism. These techniques were crucial for scale development, validation, theory testing, and theory building. For example, Choi and Sirakaya (2005) created the Sustainable Tourism Attitude Scale (SUS-TAS) and Woosnam and Norman (2010) developed the Emotional Solidarity Scale (ESS), both of which were developed using SEM techniques. These scales reflect the growing importance of considering how residents' interactions and relationships with tourists factor into their perspectives on tourism development. Furthermore, the proliferation of the use of SEM at this stage was also marked by a greater application of theoretical frameworks, such as social exchange theory, as seen in the works of Andereck et al. (2005), Choi and Murray (2010), Jurowski and Gursoy (2004), and McGehee and Andereck (2004). Nevertheless, SEM is based on the fundamental assumption that theory should guide every aspect of the modeling process. With this assumption in mind, Duncan (1975) provided a useful reflection on the use of structural models: "Do not undertake the study of structural equation models (or, for that matter, any other topic in sociological methods) in the hope of acquiring a technique that can be applied mechanically to a set of numerical data with the expectation that the result will automatically be 'research'" (p. 150). Additionally, Duncan remarked that if research using structural models "are contributions to science (and not merely exercises in quantitative technique), it is because the models rest on creative, substantial, and sound (sociological) theory" (p. 151).

In the past decade, we have entered a new stage marked by even greater development and testing of theoretical models in tourism research, with the top four tourism journals either explicitly or implicitly requiring theoretical application as a basis for publication. Much of the current work in this field involves the use of advanced statistical analysis, such as SEM, as seen in the works of Boley et al. (2014), Látková and Vogt (2012), and Nunkoo and Ramkissoon (2012). To effectively develop and modify a model, it is crucial for theory to guide every aspect of SEM modeling. However, some authors (Hair et al. 2019; Hoyle, 2023) have noted that SEM can be misused when data are solely fitted to a suitable model without a theoretical basis, which can lead to misleading results and potentially harm the validity of scientific research. Therefore, it is essential to ensure that a SEM model is based on sound theoretical foundations and not solely reliant on fitting the data. This idea was also emphasized by Nunkoo and Ramkissoon (2012) in their paper on SEM and regression analysis in tourism research:

> Researchers should be clear about the nature of 'theory' and its meaning in SEM analysis. The word "theory" means different things to different people, and it has been applied in very broad contexts. Researchers using SEM should distinguish between the common notion of "theory" and the formal definition of a "theory" and we argue that SEM analysis should be based on the formal definition of a theory (p. 782).

In fact, it is important for researchers to understand that even though SEM modeling is a complex and advanced statistical tool, it cannot be used in isolation without any theoretical basis (Hoyle, 2023). In other words, SEM modeling should be integrated with a solid theoretical framework to make meaningful contributions to the advancement of social impacts of tourism research. Within this stage, we also see more research using SEM and sound theoretical frameworks to treat social impacts of tourism as an antecedent of outcome variables, such as emotional solidarity (Munanura et al., 2023; Phuc & Nguyen, 2023, receptiveness toward tourists during COVID-19 (Tilaki et al., 2021; Woosnam et al., 2022), and pro-tourism behavioral support (Erul, Woosnam, & McIntosh, 2020; Ribeiro et al., 2017).

To pave the way for further research on the social impacts of tourism, this work aims to achieve several objectives. Firstly, we will highlight the methodological advancements in this field that date back to the late 1970s. Secondly, we will showcase the valuable contributions of the eight chapters that comprise this book. Lastly, we will provide various methodological suggestions for future studies on the social impacts of tourism to conclude this introductory chapter.

Highlighting methodological advancements in assessing social impacts of tourism

Many of us owe a debt of gratitude to Valene Smith for her seminal editorial work, *Hosts and guests: The anthropology of tourism*. The ethnographic work compiled by Smith (1977) truly set our work in motion concerning social impacts of tourism research from the residents' perspectives. For it was in this work that we were introduced to definitions of tourism as well as the beginnings of sociocultural impacts of tourism research, covering economic contributions as well as social and cultural changes to local communities. By the close of the work, Nunez (1977) introduced us to the notion of acculturation theory and the realization that acculturation effect can be more pronounced among local communities and their residents than among those visiting. From a predominantly qualitative standpoint, we began to understand that it is extremely difficult to disregard the interaction and relationships with tourists in understanding how residents and their communities are impacted from tourism and its development.

Following Smith's (1977) text, we see a continued focus on conceptual, descriptive, and exploratory work in the early 1980s, with much of this research being quantitative (Nunkoo, Smith, & Ramkissoon, 2013) and helping to pave the way for much of the post-positivistic research to follow (Sharpley, 2014). It should be noted that at that time, we witnessed the creation and development of the top tourism journals—*Western Council for Travel Research* (currently *Journal of Travel Research*) launching in 1968, *Annals of Tourism Research* beginning in 1973; and *The International Journal of Tourism Management* (currently *Tourism Management*) starting in 1980—to disseminate this burgeoning body of research. To help catapult methodological advancements in social impacts of tourism forward, Mathieson and Wall's (1982) work *Tourism: Economic, physical and social impacts* ushered in a greater focus on quantitative measures to assess impacts. Echoing what others claimed at the time, Mathieson and Wall (1982) argued that although the focus of tourism impacts began as the positive economic impacts for destinations, more research was needed that focused on the social changes brought about by tourism, even if it highlights the negative damages. As Henderson (2018) alluded to in her review of the text, the notion of sustainability is implicit in the work, especially now considering the widespread acceptance of considering residents' perspectives of tourism in planning and managing for sustainable tourism.

Though Mathieson and Wall (1982) review the current state of research undertaken on residents' perceptions of tourism impacts, it is rather obvious that much of the quantitative work in this line of research began with rather simplistic analytical techniques in describing impacts or comparing means of perceived impacts across various groups (i.e., using t-tests, ANOVAs, etc.) (Nunkoo, Smith, & Ramkissoon, 2013). Over time, we see a progression of more predictive models employing linear regression analyses to explain perceived social impacts of tourism (Sharpley, 2014). With the advent of multiple-item measures to gauge social impacts of tourism, we began to see segmentation studies employing principal components analysis (PCA)/exploratory factor analysis (EFA) and cluster analysis (see Andriotis & Vaughan, 2003; Fredline & Faulkner, 2000; Madrigal, 1995; Ribeiro, Valle, & Silva, 2013; Williams & Lawson, 2001). Then we see formal scales developed and tested to measure social impacts of tourism (Ap & Crompton, 1998; Howard & Lankford). As we moved into the 2010s and 2020s, it has become more commonplace to see more advanced predictive models employing Anderson and Gerbing's (1988) two-stage analysis involving confirmatory factor analysis (CFA) and SEM (Nunkoo, Smith, & Ramkissoon, 2013; Sharpley, 2014). Such work tests empirically and/or theoretically deduced models, considering a host of predictive constructs in explaining perceived social impacts of tourism as well as residents' support for tourism. To echo these points above, Nunkoo, Smith, and Ramkissoon (2013) reviewed statistical techniques used in 140 articles (between 1984 and 2010) concerning social impacts of tourism within *Journal of Travel Research*, *Tourism Management*, and *Annals of Tourism Research* and reported that the primary techniques (some of which were used in tandem within the same study) used were: descriptives (65.9%), t-tests/ANOVA (31.2%), PCA/EFA (29.7%), regression/correlation (26.8%), and CFA/SEM/path analysis (24.6%).

Looking back at the 40-plus years of research surrounding social impacts of tourism research, we can categorize many of the predictors of residents' perceptions of social impacts of tourism into internal, interactive, and external variables (see Deery Jago, & Fredline, 2012; Gursoy et al., 2019; Sharpley, 2014) (see Table I.1). Internal variables are those that are unique to the residents, oftentimes represented by psychological measures and sociodemographic concepts. Those that have been used with some regularity within the literature to predict residents' social impacts of tourism include community

Table 1. Predictor variables in explaining residents' perceptions of social impacts of tourism

Classification (*Variable*)		Key Studies
Internal		
	Community attachment	Andereck et al., 2005; Chen & Chen, 2010; Eslami et al., 2019; Gannon, Rasoolimanesh, & Taheri, 2021; Rasoolimanesh et al., 2017; McCool & Martin, 1994; Ribeiro, Kim, & Woosman, 2020.
	Economic dependence	Brougham & Butler, 1981; Smith & Krannich, 1998; Snaith & Haley, 1999; Wang & Pfister, 2008
	Empowerment	Aleshinloye et al., 2021; Boley et al., 2014; Joo et al., 2020; Strzelecka, Boley, & Strzelecka, 2017.
	Knowledge of tourism	Davis, Allen, & Cosenza, 1988; Joo et al., 2020; Nunkoo, 2015
	Perceived benefits/costs of tourism	Ap, 1992; Látková & Vogt, 2012; McGehee & Andereck, 2004; Wang & Pfister, 2008;
	Personal values	Choi & Murray, 2010; Nunkoo & Ramkissoon, 2009; Phuc & Nguyen, 2023;
	Place attachment	Chen & Dwyer, 2018; Eusébio, Vieira, & Lima, 2018; Lee et al., 2020; Stylidis, 2018;
	Socio-demographics (age, gender, education, etc.)	Fredline & Faulkner, 2000; Huh & Vogt, 2008; Mason & Cheyne, 2000; Sharma & Dyer, 2009; Snaith & Haley, 1999; Tosun, 2002
	Trust	Gursoy et al. 2017; Mody et al., 2023; Nunkoo & Smith, 2013; Nunkoo et al., 2018 Ouyang, Gursoy, & Sharma, 2017; Woosnam et al., 2022
Interactive	*Emotional solidarity with tourists*	Munanura et al., 2023; Ribeiro et al., 2017; Woosnam, 2012
	Degree of interaction with tourists	Aleshinloye et al., 2020; Andereck et al., 2005; Eusébio, Vieira, & Lima, 2018; Lawson et al., 1998; Teye, Sirakaya, & Sönmez, 2002
	Social distance with tourists	Joo et al., 2018; Thyne, Watkins, & Yoshida, 2018; Thyne et al., 2022
External	*Density of tourists/tourism development*	Vargas-Sánchez, Porras-Bueno, & de los Ángeles Plaza-Mejía, 2011; Vargas-Sánchez et al., 2014
	Destination stage of tourism development	Allen et al., 1988; Lepp, 2008; Long, Perdue, & Allen, 1990; Vargas-Sánchez, Plaza-Mejia, & Porras-Bueno, 2009
	Distance from tourism zone	Belisle & Hoy, 1980; Jurowski & Gursoy, 2004; Raymond & Brown, 2007; Sharma & Dyer, 2009
	Nature/type of tourists/ tourism	Johnson, Snepenger, & Akis, 1994; Sheldon & Var, 1984
	Seasonality	Belisle & Hoy, 1980; Sheldon & Var, 1984; Vargas-Sanches et al., 2014

attachment, economic dependence, empowerment, knowledge of tourism, perceived benefits/costs of tourism, personal values, place attachment, sociodemographics, and trust. Interactive variables are those that involve relationship aspects with tourists. Though these variables (i.e., emotional solidarity, degree of interaction, and social distance) have received less attention, they are gaining some momentum lately. External variables involve measures that are specific to the destination and include density of tourists/tourism development, destination stage of tourism development, distance from the tourism zone, nature/type of tourists/tourism, and seasonality.

Though quantitative research methods and analysis have received the lion's share of attention within social impacts of tourism research, such a trajectory may be slowing down, in favor of a mixed methods approach. As Hadinejad et al. (2019) argue, "There is also a lack of use of mixed methods to collect and analyse data on residents' attitudes literature based on a pragmatic approach…mixed methods can provide both subjective and objective analyses of residents' perceptions" (p. 159). We see a mixed methods approach growing in support as *Tourism Management* has made it a requirement that publications now incorporate multiple forms of data. On their "guide for authors" webpage, *Tourism Management* (2021) states that,

> In the case of manuscripts based on a quantitative analysis using data derived from a single survey, such papers should be accompanied by supplementary materials such as interviews, data from another survey to permit comparative or longitudinal analysis, or similar supplementary datasets to provide further insights into the quantitative data analysis. This also applies to panel data unless evidence is provided that the panel data are representative of pertinent populations.

Evidence of mixed methods research in social impacts of tourism research is highlighted through the review of papers employing such a design within the leading tourism journals between 1998 and 2019 (Molina-Azorín & Font, 2016; Truong, Liu, & Yu, 2020) as well as throughout the work within this book (Martín-Martín et al., 2023). From a pragmatic standpoint, the methods one employs are largely contingent upon whether the intent of the research is on theoretical development or theoretical testing. In either case, advancing theory is crucial for the continued evolution of social impacts of tourism research.

Overview of the chapters within this book

Considering the significant advancements in tourism's social impact research over the past three decades, we present the eight chapters that make up this book. Each chapter delves into methodological aspects, encompassing both general strategies and tailored, case-specific techniques. In one chapter, Olya (2023), highlighting a very pragmatic approach, argues that methods and theory are inextricably linked, and we see this through his employment of PLS-SEM and fuzzy set Qualitative Comparative Analysis (fsQCA) and the theory of change to ultimately develop a framework leading to greater effects of our social impacts of tourism research. Within this book, advancements are made in the way of relationships between community residents and tourists, social impacts and residents' impacts, residents' empowerment, overtourism, and methodologies—both qualitative and quantitative in nature. Each will be addressed in turn.

Resident–tourist relationships, social impacts, and residents' attitudes

Two other chapters within this book develop emotional solidarity theory by testing models whereby the emotional solidarity construct acts as a mediator between beliefs or values and residents' attitudinal support for tourism. In Japan, Maruyama, Ribeiro, and Woosnam (2023) demonstrate that both Korean and Brazilian residents' (as minority ethnic groups in two distinct cities) ethnic attitudes toward and emotional solidarity with Japanese explained roughly 70% of the variance in support for tourism focused on the minority groups' culture.

One chapter within this book advance research concerning social impacts and residents' attitudes toward tourism in general. Mair et al. (2022) provide us with one of the first systematic reviews of social impacts of mega-events, highlighting direct impacts on residents (e.g., volunteering, education, and skills, social cohesion, civic pride, and social capital, inclusion and diversity, and sport participation, infrastructure, and health) and indirect impacts on the destination (e.g., business and government networks, destination branding, destination preparedness, and accessibility). Mair et al. (2022) leave us with ample research opportunities to pursue as destinations move closer to welcoming back mega-events.

The theme of examining residents' perceptions of impacts of peer-to-peer accommodations continues in the work by Jordan et al. (2023) along with considerations of general tourism and cruise tourism impacts in Portugal. Jordan et al. report that residents, with relative ease, differentiated between the various forms of tourism. Koščak et al. (2023) present descriptive findings from their study across six European destinations concerning residents' attitudes about tourism. The uniqueness of this work lies in the inclusivity of oft-neglected study participants. Koščak et al. (2023) garners perspectives of youth—both concerning perspectives of tourism impacts. In essence, understanding such voices is a step in the direction toward empowering residents to actively participate in planning for sustainable tourism.

Overtourism

Nearly as hot a topic as residents' perceptions of peer-to-peer accommodations is that concerning overtourism in destinations. Martín-Martín et al. (2022) consider the nexus of the two using the human-scale development approach to best capture residents' needs (both in negative and axiological contexts). Martín-Martín et al. (2022) undertake their research using a mixed methods approach.

Methodologically focused

In addition to some interesting methodological advancements noted by Olya (2023), two others stand out within this book. The first is the article by Tse and Tung (2022) that examines implicit stereotypes of mainland Chinese residents among Hong Kong, Malaysia, Singapore, and Thailand using the implicit association test or IAT. The authors use the computerized program to measure study participants' responses and response times to cued images of mainland Chinese residents (as implicit stereotypes), and then compare these with explicit stereotypes (measured through 12 behavioral likelihood items). The second chapter, by Chamarro, Cobo-Benita, and Amo (2023), focuses on how Madrid residents' and its visitors' perceptions of the SUS-TAS (in the context of a destination with a high percentage of peer-to-peer accommodations) may be potentially different (using a multi-group invariance analysis). Such an analytical technique was also employed by Maruyama et al. (2023) within this book.

Moving forward

From the work presented in these proceeding eight chapters, various future research opportunities exist. Let us first start with one of the most obvious. Many of the articles in this issue focused on the relationship between residents and tourists, providing greater support for the idea that representatives of each group are indissolubly connected (Stylidis, 2022). Work that continues to operationalize measures of social contact and the contact theory between residents and tourists (building on the work of Fan, 2022 and Fan et al., 2022) will be most timely. With more research focusing on emotional solidarity between residents and tourists, more work is needed to examine additional outcomes of the construct beyond attitudinal measures. For instance, how does emotional solidarity prompt an individual's likelihood to behave in a certain way (behavioral support or withdrawal) (Erul et al., 2020) or better yet, actual behavior. Future research should also shift slightly to consider the construct from tourists' perspectives (like the work by Patwardhan et al., 2020 and Woosnam & Aleshinloye,

2013). Finally, the original emotional solidarity theoretical framework should be revisited to determine if inclusion of additional constructs may provide a more robust perspective of the role such relationships play in explaining outcome variables like behavioral support, engagement, or involvement. Considering solidarity in tandem with complementary theories (e.g., cognitive appraisal theory, interaction ritual theory, mere exposure theory, protection motivation theory, social cognition theory, theory of formal and substantive rationality, etc.) may go even further.

Another area of opportunity for research in social impacts of tourism surrounds greater theoretical model testing that encompasses multiple forms of data. We suggest that future researchers should design research utilizing multiple case studies (involving quantitative and qualitative data) and/or panel data. This will help to allay any potential concerns of findings being more idiographic in nature. Future testable models need to be driven by theoretical frameworks to further advance our field and address concerns stated by multiple authors regarding limited theoretical development and testing (Gursoy & Nunkoo, 2019; Hadinejad et al., 2019; and Nunkoo, Smith, & Ramkissoon, 2013). Furthermore, proposed models should include outcome variables measuring either behavioral intentions or actual behavior. This approach will help to ensure greater actionable research with strong practical implications. Additionally, since fsQCA is growing in popularity, ways in which to undertake qualitative analysis alongside more advanced quantitative approaches would surely help to triangulate findings.

Further longitudinal research is also needed. This involves data collection at multiple points throughout time from the same study population using the same measures or constructs. All too often we see authors indicating in final sections of their articles that their findings would benefit from longitudinal research designs. Works most recently by Bimonte, D'Agostino, Grilli, and Pagliuca (2019) and Liang, Luo, and Bao (2021) in the context of residents' perceptions of tourism provide strong examples to follow. Given the ever-changing landscape of travel, now is an ideal time to undertake research that monitors residents' perspectives and how they change over time—something authors conducting social impacts research have suggested for years. The work by Kamata (2022) is an invaluable guide. Though financial and time constraints often prohibit engaging in longitudinal research, how much more powerful would results be if researchers presented consistent (or even contrary) findings based on the sample!

As our field makes a push toward realizing many of the United Nation's sustainable development goals (SDGs) (UNWTO, 2017) by 2030, we need to be deliberate in employing the most suitable methodological approaches, even if that means embracing those that are rather novel in nature. In fact, we would argue that a field progresses by tackling difficult questions and by shifting the pendulum forward one new method at a time. We implore tourism researchers to be bold in implementing new methods—the future of social impacts of tourism research may very well depend on it.

References

Aguinis, H., Bergh, D., & Molina-Azorin, J. F. (2022). Methodological challenges and insights for future international business research. *Journal of International Business Studies*, *54*(2), 219–232. https://doi.org/10.1057/s41267-022-00578-8

Aleshinloye, K. D., Fu, X., Ribeiro, M. A., Woosnam, K. M., & Tasci, A. D. (2020). The influence of place attachment on social distance: Examining mediating effects of emotional solidarity and the moderating role of interaction. *Journal of Travel Research*, *59*(5), 828–849.

Aleshinloye, K. D., Woosnam, K. M., Tasci, A. D. A., & Ramkissoon, H. (2021). Antecedents and outcomes of resident empowerment through tourism. *Journal of Travel Research*, *61*(3), 656–673. https://doi.org/10.1177/0047287521990437

Allen, L. R., Long, P. T., Perdue, R. R., & Kieselbach, S. (1988). The impact of tourism development on residents' perceptions of community life. *Journal of Travel Research*, *27*(1), 16–21.

Andereck, K. L., Valentine, K. M., Knopf, R. C., & Vogt, C. A. (2005). Residents' perceptions of community tourism impacts. *Annals of Tourism Research*, *32*(4), 1056–1076.

Anderson, J. C., & Gerbing, D. W. (1988). Structural equation modeling in practice: A review and recommended two-step approach. *Psychological Bulletin*, *103*(3), 411–423.

Andriotis, K., & Vaughan, R. D. (2003). Urban residents' attitudes toward tourism development: The case of Crete. *Journal of Travel Research*, *42*(2), 172–185.

Ap, J. (1992). Residents' perceptions on tourism impacts. *Annals of Tourism Research*, *19*(4), 665–690.

Ap, J., & Crompton, J. L. (1998). Developing and testing a tourism impact scale. *Journal of Travel Research*, *37*(2), 120–130.

Belisle, F. J., & Hoy, D. R. (1980). The perceived impact of tourism by residents: A case study in Santa Marta, Colombia. *Annals of Tourism Research*, *7*(1), 83–101.

Bimonte, S., D'Agostino, A., Grilli, G., & Pagliuca, M. (2019). Tourist season and residents' life satisfaction: Empirical evidence from a longitudinal design in a Mediterranean destination. *International Journal of Tourism Research*, *21*(3), 323–333.

Boley, B. B., McGehee, N. G., Perdue, R. R., & Long, P. (2014). Empowerment and resident attitudes toward tourism: Strengthening the theoretical foundation through a Weberian lens. *Annals of Tourism Research*, *49*, 33–50.

Brougham, J. E., & Butler, R. W. (1981). A segmentation analysis of resident attitudes to the social impact of tourism. *Annals of Tourism Research*, *8*(4), 569–590.

Butler, R. W. (1980). The concept of a tourist area cycle of evolution: Implications for management of resources. *Canadian Geographer*, *24*(1), 5–12.

Chamarro, A., Cobo-Benita, J., & Herrero Amo, M. D. (2023). Towards sustainable tourism development in a mature destination: Measuring multi-group invariance between residents and visitors' attitudes with high use of accommodation-sharing platforms. *Journal of Sustainable Tourism*, *31*(2), 520–537. https://doi.org/10.1080/09669582.2020.1870988

Chen, C. F., & Chen, P. C. (2010). Resident attitudes toward heritage tourism development. *Tourism Geographies*, *12*(4), 525–545.

Chen, N., & Dwyer, L. (2018). Residents' place satisfaction and place attachment on destination brand-building behaviors: Conceptual and empirical differentiation. *Journal of Travel Research*, *57*(8), 1026–1041.

Choi, H. C., & Murray, I. (2010). Resident attitudes toward sustainable community tourism. *Journal of Sustainable Tourism*, *18*(4), 575–594.

Choi, H. S. C., & Sirakaya, E. (2005). Measuring residents' attitude toward sustainable tourism: Development of sustainable tourism attitude scale. *Journal of Travel Research*, *43*(4), 380–394.

Crowley, S. L., & Fan, X. (1997). Structural equation modeling: Basic concepts and applications in personality assessment research. *Journal of Personality Assessment*, 68(3), 508–531.

Davis, D., Allen, J., & Cosenza, R. M. (1988). Segmenting local residents by their attitudes, interests, and opinions toward tourism. *Journal of Travel Research*, *27*(2), 2–8.

Deery, M., Jago, L., & Fredline, L. (2012). Rethinking social impacts of tourism research: A new research agenda. *Tourism Management*, *33*(1), 64–73.

Doxey, G. V. (1975). A causation theory of visitor-resident irritants, methodology, and research inferences. *Sixth annual conference proceedings of the Travel Research Association*, San Diego, CA: Travel and Tourism Research Association, 195–198.

Duncan, O. D. (1975). *Introduction to structural equation models*. New York: Academic Press.

Erul, E., Woosnam, K. M., & McIntosh, W. A. (2020). Considering emotional solidarity and the theory of planned behavior in explaining behavioral intentions to support tourism development. *Journal of Sustainable Tourism*, *28*(8), 1158–1173.

Eslami, S., Khalifah, Z., Mardani, A., Streimikiene, D., & Han, H. (2019). Community attachment, tourism impacts, quality of life and residents' support for sustainable tourism development. *Journal of Travel & Tourism Marketing*, *36*(9), 1061–1079.

Eusébio, C., Vieira, A. L., & Lima, S. (2018). Place attachment, host–tourist interactions, and residents' attitudes towards tourism development: The case of Boa Vista Island in Cape Verde. *Journal of Sustainable Tourism*, *26*(6), 890–909.

Evans-Pritchard, D. (1989). How "they" see "us": Native American images of tourists. *Annals of Tourism Research*, *16*(1), 89–105.

Fan, D. X. (2022). Understanding the tourist-resident relationship through social contact: progressing the development of social contact in tourism. *Journal of Sustainable Tourism*, https://doi.org/10.1080/09669582.2020.1852409

Fan, D. X., Qiu, H., Jenkins, C. L., & Lau, C. (2022). Towards a better tourist-host relationship: the role of social contact between tourists' perceived cultural distance and travel attitude. *Journal of Sustainable Tourism*, https://doi.org/10.1080/09669582.2020.1783275

Fredline, E., & Faulkner, B. (2000). Host community reactions: A cluster analysis. *Annals of Tourism Research*, *27*(3), 763–784.

Gannon, M., Rasoolimanesh, S. M., & Taheri, B. (2021). Assessing the mediating role of residents' perceptions toward tourism development. *Journal of Travel Research*, *60*(1), 149–171.

Gursoy, D., & Nunkoo, R. (Eds) (2019). *The Routledge handbook of tourism impacts: Theoretical and applied perspectives*. New York: Routledge.

Gursoy, D., Ouyang, Z., Nunkoo, R., & Wei, W. (2019). Residents' impact perceptions of and attitudes towards tourism development: A meta-analysis. *Journal of Hospitality Marketing & Management*, *28*(3), 306–333.

Gursoy, D., Yolal, M., Ribeiro, M. A., & Panosso Netto, A. (2017). Impact of trust on local residents' mega-event perceptions and their support. *Journal of Travel Research*, *56*(3), 393–406. doi:10.1177/0047287516643415

Hadinejad, A., D. Moyle, B., Scott, N., Kralj, A., & Nunkoo, R. (2019). Residents' attitudes to tourism: A review. *Tourism Review*, *74*(2), 150–165. https://doi.org/10.1108/tr-01-2018-0003.

Hahn, E. D., & Ang, S. H. (2017). From the editors: New directions in the reporting of statistical results in the Journal of World Business. *Journal of World Business, 52*(2), 125–126. https://doi.org/10.1016/j.jwb.2016.12.003

Hair, J., Black, W., Babin, B., & Anderson, R. (2019). *Multivariate data analysis* (8th ed.). Hampshire, UK: Cengage Learning EMEA.

Henderson, J. C. (2018). Classic book review of *Tourism: Economic, physical and social impacts*, by A. Mathieson and G. Wall. *Annals of Leisure Research*, *21*(4), 522–524.

Hoyle, R. H. (Ed.). (2023). *Handbook of structural equation modeling* (2nd Ed.). New York: Guilford Press.

Huh, C., & Vogt, C. A. (2008). Changes in residents' attitudes toward tourism over time: A cohort analytical approach. *Journal of Travel Research*, *46*(4), 446–455.

Johnson, J. D., Snepenger, D. J., & Akis, S. (1994). Residents' perceptions of tourism development. *Annals of Tourism Research*, *21*(3), 629–642.

Joo, D., Tasci, A. D., Woosnam, K. M., Maruyama, N. U., Hollas, C. R., & Aleshinloye, K. D. (2018). Residents' attitude towards domestic tourists explained by contact, emotional solidarity and social distance. *Tourism Management*, *64*, 245–257.

Joo, D., Woosnam, K. M., Strzelecka, M., & Boley, B. B. (2020). Knowledge, empowerment, and action: Testing the empowerment theory in a tourism context. *Journal of Sustainable Tourism*, *28*(1), 69–85.

Jordan, E. J., Vieira, J. C., Santos, C. M., & Huang, T. Y. (2023). Do residents differentiate between the impacts of tourism, cruise tourism, and Airbnb tourism?. *Journal of Sustainable Tourism*, https://doi.org/10.1080/09669582.2020.1833894

Jurowski, C., & Gursoy, D. (2004). Distance effects on residents' attitudes toward tourism. *Annals of Tourism Research*, *31*(2), 296–312.

Kamata, H. (2022). Tourist destination residents' attitudes towards tourism during and after the COVID-19 pandemic. *Current Issues in Tourism*, https://doi.org/10.1080/13683500.2021.1881452

Koščak, M., Knežević, M., Binder, D., Pelaez-Verdet, A., Işik, C., Mićić, V., ... & Šegota, T. (2023). Exploring the neglected voices of children in sustainable tourism development: A comparative study in six European tourist destinations. *Journal of Sustainable Tourism*, https://doi.org/10.1080/09669582.2021.1898623

Lankford, S. V., & Howard, D. R. (1994). Developing a tourism impact attitude scale. *Annals of Tourism Research*, *21*(1), 121–139.

Látková, P., & Vogt, C. A. (2012). Residents' attitudes toward existing and future tourism development in rural communities. *Journal of Travel Research*, *51*(1), 50–67.

Lawson, R. W., Williams, J., Young, T. A. C. J., & Cossens, J. (1998). A comparison of residents' attitudes towards tourism in 10 New Zealand destinations. *Tourism Management*, *19*(3), 247–256.

Lee, S., Joo, D., Lee, C. K., & Woosnam, K. M. (2020). Korean DMZ tourists' perceived similarity and shared beliefs in predicting place attachment and support for tourism development. *Journal of Destination Marketing & Management*, *18*, 100467.

Lepp, A. (2008). Attitudes towards initial tourism development in a community with no prior tourism experience: the case of Bigodi, Uganda. *Journal of Sustainable Tourism, 16*(1), 5–22.

Liang, Z., Luo, H., & Bao, J. (2021). A longitudinal study of residents' attitudes toward tourism development. *Current Issues in Tourism, 24*(3), 3309–3323.

Liu, J. C., Sheldon, P. J., & Var, T. (1987). Resident perception of the environmental impacts of tourism. *Annals of Tourism Research, 14*(1), 17–37. doi:10.1016/0160-7383(87)90045-4

Liu, J. C., & Var, T. (1986). Resident attitudes toward tourism impacts in Hawaii. *Annals of Tourism* Research, 13(2), 193–214.

Loehlin, J. C., & Beaujean, A. A. (2017*). Latent variable models: An introduction to factor, path, and structural equation analysis.* (5th ed). New York. Taylor & Francis.

Long, P. T., Perdue, R. R., & Allen, L. (1990). Rural resident tourism perceptions and attitudes by community level of tourism. *Journal of Travel Research, 28*(3), 3–9.

Madrigal, R. (1995). Residents' perceptions and the role of government. *Annals of Tourism Research, 22*(1), 86–102.

Mair, J., Chien, P. M., Kelly, S. J., & Derrington, S. (2022). Social impacts of mega-events: A systematic narrative review and research agenda. *Journal of Sustainable Tourism*, https://doi.org/10.1080/09669582.2020.1870989

Martín-Martín, J. M., Prados-Castillo, J. F., Jimenez Aguilera, J. D. D., & Porras González, E. (2023). Interferences generated on the well-being of local communities by the activity of online platforms for tourist accommodation. *Journal of Sustainable Tourism, 31*(2), 483–503. https://doi.org/10.1080/09669582.2020.1861455

Maruyama, N. U., Ribeiro, M. A., & Woosnam, K. M. (2023). The effect of minority residents' attitudes and emotional solidarity on ethnic neighborhood tourism: A multigroup invariance analysis. *Journal of Sustainable Tourism, 31*(2), 383–405. https://doi.org/10.1080/09669582.2020.1849238

Mason, P. (2016). *Tourism impacts, planning and management* (3rd Ed.). New York: Routledge.

Mason, P., & Cheyne, J. (2000). Residents' attitudes to proposed tourism development. *Annals of Tourism Research, 27*(2), 391–411.

Mathieson, A., & Wall, G. (1982). *Tourism: Economic, physical and social impacts*. Harlow, UK: Longman.

McCool, S. F., & Martin, S. R. (1994). Community attachment and attitudes toward tourism development. *Journal of Travel Research, 32*(3), 29–34.

McGehee, N. G., & Andereck, K. L. (2004). Factors predicting rural residents' support of tourism. *Journal of Travel Research, 43*(2), 131–140.

Meyer, K. E., van Witteloostuijn, A., & Beugelsdijk, S. (2017). What's in a p? Reassessing best practices for conducting and reporting hypothesis-testing research. *Journal of International Business Studies, 48*(5), 535–551. https://doi.org/10.1057/s41267-017-0078-8

Milman, A., & Pizam, A. (1988). Social impacts of tourism on central Florida. *Annals of Tourism Research, 15*(2), 191–204.

Mody, M., Woosnam, K. M., Suess, C., & Dogru, T. (2023). Hapless victims or empowered citizens? Understanding residents' attitudes towards Airbnb using Weber's Theory of Rationality and Foucauldian concepts. *Journal of Sustainable Tourism, 31*(2), 284–306. https://doi.org/10.1080/09669582.2020.1834567

Molina-Azorín, J. F., & Font, X. (2016). Mixed methods in sustainable tourism research: An analysis of prevalence, designs and application in JOST (2005–2014). *Journal of Sustainable Tourism, 24*(4), 549–573.

Munanura, I. E., Needham, M. D., Lindberg, K., Kooistra, C., & Ghahramani, L. (2023). Support for tourism: The roles of attitudes, subjective wellbeing, and emotional solidarity. *Journal of Sustainable Tourism, 31*(2), 581–596. https://doi.org/10.1080/09669582.2021.1901104

Nunez, T. (1977). Touristic studies in anthropological perspective. In V. Smith (Ed.) *Hosts and guests: The anthropology of tourism* (pgs. 265–279). Philadelphia: University of Pennsylvania Press.

Nunkoo, R. (2015). Tourism development and trust in local government. *Tourism Management, 46*(0), 623–634. https://doi.org/10.1016/j.tourman.2014.08.016

Nunkoo, R., & Gursoy, D. (2019). Introduction to tourism impacts. In R. Nunkoo and D. Gursoy (Eds). *The Routledge handbook of tourism impacts: Theoretical and applied perspectives* (pgs. 1–20). New York: Routledge.

Nunkoo, R., & Ramkissoon, H. (2009). Influence of values on residents' attitudes toward tourism. *Tourism Analysis, 14*(2), 241–244.

Nunkoo, R., & Ramkissoon, H. (2012). Power, trust, social exchange and community support. *Annals of Tourism Research, 39*(2), 997–1023.

Nunkoo, R., Ribeiro, M. A., Sunnassee, V., & Gursoy, D. (2018). Public trust in mega event planning institutions: The role of knowledge, transparency and corruption. *Tourism Management, 66*, 155–166. doi:10.1016/j.tourman.2017.11.010

Nunkoo, R., & Smith, S. L. (2013). Political economy of tourism: Trust in government actors, political support, and their determinants. *Tourism Management, 36*, 120–132.

Nunkoo, R., Smith, S. L., & Ramkissoon, H. (2013). Residents' attitudes to tourism: A longitudinal study of 140 articles from 1984 to 2010. *Journal of Sustainable Tourism, 21*(1), 5–25.

Olya, H. G. (2023). Towards advancing theory and methods on tourism development from residents' perspectives: Developing a framework on the pathway to impact. *Journal of Sustainable Tourism*, https://doi.org/10.1080/09669582.2020.1843046

Ouyang, Z., Gursoy, D., & Sharma, B. (2017). Role of trust, emotions and event attachment on residents' attitudes toward tourism. *Tourism Management, 63*, 426–438.

Patwardhan, V., Ribeiro, M. A., Payini, V., Woosnam, K. M., Mallya, J., & Gopalakrishnan, P. (2020). Visitors' place attachment and destination loyalty: Examining the roles of emotional solidarity and perceived safety. *Journal of Travel Research, 59*(1), 3–21.

Perdue, R. R., Long, P. T., & Allen, L. (1990). Resident support for tourism development. *Annals of Tourism Research, 17*(4), 586–599.

Phuc, H. N., & Nguyen, H. M. (2023). The importance of collaboration and emotional solidarity in residents' support for sustainable urban tourism: Case study Ho Chi Minh City. *Journal of Sustainable Tourism, 31*(2), 245–264. https://doi.org/10.1080/09669582.2020.1831520

Rasoolimanesh, S. M., Roldán, J. L., Jaafar, M., & Ramayah, T. (2017). Factors influencing residents' ptoward tourism development: Differences across rural and urban World Heritage Sites. Journal of Travel Research, 56(6), 760–775. https://doi.org/10.1177/0047287516662354

Raymond, C., & Brown, G. (2007). A spatial method for assessing resident and visitor attitudes towards tourism growth and development. *Journal of Sustainable Tourism, 15*(5), 520–540.

Ribeiro, M. A., Kim, Y. H., & Woosnam, K. M. (2020). Residents' perception and their support for tourism development: The case of South Korea. In P. Pinto & M. Guerreiro (Eds.), *Handbook of research on resident and tourist perspectives on travel destinations* (pgs. 140–165). Hershey, PA, USA: IGI Global.

Ribeiro, M. A., Pinto, P., Silva, J. A., & Woosnam, K. M. (2017). Residents' attitudes and the adoption of pro-tourism behaviours: The case of developing island countries. *Tourism Management, 61*, 523–537.

Ribeiro, M. A., Valle, P. O. d., & Silva, J. A. (2013). Residents' attitudes towards tourism development in Cape Verde Islands. *Tourism Geographies, 15*(4), 654–679. https://doi.org/10.1080/14616688.2013.769022

Rothman, R. A. (1978). Residents and transients: Community reaction to seasonal visitors. *Journal of Travel Research, 16*(3), 8–13.

Ruhanen, L., Moyle, C. L., & Moyle, B. (2019). New directions in sustainable tourism research. *Tourism Review, 74*(2), 138–149.

Shah, R., & Goldstein, S. M. (2006). Use of structural equation modeling in operations management research: Looking back and forward. *Journal of Operations Management, 24*(2), 148–169.

Sharma, B., & Dyer, P. (2009). An investigation of differences in residents' perceptions on the Sunshine Coast: Tourism impacts and demographic variables. *Tourism Geographies, 11*(2), 187–213.

Sharpley, R. (2014). Host perceptions of tourism: A review of the research. *Tourism Management, 42*, 37–49.

Sheldon, P. J., & Var, T. (1984). Resident attitudes to tourism in North Wales. *Tourism Management, 5*(1), 40–47.

Smith, M. D., & Krannich, R. S. (1998). Tourism dependence and resident attitudes. *Annals of Tourism Research, 25*(4), 783–802.

Smith, V. (Ed.) (1977). *Hosts and guests: The anthropology of tourism*. Philadelphia: University of Pennsylvania Press.

Snaith, T., & Haley, A. (1999). Residents' opinions of tourism development in the historic city of York, England. *Tourism Management, 20*(5), 595–603.

Strzelecka, M., Boley, B. B., & Strzelecka, C. (2017). Empowerment and resident support for tourism in rural Central and Eastern Europe (CEE): The case of Pomerania, Poland. *Journal of Sustainable Tourism, 25*(4), 554–572.

Stylidis, D. (2018). Place attachment, perception of place and residents' support for tourism development. *Tourism Planning & Development*, *15*(2), 188–210.

Stylidis, D. (2022). Exploring resident–tourist interaction and its impact on tourists' destination image. *Journal of Travel Research*, https://doi.org/10.1177/0047287520969861

Teye, V., Sirakaya, E., & Sönmez, S. F. (2002). Residents' attitudes toward tourism development. *Annals of Tourism Research*, *29*(3), 668–688.

Thyne, M., Watkins, L., & Yoshida, M. (2018). Resident perceptions of tourism: The role of social distance. *International Journal of Tourism Research*, *20*(2), 256–266.

Thyne, M., Woosnam, K. M., Watkins, L., & Ribeiro, M. A. (2022). Social distance between residents and tourists explained by residents' attitudes concerning tourism. *Journal of Travel Research*, *61*(1), 150–169. doi:10.1177/0047287520971052

Tilaki, M. J. M., Abooali, G., Marzbali, M. H., & Samat, N. (2021). Vendors' attitudes and perceptions towards international tourists in the Malaysia Night Market: Does the COVID-19 outbreak matter? *Sustainability*, *13*(3), 1553.

Tosun, C. (2002). Host perceptions of impacts: A comparative tourism study. *Annals of Tourism Research*, *29*(1), 231–253.

Tourism Management (2021). Guide for authors. Retrieved on 6 October 2021 from https://www.elsevier.com/journals/tourism-management/0261-5177/guide-for-authors

Truong, D., Liu, R. X., & Yu, J. J. (2020). Mixed methods research in tourism and hospitality journals. *International Journal of Contemporary Hospitality Management*, *32*(4), 1563–1579.

Tse, W. T. S., & Tung, V. W. S. (2022). Assessing explicit and implicit stereotypes in tourism: Self-reports and implicit association test. *Journal of Sustainable Tourism*, https://doi.org/10.1080/09669582.2020.1860995

UNWTO. (2017). *Tourism and the sustainable development goals—Journey to 2030*. Retrieved on 1 January 2022 from www.e-unwto.org/doi/epdf/10.18111/9789284419401

Vargas-Sánchez, A., Plaza-Mejia, M. D. L. Á., & Porras-Bueno, N. (2009). Understanding residents' attitudes toward the development of industrial tourism in a former mining community. *Journal of Travel Research*, *47*(3), 373–387.

Vargas-Sánchez, A., Porras-Bueno, N., & de los Ángeles Plaza-Mejía, M. (2011). Explaining residents' attitudes to tourism: Is a universal model possible? *Annals of Tourism Research*, *38*(2), 460–480.

Vargas-Sánchez, A., Porras-Bueno, N., & Plaza-Mejía, M. d. l. Á. (2014). Residents' attitude to tourism and seasonality. *Journal of Travel Research*, *53*(5), 581–596. https://doi.org/10.1177/0047287513506295

Wang, Y., & Pfister, R. E. (2008). Residents' attitudes toward tourism and perceived personal benefits in a rural community. *Journal of Travel Research*, *47*(1), 84–93.

Williams, J., & Lawson, R. (2001). Community issues and resident opinions of tourism. *Annals of Tourism Research*, *28*(2), 269–290.

Woosnam, K. M. (2012). Using emotional solidarity to explain residents' attitudes about tourism and tourism development. *Journal of Travel Research*, *51*(3), 315–327.

Woosnam, K. M., & Aleshinloye, K. D. (2013). Can tourists experience emotional solidarity with residents? Testing Durkheim's model from a new perspective. *Journal of Travel Research*, *52*(4), 494–505.

Woosnam, K. M., & Norman, W. C. (2010). Measuring residents' emotional solidarity with tourists: Scale development of Durkheim's theoretical constructs. *Journal of Travel Research*, *49*(3), 365–380.

Woosnam, K. M., & Ribeiro, M. A. (2023). Methodological and theoretical advancements in social impacts of tourism research. *Journal of Sustainable Tourism*, *31*(2), 187–203.

Woosnam, K. M., Russell, Z., Ribeiro, M. A., Denley, T. J., Rojas, C., Hadjidakis, E., Barr, J., & Mower, J. (2022). Residents' pro-tourism behaviour in a time of COVID-19. *Journal of Sustainable Tourism,*, *30*(8), 1858–1877. https://doi.org/10.1080/09669582.2021.1985128

Do residents differentiate between the impacts of tourism, cruise tourism, and Airbnb tourism?

Evan J. Jordan, José Cabral Vieira, Carlos M. Santos and Ting-Yen (Tim) Huang

ABSTRACT

Tourism impacts the quality of life of host community residents in myriad ways. Most research seeking to understand those impacts has either focused on tourism in general or a single specific type of tourism. Little is understood about how residents attribute perceived impacts to different forms of tourism when they are present in the same community. We surveyed 995 residents of the island of São Miguel in the Azores archipelago to examine perceptions of tourism's quality of life impacts using Andereck and Nyaupane's Tourism Quality of Life Impacts scale in relation to tourism in general, cruise tourism, and Airbnb tourism. While confirmatory factor analysis (CFA) revealed poor fit of the measurement model across all forms of tourism, repeated measures ANOVA analysis of individual impact questions revealed that residents did differentiate between the impacts of various types of tourism in their community. Residents thought that tourism in general was the most detrimental to urban issues like traffic and crowding and that cruise tourism provided the least positive impacts to their community across all other TQOL variables. These findings indicate that future research on resident perceptions of tourism impacts should seek to measure impacts related to the specific forms of tourism.

1. Introduction

The differences in residents' perceptions of the impacts of different types of tourism are poorly understood. Most studies of tourism impacts have focused on either a single type of tourism or tourism in general, but rarely both. However, many tourism host communities host various types of tourism, which are likely to have their own unique impacts within the community. As the tourism industry recovers from COVID-19 related downturn, tourism planners need accurate information about community tourism impacts in order to make informed decisions for the future. We aim to fill this gap in the understanding of perceived tourism impacts by measuring residents' perceived impacts related to cruise tourism, short term vacation rental-based tourism (known colloquially as Airbnb tourism), and tourism in general as they have not been measured before. Specifically, we seek to measure residents' perceived impacts of these three types of tourism a) in a single community b) at the same point in time c) with the same measurement tool, effectively eliminating three confounding variables that have hampered researchers' ability to make direction comparisons of the impacts of different types of tourism. We believe that findings from this study will advance the understanding of the measurement of resident's perceived tourism impacts and provide a framework for future studies to differentiate between tourism impacts in tourism host communities.

The impacts of tourism on host community residents are one of the most examined topics in the tourism literature over the last 50 years (Sharpley, 2014). From traffic and crowding to cost of living to opportunities for recreation and leisure, tourism impacts have long been studied from many angles. Residents' perceptions are influenced by a variety of intrinsic (e.g. personal benefit from tourism or employment in the tourism industry) and extrinsic variables (e.g., level of tourism development, type of community, stakeholder groups) (Cheung & Li, 2019; Gursoy et al., 2019; Ribeiro et al., 2018; Sharma & Dyer, 2009; Sharpley, 2014; Suess et al., 2018). Cruise tourism and Airbnb tourism were selected for comparison to tourism in general specifically because they are two of the most rapidly growing segments within the tourism industry. According to the Cruise Line Industry Association, there were 28.5 million cruise passengers in 2018, compared to just 17.8 million in 2009 (CLIA, 2019). In terms of short term vacation rentals, Airbnb recently eclipsed the 500 million guest mark after the company started in 2007. Furthermore, recent research has shown that cruises and short term vacation rentals have impacts that extend beyond traditionally recognized tourism impacts into the realm of health and wellness (Jordan, Lesar, et al., 2019). Cruise ships are in ports of call for a very short period of time, and while they do have some economic benefits, leakage and other social and environmental impacts are often more substantial (Chase & McKee, 2003; Jordan & Vogt, 2017). In contrast, Airbnb based tourism brings travelers into residential areas for longer periods of time, leading to a much different set of impacts (Jordan & Moore, 2018; Suess et al., 2020).

To compare residents' perceived impacts of cruise tourism, Airbnb tourism, and tourism in general, we surveyed a systematic random sample of 995 residents of the island of São Miguel, a mid-Atlantic Portuguese island in the Azores archipelago. Resident study participants responded to a battery of tourism quality of life impact questions developed by Andereck and Nyaupane (2011) and subsequently used as a tourism impacts scale by Jordan, Spencer, and Prayag (2019). These questions were asked three times in the same survey – once about tourism in general, once about cruise tourism, and once about Airbnb based tourism. In the following paper, we first conduct a series of confirmatory factor analyses (CFA) to examine the reliability and validity of Andereck and Nyaupane's (2011) measurement tool in measuring different types of tourism impacts. Then, we conduct a repeated measure analysis of variance (ANOVA) to compare residents' perceptions of the impacts of each type of tourism.

Theoretically, this study represents an important step forward in the understanding of how residents differentiate between perceived impacts of multiple types of tourism, which have often been lumped together in previous research. Methodologically, we provide guidance for those studying residents' perceived tourism impacts in the future about how to measure perceived impacts in communities with multiple tourism segments. Practically, our findings should help inform community planners and developers about the specific impacts brought about by various types of tourism in the same community, which can be used to make informed development decisions when multiple options are available.

2. Literature review

In this literature review, we first examine the tourism quality of life impacts framework used as the foundation of our study (Andereck & Nyaupane, 2011). Then, we examine the types of tourism impacts that residents of tourism host communities have identified in previous tourism research. We start by examining impacts attributed to tourism in general, then impacts attributed specifically to cruise tourism, and final impacts attributed specifically to Airbnb tourism.

Throughout this paper, we refer to any short term vacation rental or peer to peer accommodation based tourism as 'Airbnb tourism'. Finally, based on findings in the literature, we put for specific hypotheses about the differences between perceived impacts across these three types of tourism which will later be tested through repeated measures ANOVA.

2.1. Tourism QOL impacts theoretical framework

Tourism influences resident QOL in numerous ways. Perhaps the greatest focus in the tourism literature to date has been on the relationship between tourism and resident income (Croes & Rivera, 2010). Certainly an increase in income both directly and indirectly as a result of tourism for local residents does impart some elements of QOL, but this approach has been criticized as far too narrow by many scholars (Sen, 2001). While previous research examining the impacts of tourism on host community residents had often been framed by economically minded theories such as Social Exchange Theory (Ward & Berno, 2011), Andereck and Nyaupane (2011) began the process of shifting the theoretical lens through such impacts were examined to more holistic framework. The holistic QOL framework posits that tourism influences many domains related to residents' QOL, and some have even argued that the primary goal of tourism development is to increase resident QOL (Chancellor et al., 2011). Andereck and Nyaupane (2011) created the tourism and quality of life scale (TQOL) as a way of integrating traditional tourism impacts measures with the implicit assumption that tourism impacts QOL. Since early work on the intersection of tourism and QOL, a quickly growing body of research has greatly expanded the QOL framework, and it has become one of the most predominant theoretical underpinnings for tourism impacts studies today (Biagi et al., 2020; Eusébio & Carneiro, 2014; Kaplanidou et al., 2013; Kim et al., 2013; Su et al., 2019; Uysal et al., 2016). Ultimately, this body of research seeks to understand how and why tourism influences the lives of those who live in tourism host communities.

There is some disagreement among scholars about what makes life 'good.' Although most agree that QOL is a subjective phenomenon, it has been conceptualized and measured in a variety of ways. Within the tourism context, researchers have long examined the influence of tourism on residents' lives. While QOL is about more than simply impacts, many tourism impacts included in past literature are now considered to be a part of QOL (e.g., cost of living, good jobs, recreation opportunities, clean air and water, etc.). As the tourism QOL body of literature has evolved, researchers have approached the measurement of tourism's impacts on elements of QOL both objectively (e.g., number of tourists visiting the community) and subjectively (e.g., perceived crowding from those tourists) (Uysal et al., 2016). In this study, we focus on the measurement of subjective QOL, as often individuals' perceptions of a phenomenon play an even greater role in their lives than reality (i.e., if someone believes there are too many tourists in their community, the actual number of tourists who visit are largely irrelevant). Furthermore, it is important to note that although the two subjects began as unique bodies of research, recent conceptual work has determined that the concept of QOL is virtually synonymous with subjective wellbeing (Camfield & Skevington, 2008; Skevington & Böhnke, 2018). For the sake of continuity in this body of literature, we use the term QOL in this study.

In their conceptualization of QOL in the tourism context, Andereck and Nyaupane (2011) based their creation of the TQOL scale on the framework provided by Schalock (1996). Their conceptualization includes eight overarching categories of TQOL: Community well-being, urban issues, way of life, community pride and awareness, natural/cultural preservation, economic strength, recreation amenities, and crime and substance abuse. A summary and description of each element of the TQOL scale is included in Table 1. These elements of QOL overlap significantly with other research that has measured QOL, and has been effectively utilized in previous research examining the relationships between perceived quality of life impacts, tourism related stress, and emotions (Jordan, Lesar, et al., 2019; Jordan, Spencer, et al., 2019).

Table 1. Factors and items of the tourism quality of life scale.

TQOL Factor	Items
Community well-being	Peace and quiet
	Feeling safe
	Clean air and water
	City services like police and fire protection
	A stable political environment
	Good public transportation
	The beauty of my community
	Quality of roads, bridges, and utility services
Urban issues	Crowding and congestion
	Traffic
	Urban sprawl and population growth
	Litter
	Zoning/land use violations
Way of life	My personal life quality
	The preservation of my way of life
	A feeling of belonging in my community
	Resident participation in local government
	Respect for my way of life
Community pride and awareness	The image of my community to others
	An understanding of different cultures
	Awareness of natural and cultural heritage
	Community pride
	Opportunities to participate in local culture
Natural/cultural preservation	Preservation of wildlife habitats
	Preservation of natural areas
	Preservation of cultural/historical sites
Economic strength	Strong and diverse economy
	Stores and restaurants owned by local residents
	The value of my house and/or land
	Enough good jobs for residents
	Plenty of retail shops and restaurants
	Fair prices for goods and services
Recreation amenities	Plenty of festivals, fairs, museums
	Live sports to watch in my community
	Quality recreation opportunities
Crime and substance abuse	Crime and vandalism
	Drug and alcohol abuse

2.2. Tourism impacts

The measurement of tourism impacts has received a great deal of attention over the past 50 years. Many studies examining tourism impacts have focused on tourism in general, rather than any one specific type of tourism. The breadth and depth of literature assessing residents' perceived impacts means that most TQOL factors have received at least some attention (Sharpley, 2014). In the next two paragraphs, we discuss each TQOL impact factor in the context of tourism in general and present past tourism literature as evidence of each impact area.

Residents have indicated that their *community well-being* has been both positively and negatively impacted in a variety of ways across destinations. For instance, Stylidis et al. (2014) found specifically that residents of Kavala, Greece perceived that the availability of community services was positively impacted by tourism, while peace and quiet was negatively affected by tourism. *Urban issues* are one of the most cited impacts in tourism communities, usually for the worse. For example, Lundberg (2017) found that residents of the island community of Marstrand, Sweden perceived a great deal of crowding and congestion from tourists who visited during the summer season. Residents' perceived impacts on *way of life* have seen increased attention recently with the shift toward focusing on QOL. Lin et al. (2017) found that residents' satisfaction with life in various cities across China was positively influenced by economic and socio-cultural benefits from tourism. *Community pride and awareness* is also dually affected by tourism, with some studies finding decreases and others increases. Sinclair-Maragh et al. (2015) found that some residents of communities in Turkey perceived that tourism negatively impacted cultural identity, while Choi and Murray (2010) found that residents of the U.S. state of Texas perceived that tourism negatively influenced the character of their community. *Natural and cultural preservation* follows the trend of bi-directional impacts. Amuquandoh (2010) found that residents in living in the Lake Bsomotwe Basin in Ghana perceived that tourism positively influenced cultural and historical sites while it simultaneously accelerated the loss of forest lands. Similarly, Stylidis et al. (2014) found that residents of Kavala, Greece perceived that tourism positively impacted them sociocultural and negatively impacted them environmentally.

Economic strength is an impact area where many residents agree that tourism is a boon rather than a curse. However, while residents of Salford, England thought tourism was a positive influence on jobs and income, they also worried about the rising cost of living due to tourism development (Schofield, 2011). Tourism community residents also have mixed perceptions of tourism's influence of *recreation amenities*. While Brida et al. (2010) found that residents of a northern Italian community perceived that tourism brought greater access to recreational facilities, Vareiro et al. (2013) found that residents of Guimarães, Portugal thought tourism decreased their access to leisure sites. Finally, research has borne out that residents generally agree that tourism has a negative influence on *crime and substance abuse*. For instance, Postma and Schmuecker (2017) recently found that residents of Hamburg, Germany perceived that tourism negatively influenced the consumption of alcohol and illegal drugs. While residents perceived QOL impacts from tourism in general are often bi-directional, perceived impacts related to cruise tourism and Airbnb tourism tend to be skewed more toward the negative, as we will discuss in the following sections of the literature review.

2.3. Cruise tourism impacts

While cruising has been a popular form of tourism for a number of years, the recent increase in the number of ports, ships, and passengers traversing the world's oceans have led to a recent boom in research examining how residents are affected by cruise ships in their ports and passengers in their towns. Similar to tourism in general, host community residents often perceive both positive and negative impacts to their quality of life. In the next two paragraphs, we will discuss each TQOL factor in the context of cruise tourism, and present past tourism literature as evidence of each impact area.

Residents' perceptions of the *community well-being* impacts related to cruise tourism are both positive and negative. Del Chiappa and Abbate (2016) found that cruise tourism development on the island of Sicily, Italy resulted in greater public and private investment in infrastructure as well as enhanced the quality of roads and other public services, while MacNeill and Wozniak (2018) found that residents perceived cruise tourism resulted in increased corruption in the community of Trujillo, Honduras. As with tourism in general, *urban issues* are almost always perceived in a negative light in relation to cruise tourism. Jordan and Vogt (2017) found that residents of the island community of Falmouth, Jamaica perceived that cruise tourism increased crowding and traffic in their community. Residents' perceptions of *way of life* benefits can be both positive and negative, and several studies have found that residents perceptions of quality of life (one element of the way of life construct) improved due to cruise tourism (Del Chiappa & Abbate, 2016; MacNeill & Wozniak, 2018). Despite some studies showing a positive impact on way of life, Gibson and Bentley (2007) found that residents of Falmouth and Cornwall, England perceived that cruise tourism interfered with their local communities' way of life in a negative way. *Cultural pride and awareness* skews toward the negative, as cruise tourism has often been viewed in the literature as exploitative of local culture. For instance, Gutberlet (2016) found that cruise tourism resulted in a loss of cultural identity for residents of Muscat, Oman. Furthermore, *natural and cultural preservation* has received a great deal of attention in relation to cruise tourism, and the majority of studies have determined that it is detrimental, particularly to the natural environment. A growing number of studies have found that residents perceive a great variety of negative environmental impacts, such as reduction in water quality and biodiversity, erosion, and destruction of reef areas for the construction of ports related the cruise tourism in their communities (Carić & Mackelworth, 2014; Johnson, 2002; Moscovici, 2017).

While *economic strength* was an area that mostly positive impacts were perceived in relation to tourism in general, the cruise tourism literature is more of a mixed bag. Del Chiappa and Abbate (2016) found that residents of the island of Sicily, Italy perceived that cruise tourism positively influenced the income of local people, but that there was a corresponding cost of living increase and the real wealth generated by the cruise tourism industry was leaked out of the local economy to other places. Residents have infrequently identified *recreation amenity* impacts in relation to cruise tourism development. However, Jordan and Vogt (2017) found while residents thought they would be allowed access to the new cruise port in the island community of Falmouth, Jamaica, they faced the opposite reality once port construction was finished. Similar to tourism in general, studies have generally found that residents perceive that cruise tourism has a negative impact on *crime and substance abuse* in their communities (MacNeill & Wozniak, 2018).

2.4. Airbnb tourism impacts

While the body of literature on cruise tourism impacts remains small, it has steadily grown over the last several decades. In contrast, the body of literature on resident perceptions of Airbnb tourism impacts was non-existent until the mid-2010s. This is simply because Airbnb did not exist as a company (although traditional bed and breakfasts were one of the first forms of tourism) until 2008 and did not explode in popularity until a few years later. In the next two paragraphs, we will discuss each TQOL factor in the context of Airbnb tourism, and present past tourism literature as evidence of each impact area.

Community well-being is an area where research has often focused in relation to residents' perceived impacts of Airbnb tourism, particularly peace and quiet and safety. In most cases, residents have perceived negative impacts. Stergiou and Farmaki (2019) found that residents of Athens, Greece thought Airbnb tourism brought strangers into their community and disrupted the peace and quiet of their neighborhoods. As with tourism in general and cruise tourism, *urban issues* impacts are viewed almost exclusively negatively in relation to Airbnb tourism. Yeager et al. (2020) found that residents of the city of Savannah, Georgia in the Unite States perceived that Airbnb tourism increased traffic and crowding in their communities. Further, residents of host communities most often perceive negative *way of life* impacts in relation to Airbnb tourism. For instance, Richards et al. (2019) found that residents of Barcelona, Spain perceived that Airbnb tourism resulted in gentrification and loss of community identity. Departing from the trend of residents perceiving mostly negative QOL impacts, residents often perceive both positive end negative *community pride and awareness* impacts in relation to Airbnb tourism. While Petruzzi, et al. (2020) found that residents of Lisbon, Portugal thought Airbnb tourism increased cultural exchange with tourists, it also resulted in loss of tradition. *Natural and cultural preservation* is an area where most residents believe Airbnb tourism results in positive impacts, contrary to the mixed perceptions about tourism in general and negative perceptions about cruise tourism. For instance, Mody (2018) found that residents thought Airbnb tourism provided incentives for the preservation and restauration of local historical buildings.

Residents' perceptions of *economic strength* are a mix of both positive and negatives. Jordan and Moore (2018) found that residents of the island of Oahu, Hawaii in the United States thought that Airbnb tourism resulted in the creation of jobs for some, but that ultimately it resulted in an increased cost of living and rising property values for the local community.

Contrary to the bodies of literature examining residents' perceived impacts related to tourism in general and cruise tourism, there is very little evidence available about residents' perceptions of *recreation amenities* in relation to Airbnb tourism. Similarly, there is little evidence of residents' perceived impacts of Airbnb tourism on *crime and substance abuse*; however, the little evidence that does exist suggests that resident perceive negative impacts in this area (Mody, 2018; Yeager et al., 2020).

2.5. Hypotheses

Based on the previously reviewed literature, it is likely that residents perceive that different types of tourism play varying roles in the tourism related QOL impacts in their communities. Therefore, we propose the following set of research hypotheses based on those perceptions:

Hypothesis 1a-h: Residents will perceive each **community well-being** impact significantly differently in relation to tourism in general, cruise tourism, and Airbnb tourism

Hypothesis 2a-e: Residents will perceive each **urban issues** impact significantly differently in relation to tourism in general, cruise tourism, and Airbnb tourism

Hypothesis 3a-e: Residents will perceive each **way of life** impact significantly differently in relation to tourism in general, cruise tourism, and Airbnb tourism

Hypothesis 4a-e: Residents will perceive each **community pride and awareness** impact significantly differently in relation to tourism in general, cruise tourism, and Airbnb tourism

Hypothesis 5a-c: Residents will perceive each **natural/cultural preservation** impact significantly differently in relation to tourism in general, cruise tourism, and Airbnb tourism

Hypothesis 6a-f: Residents will perceive each **economic strength** impact significantly differently in relation to tourism in general, cruise tourism, and Airbnb tourism

Hypothesis 7a-c: Residents will perceive each **recreation amenities** impact significantly differently in relation to tourism in general, cruise tourism, and Airbnb tourism

Hypothesis 8a-b: Residents will perceive each **crime and substance abuse** impact significantly differently in relation to tourism in general, cruise tourism, and Airbnb tourism

3. Methods

3.1. Study site

Data for this study were collected on the Portuguese island of São Miguel, in the Azores archipelago. The island of São Miguel is an appropriate site for a study examining residents' differentiation between several types of tourism impacts, as it has recently experienced a rapid growth in both cruise visitors and visitors staying in Airbnb type vacation rentals. These two specific forms of tourism are highly visible in the Azores, as cruise passengers recently reach an all-time high of 170,000 passengers (compared to a population of 140,000 in São Miguel) and Airbnb stays tripled between 2015 and 2017 (Invest in Azores, 2019). These two changes followed a process of airline liberalization in two corridors between the Azores and the Portuguese mainland (allowing the entrance of low-cost airlines) in 2015 as well as the construction and marketing of a new cruise port in the city of Ponta Delgada in 2008. São Miguel serves mainly as a port-of-call rather than a home port for cruise ships, and is often used as a stop on transatlantic repositioning cruises in the spring and fall as ships are transferred between the Caribbean and Mediterranean cruise markets. The result was not only a significant expansion of the number of visitors, but likely also a new type of visitor skewed towards Airbnb type accommodation . While these are not the *only* types of tourism that occur in São Miguel, it stands to reason that the recency and rapidity of the increase in both likely allows residents to clearly delineate the changes they have experienced during the rise of both forms of tourism.

3.2. Data collection

Cross-sectional data were collected in a systematic random sample of residents of São Miguel between June and September 2019. All data collection activities followed ethics guidelines established by the local university's code of ethics regarding human subjects research. Data were collected via a pencil and paper survey, which were filled out by participants in the presence of trained researchers. A total of 14 researchers participated in data collection activities. The 13-page survey instrument contained approximately 150 questions about tourism impacts and other tourism related variables. The survey instrument was originally written in English and translated to Portuguese by a bi-lingual native Azorean resident with tourism subject knowledge. The instrument was pilot tested on a small sample of Azorean residents to ensure understanding of the translated questions, and minor changes were made based on feedback from those who completed the survey. Trained researchers approached individuals residing in every fifth house across the island of São Miguel. A total of 1,875 households were approached to participate in the survey. A total of 225 individuals started the survey but did not complete it. A total of 995 individuals completed the survey fully, resulting in an overall responses rate of 53.1%. An incentive was not offered for participation in the survey.

3.3. Measurement and analysis

Data reported in this paper were measured using the 36 item TQOL scale developed by Andereck and Nyaupane (2011). Participants were asked to indicate how they thought each TQOL item was affected by tourism, cruise tourism, and Airbnb tourism in their community from 0 (very negatively) to 4 (very positively. The TQOL scale was present in the survey instrument three times – first in reference to tourism, second in reference to cruise tourism, and third in reference to Airbnb tourism. The questions were asked first in reference to tourism in general in an effort to remove any priming effect asking questions about specific type so of tourism might have on survey respondents. While there are several other scales that have been recently developed for the measurement of tourism quality of life impacts, this scale was chosen to maintain consistency in the measurement of resident's perceived impacts across island destinations (Jordan, Spencer, et al., 2019). Analysis of the data collected via the TQOL instrument were analyzed in a two-step process. First, confirmatory factor analysis (CFA) was conducted to assess the reliability and validity of the scale in measuring the eight factors of tourism related quality of life impacts in reference to tourism, cruise tourism, and Airbnb tourism. Then, repeated measures analysis of variance (ANOVA) was conducted to compare resident's perceptions of each scale item across each type of tourism.

Study CFAs were estimated using the lavaan package in the R Studio software environment. A battery of absolute and relative fit indices were utilized to determine 'good' model fit. Values for each fit index used to determine 'good' fit are: include comparative fit index (CFI) < 0.9, Tucker Lewis index (TLI) > 0.9, root mean square error of approximation (RMSEA) < 0.08, and standardized root mean square residual (SRMR) < 0.08 (Bentler, 1990; Hu & Bentler, 1999). Utilizing this combination of fit indices reduces the likelihood of Type I and Type II error (Schmidt et al., 2003). Construct validity was assessed by examining average variance extracted,

Table 2. Sample characteristics.

Demographic Variable	Value
Age	
18-25 years	20.4%
26-35 years	23.3%
36-45 years	22.3%
46-55 years	15.0%
56-65 years	10.2%
66 or older	8.7%
Gender	
Male	36.5%
Female	63.5%
Highest education level	
Primary school	27.7%
Secondary school	38.9%
Higher education	33.4%
Employment	
Self employed	9.9%
Employed	58.1%
Unemployed	6.6%
Student	9.2%
Other	16.2%
Yearly household income	
Less than €25,000	73.0%
€25,000 to €34,999	16.0%
€35,000 to €49,999	5.6%
€50,000 to €74,999	2.4%
€75,000 to €99,999	1.2%
€100,000 or greater	1.6%

composite reliability. Repeated measures ANOVAS were estimated in the IBM SPSS software package.

4. Results

4.1. Sample characteristics

The age of survey respondents was evenly distributed across age groups and closely matches the age distribution of São Miguel Island according to the latest census data from 2011 (Table 2). A much greater portion of survey respondents were female (64.5%) than male (36.5%), indicating an overrepresentation of women compared to the São Miguel population (50.7% female). A fairly even distribution of respondents had completed primary school, secondary school, and higher education as their highest level of education attained. Most respondents were employed (58.1%), and the majority of participants had a household income under €25,000. Data about education and income were not available from the Azorean census.

4.2. CFA results

Individual CFAs were estimated for the TQOL scale for tourism in general, cruise tourism, and Airbnb tourism (Table 3). Initial model fit for all three models was poor. Model fit for the CFA measuring tourism quality of life impacts from tourism in general were: CFI=.821, TLI=.802, RMSEA=.073, SRMR=.083. Model fit statistics for the CFA measuring tourism quality of life impacts from cruise tourism were CFI=.800, TLI=.779, RMSEA=.088, SRMR=.116. Model fit statistics for the CFA measuring tourism and quality of life impacts for Airbnb tourism were CFI=.813, TLI=.793, RMSEA=.083, SRMR=.085. The only fit statistic that met the threshold for 'good' model fit was the RMSEA value for tourism in general. All other fit statistics for all three models were outside established thresholds for 'good' fit.

Table 3. Summary of study CFAs.

TQOL Factor	Items	Tourism*	Cruise Tourism**	Airbnb Tourism***
Community well-being		AVE=.342 CR=.795	AVE=.343 CR=.797	AVE=.398 CR=.821
	Peace and quiet	.666	.714	.715
	Feeling safe	.674	.744	.530
	Clean air and water	.620	.561	.662
	City services like police and fire protection	.562	.694	.745
	A stable political environment	.521	.235	.710
	Good public transportation	.502	.530	.613
	The beauty of my community	.551	.679	.636
	Quality of roads, bridges, and utility services	.545	.646	.551
Urban issues		AVE=.371 CR=.736	AVE=.521 CR=.816	AVE=.501 CR=.821
	Crowding and congestion	.585	.896	.882
	Traffic	.681	.851	.881
	Urban sprawl and population growth	.444	.385	.469
	Litter	.704	.662	.649
	Zoning/land use violations	.575	.342	.473
Way of life		AVE=.477 CR=.796	AVE=.675 CR=.897	AVE=.529 CR=.836
	My personal life quality	.907	.939	.849
	The preservation of my way of life	.909	.952	.762
	A feeling of belonging in my community	.562	.570	.643
	Resident participation in local government	.315	.322	.393
	Respect for my way of life	.518	.843	.821
Community pride and awareness		AVE=.493 CR=.827	AVE=.369 CR=.739	AVE=.573 CR=.869
	The image of my community to others	.536	.679	.712
	An understanding of different cultures	.720	.809	.732
	Awareness of natural and cultural heritage	.775	.871	.817
	Community pride	.767	.414	.791
	Opportunities to participate in local culture	.693	.652	.721
Natural/cultural preservation		AVE=.869 CR=.951	AVE=.827 CR=.935	AVE=.881 CR=.956
	Preservation of wildlife habitats	.928	.909	.939
	Preservation of natural areas	.974	.962	.969
	Preservation of cultural/historical sites	.886	.851	.906
Economic strength		AVE=.363 CR=.761	AVE=.432 CR=.789	AVE=.315 CR=.726
	Strong and diverse economy	.701	.733	.676
	Stores and restaurants owned by local residents	.707	.768	.727
	The value of my house and/or land	.512	.421	.420
	Enough good jobs for residents	.633	.670	.623
	Plenty of retail shops and restaurants	.740	.886	.675
	Fair prices for goods and services	.244	.101	.289

(continued)

Table 3. Continued.

TQOL Factor	Items	Tourism*	Cruise Tourism**	Airbnb Tourism***
Recreation amenities		AVE=.596 CR=.815	AVE=.631 CR=.835	AVE=.669 CR=.859
	Plenty of festivals, fairs, museums	.730	.778	.768
	Live sports to watch in my community	.774	.765	.849
	Quality recreation opportunities	.817	.839	.846
Crime and substance abuse		AVE=.445 CR=.613	AVE=.835 CR=.910	AVE=.801 CR=.889
	Crime and vandalism	.903	.911	.877
	Drug and alcohol abuse	.516	.917	.913

*Tourism fit statistics: CFI=.821, TLI=.802, RMSEA=.073, SRMR=.083.
**Cruise tourism fit statistics: CFI=.800, TLI=.779, RMSEA=.088, SRMR=.116.
***Airbnb tourism fit statistics: CFI=.813, TLI=.793, RMSEA=.083, SRMR=.085.

Each CFA also revealed poor construct validity (Tables 4–6). In terms of convergent validity, many constructs within each CFA had factor loadings below the recommended 0.5 value, although most composite reliability values were above the recommended 0.7 threshold (Hair et al., 2010). With the notable exception of natural/cultural preservation and recreation amenities, many average variance extracted (AVE) values falling below the 0.5 threshold recommended by Hair et al. (2010). In terms of discriminant validity, several squared correlations between latent variables were greater than the AVE, revealing potential multicollinearity and poor evidence for discriminant validity (Grewal et al., 2004; Hair et al., 2010).

4.3. Repeated measures ANOVA results

A repeated measures ANOVA was conducted to examine whether there were significant differences in resident perceptions of tourism QOL impacts in relation to tourism in general, cruise tourism, and Airbnb tourism. Mauchly's test of sphericity indicated that the assumption of sphericity in all items was not violated, therefore no correction was used. In most cases, the repeated measures ANOVA showed significant differences in the majority of mean scores of TQOL items when answered in relation to tourism in general, cruise tourism, and Airbnb tourism. Results of the standard univariate F tests, including F statistics, significance level, and mean scores for each variable related to tourism in general, cruise tourism, and Airbnb tourism are reported in Table 7 below.

5. Discussion and conclusions

In this paper, we aimed to advance the understanding of the measurement of resident's perceived tourism impacts and provide a framework for future studies to differentiate between tourism impacts in tourism host communities. Our study CFAs revealed that the TQOL measurement tool performed poorly in the measurement of latent constructs; however, certain constructs were measured more reliability and validly in relation to some types of tourism rather than others. Furthermore, our repeated measures ANOVA revealed that residents *do* differentiate between individual tourism impact items across three different types of tourism – even when they are measured a) in a single community b) at the same point in time c) with the same measurement tool.

Table 4. Correlations and squared correlations between tourism model constructs.

Tourism in general	WB	UI	WL	CP	NCP	ES	RA	CSA
Community well-being (WB)	**0.342**	0.484	0.508	0.255	0.272	0.199	0.123	0.306
Urban issues (UI)	.696	**0.371**	0.387	0.032	0.184	0.024	0.009	0.125
Way of life (WL)	0.713	.622	**0.477**	0.200	0.150	0.134	0.097	0.132
Community pride and awareness (CP)	0.505	0.179	0.447	**0.493**	0.226	0.479	0.339	0.147
Natural/cultural preservation (NCP)	0.522	0.429	0.387	0.475	**0.869**	0.144	0.065	0.112
Economic strength (ES)	0.446	0.154	0.366	0.692	0.379	**0.363**	0.392	0.141
Recreation amenities (RA)	0.351	0.095	0.312	0.582	0.255	0.626	**0.596**	0.231
Crime and substance abuse (CSA)	0.553	0.353	0.363	0.383	0.334	0.376	0.481	**0.446**

*The diagonal represents the AVE for each construct.
**Numbers below the diagonal are correlations and numbers above the diagonal are squared correlations.

Table 5. Correlations and squared correlations between cruise tourism model constructs.

Cruise tourism	WB	UI	WL	CP	NCP	ES	RA	CSA
Community well-being (WB)	**0.343**	0.235	0.648	0.275	0.312	0.133	0.180	0.382
Urban issues (UI)	0.485	**0.521**	0.177	0.000	0.120	0.009	0.000	0.088
Way of life (WL)	0.805	0.421	**0.369**	0.230	0.309	0.128	0.125	0.338
Community pride and awareness (CP)	0.524	−0.019	0.480	**0.827**	0.160	0.466	0.254	0.120
Natural/cultural preservation (NCP)	0.559	0.347	0.556	0.400	**0.432**	0.099	0.100	0.210
Economic strength (ES)	0.365	−0.096	0.358	0.683	0.314	**0.631**	0.342	0.068
Recreation amenities (RA)	0.424	−0.001	0.354	0.504	0.317	0.585	**0.835**	0.141
Crime and substance abuse (CSA)	0.618	0.296	0.581	0.346	0.458	0.260	0.376	**0.487**

*The diagonal represents the AVE for each construct.
**Numbers below the diagonal are correlations and numbers above the diagonal are squared correlations.

Table 6. Correlations and squared correlations between Airbnb tourism model constructs.

Airbnb tourism	WB	UI	WL	CP	NCP	ES	RA	CSA
Community well-being (WB)	**0.398**	0.265	0.521	0.361	0.231	0.333	0.181	0.319
Urban issues (UI)	0.515	**0.501**	0.226	0.040	0.114	0.036	0.014	0.140
Way of life (WL)	0.722	0.475	**0.530**	0.248	0.183	0.213	0.098	0.198
Community pride and awareness (CP)	0.601	0.200	0.498	**0.573**	0.311	0.514	0.321	0.125
Natural/cultural preservation (NCP)	0.481	0.337	0.428	0.558	**0.881**	0.245	0.128	0.152
Economic strength (ES)	0.577	0.189	0.461	0.717	0.495	**0.315**	0.340	0.130
Recreation amenities (RA)	0.426	0.117	0.313	0.567	0.358	0.583	**0.670**	0.107
Crime and substance abuse (CSA)	0.565	0.374	0.445	0.354	0.390	0.360	0.327	**0.801**

*The diagonal represents the AVE for each construct.
**Numbers below the diagonal are correlations and numbers above the diagonal are squared correlations.

5.1. Theoretical implications

Both the CFA models and repeated measures ANOVA model reveals interesting findings related to the measurement of the social impacts of tourism. They are discussed individually in the paragraphs below.

5.1.1. Latent variable analysis of the TQOL measurement tool

There are several elements of the study CFA models that merit discussion here. First – an initial look at model fit indicates the TQOL scale performed poorly as a measure of latent impact variables across all forms of tourism. This finding differs somewhat from previous research using the scale as a latent variable measurement tool (Jordan, Spencer, et al., 2019). A closer examination of fit statistics reveals some interesting findings. The incremental fit indices (CFI and TLI) were well outside established norms for 'good' fitting models (Miles & Shevlin, 2007). However, incremental fit indices are notoriously sensitive to models with a large number of parameters, which the TQOL has (Kenny & McCoach, 2003). An examination of absolute fit indices of the model (RMSEA and SRMR) tells a slightly different story than incremental fit – for each model they are either slightly under the established value for 'good' fit, or only slightly over it (Chen et al.,

Table 7. Summary of repeated measures ANOVA.

		F	Sig.	Tourism (general) μ	Cruise tourism μ	Airbnb tourism μ
Community well-being	Peace and quiet	7.623	.001	1.86	1.98	1.92
	Feeling safe	*0.204*	*.815*	*2.09*	*2.10*	*2.08*
	Clean air and water	23.672	.000	1.80	1.70	1.88
	City services like police and fire protection	6.470	.002	2.15	2.09	2.09
	A stable political environment	*1.880*	*.153*	*2.06*	*2.10*	*2.04*
	Good public transportation	9.377	.000	2.21	2.12	2.13
	The beauty of my community	23.029	.000	2.49	2.27	2.35
	Quality of roads, bridges, and utility services	58.451	.000	2.48	2.18	2.38
Urban issues	Crowding and congestion	22.490	.000	1.32	1.57	1.53
	Traffic	77.626	.000	1.20	1.58	1.47
	Urban sprawl and population growth	6.626	.001	1.91	2.00	1.93
	Litter	18.737	.000	1.43	1.59	1.51
	Zoning/land use violations	52.297	.000	1.69	1.91	1.76
Way of life	*My personal life quality*	*0.667*	*.513*	*2.01*	*2.00*	*1.99*
	The preservation of my way of life	*0.314*	*.731*	*2.02*	*2.01*	*2.03*
	A feeling of belonging in my community	*0.629*	*.533*	*2.12*	*2.09*	*2.12*
	Resident participation in local government	15.696	.000	2.05	2.00	2.10
	Respect for my way of life	*1.566*	*.209*	*2.13*	*2.09*	*2.08*
Community pride and awareness	The image of my community to others	14.526	.000	2.45	2.33	2.45
	An understanding of different cultures	75.055	.000	2.79	2.50	2.67
	Awareness of natural and cultural heritage	131.848	.000	2.78	2.41	2.50
	Community pride	36.899	.000	2.80	2.49	2.54
	Opportunities to participate in local culture	66.817	.000	2.54	2.27	2.37
Natural/cultural preservation	Wildlife habitats	23.188	.000	2.17	2.00	2.08
	Natural areas	15.573	.000	2.18	2.00	2.08
	Cultural/historical sites	28.448	.000	2.28	2.08	2.16
Economic strength	Strong and diverse economy	144.667	.000	2.88	2.40	2.65
	Stores and restaurants owned by local residents	75.754	.000	2.80	2.44	2.57
	The value of my house and/or land	94.649	.000	2.39	2.04	2.39
	Enough good jobs for residents	30.329	.000	2.35	2.19	2.16
	Plenty of retail shops and restaurants	44.273	.000	2.74	2.51	2.56
	Fair prices for goods and services	52.028	.000	1.32	1.59	1.36
Recreation amenities	Plenty of festivals, fairs, museums	117.580	.000	2.54	2.25	2.30
	Live sports to watch in my community	61.018	.000	2.30	2.09	2.17
	Quality recreation opportunities	77.760	.000	2.38	2.15	2.22
Crime and substance abuse	Crime and vandalism	5.067	.006	2.09	2.04	2.04
	Drug and alcohol abuse	*0.706*	*.494*	*2.00*	*1.98*	*2.01*

*Measured on a 5 point Likert scale, 0 = very negatively, 4 = very positively. *Italicized rows were not significant.* Bold values are the lowest among the three types of tourism.

2008). Therefore, it is quite possible that the reason the TQOL measurement model suffers from poor fit is due to many indicator variables per construct. This is perhaps unsurprising, as the TQOL measure was not originally developed using latent variable analysis (Andereck & Nyaupane, 2011). It is certainly possible that a more parsimonious model with fewer indicator variables could improve model fit of the TQOL measurement tool when measuring tourism impacts.

While overall model fit was poor, there are some elements of the validity analysis that provide further information about *why* the model fit poorly and some potential fixes that could be used in the future. First – while all loadings were significant, a variety of individual items across all three types of tourism did not load onto the latent variable > .5, informing us that they are a poor indicator of the construct. Second, the AVE of many latent constructs were below .5, indicating that the randomness explains more of the latent variable than the indicators do. Third and finally, the correlations between latent variables were unusually high, indicating that multicollinearity may be a problem with this measurement scale (Grewal et al., 2004). These findings provide further evidence that a 'trimmed' TQOL model may provide a better measurement tool for exploring latent tourism impact variables.

5.1.2. Resident differentiation between tourism impacts

The repeated measures ANOVA reveals that residents of a tourism host community *do* differentiate between the impacts of tourism in general, cruise tourism, and Airbnb tourism. While previous studies have found differing levels of support for various types of tourism (Nunkoo & Gursoy, 2017), this is the first time that resident perceptions of the impacts of multiple types of tourism have been measured in the same community at a single point in time. The model reveals support for 29 of the 37 hypotheses – the only tourism quality of life impacts that residents did *not* find to differ across the three types of tourism were feeling safe, a stable political environment, my personal life quality, the preservation of my way of life, a feeling of belonging in my community, respect for my way of life, and drug and alcohol abuse. A total of 7 items were perceived to have negative impacts (a mean of less than 2) across all three types of tourism, meaning tourism was generally viewed in a positive light by study participants. This finding is similar to previous research utilizing Andereck and Nyaupane's (2011) TQOL scale (Andereck & Nyaupane, 2011; Jordan, Spencer, et al., 2019). It is important to note, however, that several of the items that were perceived most negatively were worded in a less positive way than other items in the scale (e.g., traffic vs. strong and diverse economy). We will discuss three specific findings from the repeated measures ANOVA that are of particular interest: (1) The way of life impact category was mostly not significantly different across three types of tourism; (2) The urban issues impacts category was perceived most negatively, and was attributed to tourism in general rather than cruise tourism or Airbnb tourism; and (3) Cruise tourism was generally perceived as having the most negative impacts across all other impact categories.

All but one item (Resident participation in local government) from the way of life impacts category was not perceived as significantly more or less impactful across different types of tourism by study participants. This finding is interesting because the way of life impacts category seems to get at the somewhat intangible overall essence of QOL, rather than more specific or visible things within their community (Fernando & Cooley, 2016). In fact, one of the questions specifically asks about personal life quality. While residents were clearly able to differentiate between other tourism and quality of life impacts – they generally thought that their overall way of life was equally affected by all of them in a way that was neither positive nor negative (the means for each item were all very close to 2, the midpoint of the Likert scale and anchored to the 'neutral' answer'). It would seem as though at this point; residents do not think their 'overall' quality of life has been all that impacted by any type of tourism, contrary to previous findings in the tourism literature (Andereck & Nyaupane, 2011; Kaplanidou et al., 2013; Kim et al., 2013; Uysal et al., 2016).

The urban issues impact category, in contrast, was perceived as categorically negatively impacted by all three types of tourism – a finding that runs parallel to much of the previous literature on tourism impacts (Andereck et al., 2005; Ap & Crompton, 1998; Jordan & Vogt, 2017; Perdue et al., 1987). Where this finding adds to the body of knowledge is that residents attributed urban issues most negatively to tourism in general, rather than specific types of tourism like cruise tourism and Airbnb tourism. It could be that it is difficult to attribute traffic in the roads and crowding on the sidewalks to a specific type of tourist, leading to residents to simply attribute them to 'tourism'. However, the face validity of this notion is questionable in an island as small as São Miguel where most residents are keenly aware of when there are cruise ships in town and which houses are being rented as Airbnb properties to tourists.

Finally, with the notable exception of peace and quiet, enough good jobs for residents, and fair prices for goods and services, all other tourism impacts were perceived to be least positive in association with cruise tourism. This finding supports previous literature, which has found cruise tourism to be one of the worst types of tourism when it comes to the benefit of residents of cruise host communities (Del Chiappa & Abbate, 2016; Gutberlet, 2016; Jordan & Vogt, 2017). This is not to say that cruise tourism was perceived overall in a negative light by residents – actually to the contrary. The means of most perceived tourism impacts skewed toward the positive, as most means were above the midpoint of 2 in the 5-point Likert scale. However, the fact remains that when asked about the impacts of tourism, cruise tourism stood out to community residents as the least beneficial to their QOL.

5.2. Practical implications

That residents differentiate between the impacts of different forms of tourism in their community indicates that communities should change the way they gather information about resident perceptions of tourism impacts. It is not enough to simply ask about perceptions of tourism any longer. Questions should be asked about residents' perceptions of ALL forms of tourism in their community to determine if one or more forms of tourism is more or less desirable for those living there. Based on that information, community leaders, tourism developers, and destination management organizations can effectively plan for future tourism with a greater level of confidence that residents' benefits from tourism are maximized and negative impacts are minimized. A comprehensive understanding of residents' perceived tourism impacts allows for a much more targeted approach to the management of tourism in host communities.

In this instance, cruise tourism was found to be perceived far more negatively than tourism in general or Airbnb tourism. For the island of São Miguel, this is a sign that perhaps the future of tourism on the island should head in a different direction. The increased visitation of tourists to the Azores indicates there is demand for the Azorean tourism product, and so it is important to make decisions about tourism development that will reduce negative impacts on those living there. However, it is important to realize that there are certain things about tourism in general and Airbnb tourism that residents also view very negatively. For instance, while residents thought that Cruise tourism was neither negative nor positive for the prices of goods and services in their community, they thought that tourism in general and Airbnb tourism were very detrimental. Community leaders in São Miguel and other tourism destinations can use such information to specifically address issues that are affecting individuals living within their communities.

While it would be naive to think that resident perceptions are the only factor considered when developing or promoting various types of tourism in any destination, it is also folly to think that they do not matter to the travel product of any given destination. Destinations seeking to strike a balance between the two can utilize resident perceptions of multiple types of tourism to become more sustainable and gain a competitive advantage in the tourism marketplace. Destination marketers and managers who truly understand how residents feel about various types of tourism in their community can work toward matching the desires of developers seeking to build cruise ports, construct new hotels, expand Airbnb offerings, etc. with the desires of those living within tourism destinations to promote a symbiosis between the two.*5.3 Limitations and recommendations for future research*

São Miguel island in the Portuguese island chain of the Azores is but one tourism community that hosts multiple types of tourism. Furthermore, women were overrepresented in our sample as compared to the general São Miguel population. Future research should explore resident perceptions of various types of tourism in various types of communities. However, researchers will likely run into the issue of survey fatigue if a community hosts, say, 10 different types of tourism. Therefore, efforts should be made to either slim down the TQOL scale to a more parsimonious number of indicators or develop a much smaller scale of tourism impacts with the purpose of measuring the impact of multiple types of tourism in multiple communities. Further research should also be undertaken to understand *why* tourists perceive certain impacts more negatively in relation to certain types of tourism. Qualitative inquiry utilizing focus groups or semi-structured, in-depth interviews would likely be more appropriate at accessing details about the reasoning behind perceived tourism impacts than the scales we used in this study.

Disclosure statement

No potential conflict of interest was reported by the authors.

ORCID

Evan J. Jordan http://orcid.org/0000-0002-6924-2628
Ting-Yen (Tim) Huang http://orcid.org/0000-0002-3497-5637

References

Amuquandoh, F. E. (2010). Residents' perceptions of the environmental impacts of tourism in the Lake Bosomtwe Basin, Ghana. *Journal of Sustainable Tourism*, *18*(2), 223–238. https://doi.org/10.1080/09669580903298531

Andereck, K. L., & Nyaupane, G. P. (2011). Exploring the nature of tourism and quality of life perceptions among residents. *Journal of Travel Research*, *50*(3), 248–260. https://doi.org/10.1177/0047287510362918

Andereck, K. L., Valentine, K. M., Knopf, R. C., & Vogt, C. A. (2005). Residents' perceptions of community tourism impacts. *Annals of Tourism Research*, *32*(4), 1056–1076. https://doi.org/10.1016/j.annals.2005.03.001

Ap, J., & Crompton, J. L. (1998). Developing and testing a tourism impact scale. *Journal of Travel Research*, *37*(2), 120–130. https://doi.org/10.1177/004728759803700203

Bentler, P. M. (1990). Comparative fit indexes in structural models. *Psychological Bulletin*, *107*(2), 238–246. https://doi.org/10.1037/0033-2909.107.2.238

Biagi, B., Ladu, M. G., Meleddu, M., & Royuela, V. (2020). Tourism and the city: The impact on residents' quality of life. *International Journal of Tourism Research*, *22*(2), 168–181. https://doi.org/10.1002/jtr.2326

Brida, J. G., Osti, L., & Barquet, A. (2010). Segmenting resident perceptions towards tourism—A cluster analysis with a multinomial logit model of a mountain community. *International Journal of Tourism Research*, *12*(5), n/a–602. https://doi.org/10.1002/jtr.778

Camfield, L., & Skevington, S. M. (2008). On Subjective Well-being and Quality of Life. *Journal of Health Psychology, 13*(6), 764–775. https://doi.org/10.1177/1359105308093860

Carić, H., & Mackelworth, P. (2014). Cruise tourism environmental impacts – The perspective from the Adriatic Sea. *Ocean & Coastal Management, 102*(Part A), 350–363. https://doi.org/10.1016/j.ocecoaman.2014.09.008

Chancellor, C., Yu, C.-P S., & Cole, S. T. (2011). Exploring quality of life perceptions in rural midwestern (USA) communities: An application of the core–periphery concept in a tourism development context. *International Journal of Tourism Research, 13*(5), 496–507. https://doi.org/10.1002/jtr.823

Chase, G. L., & McKee, D. L. (2003). The economic impact of cruise tourism on Jamaica. *Journal of Tourism Studies, 14*(2), 16–22. CABDirect2.

Chen, F., Curran, P. J., Bollen, K. A., Kirby, J., & Paxton, P. (2008). An Empirical Evaluation of the Use of Fixed Cutoff Points in RMSEA Test Statistic in Structural Equation Models. *Sociological Methods & Research, 36*(4), 462–494. https://doi.org/10.1177/0049124108314720

Cheung, K. S., & Li, L.-H. (2019). Understanding visitor–resident relations in overtourism: Developing resilience for sustainable tourism. *Journal of Sustainable Tourism, 27*(8), 1197–1216. https://doi.org/10.1080/09669582.2019.1606815

Choi, H. C., & Murray, I. (2010). Resident attitudes toward sustainable community tourism. *Journal of Sustainable Tourism, 18*(4), 575–594. https://doi.org/10.1080/09669580903524852

CLIA (2019). Cruise Lines International Association (CLIA) Reveals Growth in Global and North American Passenger Numbers and Insights. Cruise Line Industry Association. https://cruising.org:443/en/news-and-research/press-room/2019/april/clia-reveals-growth

Croes, R., & Rivera, M. A. (2010). Testing the empirical link between tourism and competitiveness: Evidence from Puerto Rico. *Tourism Economics, 16*(1), 217–234. https://doi.org/10.5367/000000010790872114

Del Chiappa, G., & Abbate, T. (2016). Island cruise tourism development: A resident's perspective in the context of Italy. *Current Issues in Tourism, 19*(13), 1372–1385. https://doi.org/10.1080/13683500.2013.854751

Eusébio, C., & Carneiro, M. J. (2014). The impact of tourism on quality of life: A segmentation analysis of the youth market. *Tourism Analysis, 19*(6), 741–757. https://doi.org/10.3727/108354214X14146846679529

Fernando, F. N., & Cooley, D. R. (2016). An oil boom's effect on quality of life (QoL): Lessons from Western North Dakota. *Applied Research in Quality of Life, 11*(4), 1083–1115. https://doi.org/10.1007/s11482-015-9422-y

Gibson, P., & Bentley, M. (2007). A Study of Impacts—Cruise Tourism and the South West of England. *Journal of Travel & Tourism Marketing, 20*(3-4), 63–77. https://doi.org/10.1300/J073v20n03_05

Grewal, R., Cote, J. A., & Baumgartner, H. (2004). Multicollinearity and Measurement Error in Structural Equation Models: Implications for Theory Testing. *Marketing Science, 23*(4), 519–529. https://doi.org/10.1287/mksc.1040.0070

Gursoy, D., Ouyang, Z., Nunkoo, R., & Wei, W. (2019). Residents' impact perceptions of and attitudes towards tourism development: A meta-analysis. *Journal of Hospitality Marketing & Management, 28*(3), 306–333. https://doi.org/10.1080/19368623.2018.1516589

Gutberlet, M. (2016). Socio-cultural impacts of large-scale cruise tourism in Souq Mutrah, Sultanate of Oman. *Fennia – International Journal of Geography, 194*(1), 46–63.

Hair, J. F., Black, W. C., Babin, B. J., & Anderson, R. E. (2010). *Multivariate data analysis: A global perspective*. Prentice Hall.

Hu, L., & Bentler, P. M. (1999). Cutoff criteria for fit indexes in covariance structure analysis: Conventional criteria versus new alternatives. *Structural Equation Modeling: A Multidisciplinary Journal, 6*(1), 1–55. https://doi.org/10.1080/10705519909540118

Invest in Azores (2019). Azores Tourism Sector. https://www.investinazores.com/index.php/en/why-azores/tourism-sector

Johnson, D. (2002). Environmentally sustainable cruise tourism: A reality check. *Marine Policy, 26*(4), 261–270. https://doi.org/10.1016/S0308-597X(02)00008-8

Jordan, E. J., Lesar, L., & Spencer, D. M. (2019). Clarifying the interrelations of residents' perceived tourism-related stress, stressors, and impacts. *Journal of Travel Research*, https://doi.org/10.1177/0047287519888287

Jordan, E. J., & Moore, J. (2018). An in-depth exploration of residents' perceived impacts of transient vacation rentals. *Journal of Travel & Tourism Marketing, 35*(1), 90–101. https://doi.org/10.1080/10548408.2017.1315844

Jordan, E. J., Spencer, D. M., & Prayag, G. (2019). Tourism impacts, emotions and stress. *Annals of Tourism Research, 75*, 213–226. https://doi.org/10.1016/j.annals.2019.01.011

Jordan, E. J., & Vogt, C. A. (2017). Residents' perceptions of stress related to cruise tourism development. *Tourism Planning & Development, 14*(4), 527–547. https://doi.org/10.1080/21568316.2017.1287123

Kaplanidou, K. (K.)., Karadakis, K., Gibson, H., Thapa, B., Walker, M., Geldenhuys, S., & Coetzee, W. (2013). Quality of Life, Event Impacts, and Mega-Event Support among South African Residents before and after the 2010 FIFA World Cup. *Journal of Travel Research, 52*(5), 631–645. https://doi.org/10.1177/0047287513478501

Kenny, D. A., & McCoach, D. B. (2003). Effect of the Number of Variables on Measures of Fit in Structural Equation Modeling. *Structural Equation Modeling: A Multidisciplinary Journal, 10*(3), 333–351. https://doi.org/10.1207/S15328007SEM1003_1

Kim, K., Uysal, M., & Sirgy, M. J. (2013). How does tourism in a community impact the quality of life of community residents? *Tourism Management*, *36*, 527–540. https://doi.org/10.1016/j.tourman.2012.09.005

Lin, Z., Chen, Y., & Filieri, R. (2017). Resident-tourist value co-creation: The role of residents' perceived tourism impacts and life satisfaction. *Tourism Management*, *61*, 436–442. https://doi.org/10.1016/j.tourman.2017.02.013

Lundberg, E. (2017). The importance of tourism impacts for different local resident groups: A case study of a Swedish seaside destination. *Journal of Destination Marketing & Management*, *6*(1), 46–55. https://doi.org/10.1016/j.jdmm.2016.02.002

MacNeill, T., & Wozniak, D. (2018). The economic, social, and environmental impacts of cruise tourism. *Tourism Management*, *66*, 387–404. https://doi.org/10.1016/j.tourman.2017.11.002

Miles, J., & Shevlin, M. (2007). A time and a place for incremental fit indices. *Personality and Individual Differences*, *42*(5), 869–874. https://doi.org/10.1016/j.paid.2006.09.022

Mody, M. (2018). *How does my neighbor feel about my Airbnb?* Boston Hospitality Review. https://www.bu.edu/bhr/files/2018/02/How-Does-My-Neighbor-Feel-About-My-Airbnb.pdf

Moscovici, D. (2017). Environmental Impacts of Cruise Ships on Island Nations. *Peace Review*, *29*(3), 366–373. https://doi.org/10.1080/10402659.2017.1344580

Nunkoo, R., & Gursoy, D. (2017). Political trust and residents' support for alternative and mass tourism: An improved structural model. *Tourism Geographies*, *19*(3), 318–339. https://doi.org/10.1080/14616688.2016.1196239

Perdue, R. R., Long, P. T., & Allen, L. (1987). Rural resident tourism perceptions and attitudes. *Annals of Tourism Research*, *14*(3), 420–429. https://doi.org/10.1016/0160-7383(87)90112-5

Petruzzi, M. A., Marques, G. S., do Carmo, M., & Correia, A. (2020). Airbnb and neighbourhoods: An exploratory study. *International Journal of Tourism Cities*, *6*(1), 72–89. https://doi.org/10.1108/IJTC-08-2019-0119

Postma, A., & Schmuecker, D. (2017). Understanding and overcoming negative impacts of tourism in city destinations: Conceptual model and strategic framework. *Journal of Tourism Futures*, *3*(2), 144–156. https://doi.org/10.1108/JTF-04-2017-0022

Ribeiro, M. A., Pinto, P., Silva, J. A., & Woosnam, K. M. (2018). Examining the predictive validity of SUS-TAS with maximum parsimony in developing island countries. *Journal of Sustainable Tourism*, *26*(3), 379–398. https://doi.org/10.1080/09669582.2017.1355918

Richards, S., Brown, L., & Dilettuso, A. (2019). The Airbnb phenomenon: The resident's perspective. *International Journal of Tourism Cities*, *6*(1), 8–26. https://doi.org/10.1108/IJTC-06-2019-0084

(1996). Schalock, R. L. (Ed.). *Quality of life: Conceptualization and measurement*. (Vol. 1). American Association on Intellectual Development.

Schmidt, F. L., Le, H., & Ilies, R. (2003). Beyond alpha: An empirical examination of the effects of different sources of measurement error on reliability estimates for measures of individual differences constructs. *Psychological Methods*, *8*(2), 206–224. https://doi.org/10.1037/1082-989X.8.2.206

Schofield, P. (2011). City resident attitudes to proposed tourism development and its impacts on the community. *International Journal of Tourism Research*, *13*(3), 218–233. https://doi.org/10.1002/jtr.814

Sen, A. (2001). *Development as Freedom*. OUP.

Sharma, B., & Dyer, P. (2009). An investigation of differences in residents' perceptions on the Sunshine Coast: Tourism impacts and demographic variables. *Tourism Geographies*, *11*(2), 187–213. https://doi.org/10.1080/14616680902827159

Sharpley, R. (2014). Host perceptions of tourism: A review of the research. *Tourism Management*, *42*, 37–49. https://doi.org/10.1016/j.tourman.2013.10.007

Sinclair-Maragh, G., Gursoy, D., & Vieregge, M. (2015). Residents' perceptions toward tourism development: A factor-cluster approach. *Journal of Destination Marketing & Management*, *4*(1), 36–45. https://doi.org/10.1016/j.jdmm.2014.10.001

Skevington, S. M., & Böhnke, J. R. (2018). How is subjective well-being related to quality of life? Do we need two concepts and both measures? *Social Science & Medicine (1982))*, *206*, 22–30. https://doi.org/10.1016/j.socscimed.2018.04.005

Stergiou, D. P., & Farmaki, A. (2019). Resident perceptions of the impacts of P2P accommodation: Implications for neighbourhoods. *International Journal of Hospitality Management*, 102411.https://doi.org/10.1016/j.ijhm.2019.102411

Stylidis, D., Biran, A., Sit, J., & Szivas, E. M. (2014). Residents' support for tourism development: The role of residents' place image and perceived tourism impacts. *Tourism Management*, *45*, 260–274. https://doi.org/10.1016/j.tourman.2014.05.006

Su, L., Huang, S. (S.)., & Nejati, M. (2019). Perceived justice, community support, community identity and residents' quality of life: Testing an integrative model. *Journal of Hospitality and Tourism Management*, *41*, 1–11. https://doi.org/10.1016/j.jhtm.2019.08.004

Suess, C., Baloglu, S., & Busser, J. A. (2018). Perceived impacts of medical tourism development on community well-being. *Tourism Management*, *69*, 232–245. https://doi.org/10.1016/j.tourman.2018.06.006

Suess, C., Woosnam, K., Mody, M., Dogru, T., & Sirakaya Turk, E. (2020). Understanding How Residents' Emotional Solidarity with Airbnb Visitors Influences Perceptions of Their Impact on a Community: The Moderating Role of Prior Experience Staying at an Airbnb. *Journal of Travel Research*, https://doi.org/10.1177/0047287520921234

Uysal, M., Sirgy, M. J., Woo, E., & Kim, H. (L.). (2016). Quality of life (QOL) and well-being research in tourism. *Tourism Management*, *53*, 244–261. https://doi.org/10.1016/j.tourman.2015.07.013

Vareiro, L. M. d C., Remoaldo, P. C., & Ribeiro, J. A. C. (2013). Residents' perceptions of tourism impacts in Guimarães (Portugal): A cluster analysis. *Current Issues in Tourism*, *16*(6), 535–551. https://doi.org/10.1080/13683500.2012.707175

Ward, C., & Berno, T. (2011). Beyond social exchange theory: Attitudes toward tourists. *Annals of Tourism Research*, *38*(4), 1556–1569. https://doi.org/10.1016/j.annals.2011.02.005

Yeager, E. P., Boley, B. B., Woosnam, K. M., & Green, G. T. (2020). Modeling Residents' Attitudes toward Short-term Vacation Rentals. *Journal of Travel Research*, *59*(6), 955–974. https://doi.org/10.1177/0047287519870255

Towards advancing theory and methods on tourism development from residents' perspectives: Developing a framework on the pathway to impact

Hossein G. T. Olya

ABSTRACT

This study argues that method should work hand-in-hand with theory to be able to generate impactful outcomes. This study provides *three* methodological, theoretical and practical contributions to the current body of knowledge on tourism development from residents' perspective. *First*, five analytical approaches—namely, covariance based-structural equation modelling (CB-SEM), partial least squares structural equation modelling (PLS-SEM), multiple regression analysis (MTA), simple regression analysis (SRA), fuzzy set Qualitative Comparative Analysis (fsQCA), and analysis of necessary conditions (ANC)—are applied to test a model to predict support for tourism development from local residents' perspective. It critically discusses the results of analytical approaches to researching support for tourism development. A guideline for conducting a symmetrical approach is provided. *Second*, this study presents a classification of theories on residents' support of tourism, which helps to tackle the complexity of this complex social phenomenon by advancing theories and highlighting the importance of a theory-method match in future research. *Third*, beyond method and theory, there is a need for an informative framework that illustrates the 'pathway to impact' of research on residents' support for sustainable tourism development. Using the theory of change, this study fills this research gap by developing a logical model demonstrating outputs, actor groups, outcomes, and impacts of residents' support for tourism development.

1. Introduction

The sustainability of tourism development depends on the support of local communities and their residents. Resident's attitude toward tourism is a well-researched area of tourism studies (Sharpley, 2014). The perceived positive and negative impacts of tourism, personal benefits from tourism, power in tourism, community attachment, community involvement, knowledge of tourism, and trust in the government are all predictors of residents' support for tourism development (Lee, 2013; Nunkoo & So, 2016; Woo et al., 2015). Despite extensive research, a more innovative approach in the application of theories and methods is needed to tackle the complexity of residents' attitude toward tourism.

Most of the relevant studies have relied on symmetrical methods (e.g., SEM) for model testing. However, the applicability of these approaches and the credibility of the results for providing a complete picture from the study phenomena have been questioned by scholars (e.g., Armstrong et al., 2001; Olya & Mehran, 2017; Woodside, 2013). Westland has indicated that symmetrical methods "were disappointingly inadequate, but the best we had at the time. Statistical power has always lagged [in terms of] the size and complexity of the networks under analysis, and as a result generated unreliable, simplistic, and inapplicable results" (Westland, 2015, p. 161). The critique of the functionality of different symmetrical methods (e.g., CB-SEM vs PLS-SEM) appears in business and management research (e.g., Guide & Ketokivi, 2015; Rönkkö et al., 2016). For example, Rönkkö et al. (2016) have argued that the use of an incorrect estimator leads to biased (inaccurate – invalid) results. The many problems that arise from using PLS can be overcome and avoided by using a less controversial estimator (e.g., AMOS, LISREL, MPLUS).

Meanwhile, Woodside (2014, p. 2502) has argued that there are 'serious problems with the near-total reliance by most researchers on symmetric statistical tests and difficulties in achieving theory advances relying on such tools. [These approaches] will end during the second decade of the 21st century'. This study contends that neither being 'the best we have at the time' nor 'know[ing] how to use it' is a decent rationale for using these methods. In fact, research methods should go hand-in-hand with theory to address a study's research questions and objectives. Olya et al. (2019b) have contended that an extension and modification of the theory of planned behaviour (TPB) may potentially contribute to the literature; however, researchers need to revisit the analytical approaches in order to present different insights from models developed based on the TPB. This could aid the justification of heterogeneous results from asymmetrical methods.

Most research on residents' behaviours and tourism development relies on a few theories (e.g., the social exchange theory and social identity theory) which are insufficient for explaining the heterogeneity of the model's testing results. In this vein, Olya and Gavilyan (2017) have found that residents who have experienced negative impacts from tourism still express their support for tourism development. This indicates that social exchange theory is insufficient in explaining complex attitudes of residents. Olya and Gavilyan (2017) have thus deduced that the complexity theory can explain such complex behaviours of residents toward tourism development. In this regard, Boley et al. (2014) have used Weber's theory of formal and substantive rationality as a complementary theory of social exchange theory to model residents' support for tourism development. To these ends, there is a paucity of research discussing the falsification or corroboration of the above theories in different contexts. Specifically, in residents' attitudes toward tourism research, there is the need for a classification of theories to guide how different theories can fit the appropriate analytical techniques.

Beyond theory and method, one must also think of how research outputs can be translated into practice to ensure benefits for the economy and the wider society. A review of the literature shows that tourism scholars have recommended such practical implications. Nonetheless, it is imperative to develop a pathway for the sustainable implementation of the proposed practical implications. Drawing on the theory of change, this study designs a framework illustrating the transformation of research outputs to their impacts. This study critically evaluates results from four symmetrical approaches and provides a guideline of standard practices for conducting symmetrical modelling. It also conducts asymmetrical modelling using fsQCA to explore solutions for high and low scores of residents' support for tourism development. This study is a first attempt to identify the necessary conditions for residents' support for tourism development. This study also develops a classification of theories that discusses the relevant theories and methods which could potentially deepen the understanding of residents' complex behaviours towards tourism development.

2. Methods

This study uses data from the research by Olya and Gavilyan (2017) to test the proposed model (Figure 1). Drawing on complexity theory, used a range of predictors, namely, personal benefits

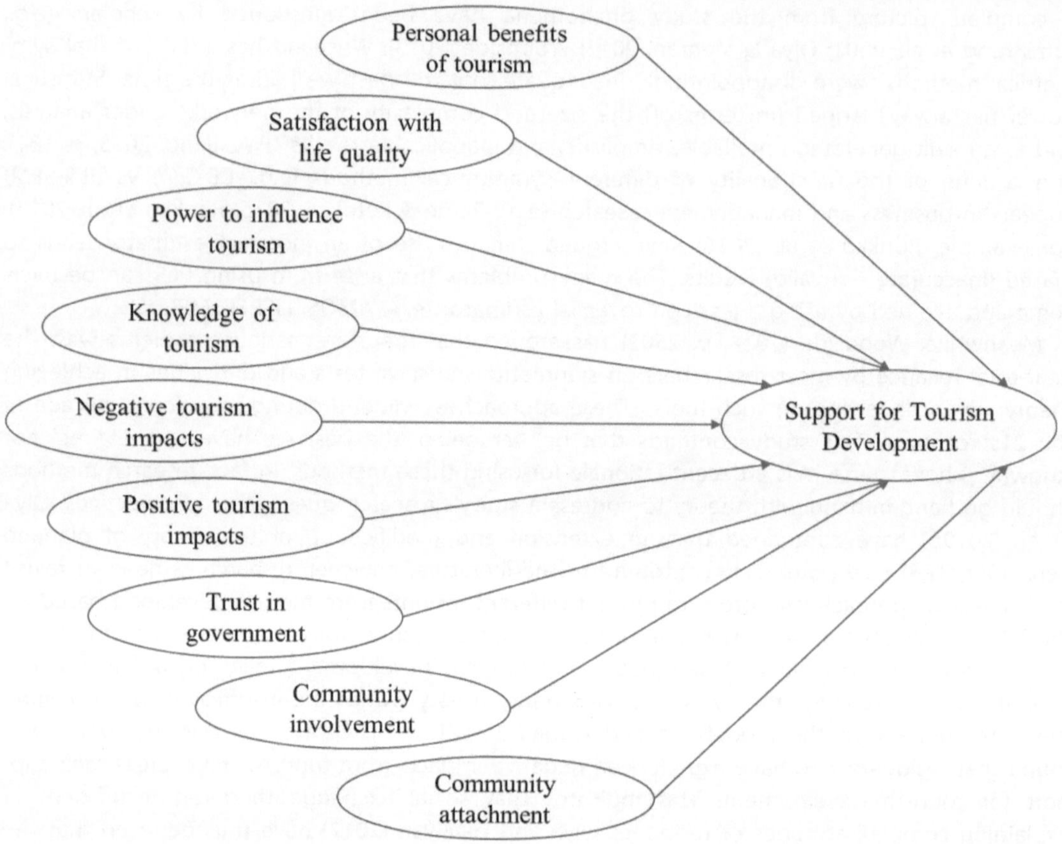

Figure 1. The research model.

of tourism, satisfaction with quality of life, power to influence tourism, knowledge of tourism, negative tourism impacts, positive tourism impacts, trust in government, community involvement, and community attachment to stimulate residents' support for tourism development. To avoid redundancy, this study refers to Olya and Gavilyan's (2017) research for detailed information on the sample, data collection procedures, measurement model testing (common method bias, reliability, and validity), and existence of contrarian cases.

This study conducted both symmetrical and asymmetrical approaches to predict support for tourism development from residents' perspectives. Symmetrical approaches (e.g., SEM and regression) aim to investigate the sufficiency of the independent variable (X) in predicting the dependent variable (Y). In symmetrical approaches, a high score of X is most likely associated with a high score of Y and a low score of X is related with a low score of Y. In asymmetrical approaches (e.g., fsQCA), a high score of X (the solution) is not necessarily linked with a high score of Y (the outcome variable). In other words, a low score of X could be associated with a high score of Y (Feng et al., 2019; Khan et al., 2018; Olya et al., 2020b). This study performed four symmetrical approaches (namely CB-SEM, PLS-SEM, MRA, and SRA) and compared their results to understand the net effect of the predictors on residents' support for tourism development. As an asymmetrical approach which functions based on Boolean Algebra, fsQCA was used to explore sufficient solutions (i.e., a combination of the predictors) to stimulate residents' support for tourism development. fsQCA helps explore solutions for low scores of residents' support, which are not opposite to solutions for high scores of residents' support for tourism development.

This study conducted an analysis of necessary conditions (ANC) to identify the necessary predictors to achieve the outcome variable. Support of tourism development is not achievable in the absence of necessary predictors. Results of the ANC are important for prioritising the plans to provide critical conditions for attaining a desired outcome (e.g., residents' support for tourism development). An overview of the aforementioned analytical approaches — including the definitions, advantages, and disadvantages of six techniques — are presented in Table 1.

The normality of the data is assessed using the two measures of skewness and kurtosis (Table A1, Appendix, Supplementary material). The values of skewness and kurtosis fell within the commonly-accepted range of ±3, which indicates the data are normally distributed (Taheri et al., 2020). As this study conducts a symmetrical analysis (e.g. SEM, MLR, and SRA), multicollinearity should not be a concern. According to the results of multicollinearity tests, the variance inflation factor (VIF) and tolerance values are less than 4 and larger than 0.25, respectively (O'brien, 2007).

3. Results

3.1. Symmetrical modelling: sufficiency of the factor

This study has conducted multiple analyses to provide a deeper insight into analytical approaches used for modelling residents' support for tourism development. The sufficiency of factors (i.e., the net effect) is investigated using four model testing methods (CB-SEM, PLS-SEM, SRA, and MRA) (Table 2). The results from CB-SEM and PLS-SEM are similar (except for the power to influence tourism); however, the results of the other two analytical approaches (MRA and SRA) are not the same, indicating the importance of using appropriate statistics to avoid decision-making based on misleading results. In four analyses, personal benefit, the positive impact of tourism, and community involvement significantly and positively affect residents' support for tourism development. Unlike personal benefit, satisfaction with the quality of life, knowledge of tourism, trust in the government, and the negative impact of tourism do not appear to be significant predictors of residents' support for tourism development (Table 2).

According to the CB-SEM results, the power to influence tourism plays a negative role in driving residents' support for tourism development. However, the results of the PLS-SEM, MRA, and SRA show that the effect of the power to influence tourism on residents' support for tourism development is not significant. These heterogeneous results have been reported in previous research. For example, Látková and Vogt (2012, p. 62) state that 'inconsistent with social exchange theory and Madrigal's (1993) findings, power was not found to be a significant predictor of tourism impacts … [and it] does not guarantee that a person will see solely the positive or negative side of the tourism industry'. Similarly, Kayat (2002) has found that the net effect of power in tourism is not sufficient for a resident to support the tourism industry, suggesting its interactive effect with other predicates—such as the level of their dependency on tourism and the willingness to adapt to changes—should be included in the model indicating residents' attitudes and behaviours. The heterogeneous roles of these factors can be addressed using fsQCA that explores solutions (recipe: a combination of factors) for understanding how power—in combination with other predictors rather than alone—behaves in predicting residents' support for tourism development.

Unlike the SEM and MRA, the results of the SRA indicate that community attachment significantly increases residents' support for tourism development ($\beta = .215$, $p < .01$). This accords with residents' attitude toward tourism research, which has found that community attachment influences support for tourism development positively (e.g., Lee, 2013). By contrast, Gannon et al. (2020) have found that community attachment has no impact on support for tourism development. Comparing the results of the four analytical methods in this study prompts scepticism about the accuracy of results obtained from the SEM and MRA approaches. Similar to this study's findings, community attachment plays a significant role in predicting support for tourism development in Gannon et al. (2020) study with the latter using SRA for hypothesis testing. The SRA results fit with the Weber's theory of substantive and formal rationality (Boley et al., 2014; Gannon et al., 2020). Furthermore, SRA is a powerful, rigorous, and valid approach that satisfies

Table 1. Overview of symmetrical and asymmetrical modelling approaches.

Approach	Technique	Definition/purpose	Benefits	Limitations
Symmetrical	SRA	It is a predictive statistical analysis to estimate the effect of one independent variable on a dependent variable.	• Assesses the net effect of the independent variable on the dependent variable • Easily Implemented (fairly simple to understand and interpret)	• Ignores data error in estimation • Requires a large sample size • Requires normality of data • Excludes the observed variables • Ignores the contrarian cases • Lacks calculation of models for low score of dependent variable
	MRA	It is an extension of SRA to investigate the impacts of two or more independent variables on a dependent variable.	• Predicts a more complicated model (effects of more than one variable responsible for a dependent variable)	• Ignores data error in estimation • Requires a large sample size • Requires normality of data and residuals • Requires absence of multicollinearity • Excludes the observed variables • Ignores the contrarian cases • Lacks calculation of models for low score of dependent variable
	CB-SEM	It uses a maximum likelihood (ML) estimation method to achieve a good model fit for a complex structural model through reproducing the covariance matrix.	• Assesses error in the structural model • Deals with the complex models • Facilitates the assessment of direct, indirect and total effects	• Requires a substantially larger sample size • Requires normality of data • Requires absence of multicollinearity • Ignores contrarian cases • Lacks calculation of models for low score of dependent variable
	PLS-SEM	It aims at predicting key target variables or investigating key "driver" variables. It uses a regression-based ordinary least squares (OLS) estimator to minimise the error terms and maximise the R^2 of the endogenous variables in a complex structural model.	• Uses composites that represent formatively-measured latent factors • Incorporates single-item measures in a structural model • Uses the latent variable scores in subsequent analyses.	• Distrusts the accuracy of results based on small sample size • Unreliable on use of correct estimator (OLS) • Doubts the ability of PLS-SEM for theory development • Requires absence of multicollinearity • Ignores the contrarian cases • Lacks calculation of models for low score of dependent variable
Asymmetrical	fsQCA	It is a set-theoretical approach that explore complex combination of the predictors (solution or recipe) explain condition leading to the outcome.	• Works efficiently with small, medium and large sample size • Computes multiple solutions to predict the outcome • Includes the contrarian cases in the model testing • Addresses the heterogeneous role of predictors in stimulating the outcome	• Ignores data error in estimation • Excludes the observed variables • Requires prior knowledge to refine the solutions • Requires software development to test the measurement model and mediation effect

(*continued*)

Table 1. Continued.

Approach	Technique	Definition/purpose	Benefits	Limitations
	ANC	It aims at identifying necessary predictors to achieve the outcome.	• Calculates solutions for low score of the outcome • Works efficiently with small to large sample size • Provides pragmatic results for practitioners to know what are the critical conditions to attain the outcome • Calculates necessary conditions for low score of the outcome	• Ignores data error in estimation • Excludes the observed variables • Requires software development to test measurement model

Note: CB-SEM: covariance-based structural equation modelling, PLS-SEM: partial-least-squares structural equation modelling, MRA: multiple regression analysis, SRA: simple regression analysis, fsQCA: fuzzy set Qualitative Comparative Analysis; ANC: analysis of necessary conditions.

Table 2. Results of symmetrical analysis to investigate sufficient predictors.

	Outcome: support for tourism development									
	CB-SEM (R^2= .18)		PLS-SEM (R^2=.30)		MRA (R^2= .22)			SRA (R^2= .38)		
Predictors	*β*	*p*	*β*	*t*	*β*	*t*	*p*	*β*	*t*	*p*
Personal benefits of tourism	**.177**	**.005**	**.226**	**1.559**	**.153**	**1.894**	**.060**	**.303**	**4.502**	**.000**
Satisfaction with life quality	.036	.575	.023	.165	.034	.400	.690	.017	.243	.808
Power to influence tourism	**−.131**	**.040**	−.141	.152	−.121	−1.597	.112	.017	.243	.808
Knowledge of tourism	−.096	.132	−.063	.470	−.087	−1.131	.259	.035	.492	.624
Negative tourism impacts	−.082	.196	−.027	.135	−.079	−1.228	.221	−.094	−1.329	.185
Positive tourism impacts	**.222**	**.000**	**.246**	**2.367**	**.220**	**3.166**	**.002**	**.320**	**4.771**	**.000**
Trust in government	−.019	.761	−.066	.881	−.011	−.163	.871	.064	.903	.368
Community involvement	**.242**	**.000**	**.240**	**1.716**	**.241**	**3.269**	**.001**	**.314**	**4.675**	**.000**
Community attachment	.097	.130	.120	.7665	.094	1.298	.196	**.215**	**3.114**	**.002**

Note: CB-SEM: covariance-based structural equation modelling, PLS-SEM: partial-least-squares structural equation modelling, MRA: multiple regression analysis, SRA: simple regression analysis. Bolded values represent sufficient predictors that significantly affect the outcome.

the research objective in terms of investigating the net effect of predictors (sufficient factors) on the model outcome (Mehran et al., 2020). Further arguments on the scepticism of SEM outputs are discussed in subsection 4.1., titled *SEM: A building with no walls or roof.*

3.2. *Analysis of necessary conditions: necessity of factors*

The necessary factors to achieve residents' support for tourism development are identified using ANC (Table 3). Negative and positive tourism impacts and community attachments are necessary to achieve support for tourism development (consistency >. 90). The role of sufficient and necessary factors in predicting support for tourism development is presented in Table 3. According to SRA and ANC results, personal benefit is sufficient but unnecessary to stimulate residents' support of tourism. Additionally, satisfaction with the quality of life, the power to influence tourism, knowledge of tourism, and trust in the government are insufficient and unnecessary predictors of support for tourism development. Community involvement is sufficient but unnecessary in predicting support for tourism development, while the positive impacts of tourism and community attachment are both sufficient and necessary predictors of support for tourism development (Table 3).

3.3. *Asymmetrical modelling: fsQCA*

While both SRA and ANC show the net effect of factors predicting the outcome, the fsQCA explores the combined effect of factors that receive different names, such as 'recipe', 'solution',

Table 3. Results of analysis of necessary conditions (ANC).

Antecedent condition	Outcome: support for tourism development		Synthesis of results of SRA and ANC (sufficiency vs necessity)
	Coverage	Consistency	
Personal benefits of tourism	.882	.201	*Sufficient* but unnecessary
Satisfaction with quality of life	.788	.227	Insufficient and unnecessary
Power to influence tourism	.840	.265	Insufficient and unnecessary
Knowledge of tourism	.809	.297	Insufficient and unnecessary
Negative tourism impacts	.904	**.231**	Insufficient but *necessary*
Positive tourism impacts	.907	**.203**	*Sufficient* and *necessary*
Trust in government	.877	.241	Insufficient and unnecessary
Community involvement	.881	.204	*Sufficient* and unnecessary
Community attachment	.904	**.207**	*Sufficient* and *necessary*

Note: bolded values represent necessary condition (consistency >.09).

Table 4. The fsQCA results to predict high and low levels of support for tourism development.

Solutions for *high* support of tourism development	Raw coverage	Unique coverage	Consistency
S1:stqulf*psim*ngim*cminv*cmatch*pbt*~trgov*knwt	0.326	0.029	0.996
S2: stqulf*psim*cminv*cmatch*pwr*pbt*~trgov*knwt	0.325	0.030	0.994
S3: stqulf*psim*ngim*cminv*cmatch*pwr*pbt*knwt	0.402	0.108	0.995
S4:~stqulf*psim*~ngim*cminv*cmatch*~pwr*pbt*trgov*knwt	0.158	0.180	0.993
Solution coverage: 0.491; *Solution consistency*: 0.992			
Solutions for *low* support of tourism development	Raw coverage	Unique coverage	Consistency
L1: ~cmatch*~trgov	0.734	0.045	0.593
L2: ~psim*~pbt	0.667	0.007	0.715
L3: ~ngim*~pbt	0.617	0.012	0.667
L4: ~stqulf*~cminv	0.728	0.041	0.633
Solution coverage: 0.875; *Solution consistency*: 0.506			

Note: stqulf: satisfaction with quality of life, psim: positive tourism impacts, ngim: negative tourism impacts, cminv: community involvement, cmatch: community attachment, pwr: power to influence tourism, pbt: personal benefits of tourism, trgov: trust in government, knwt: knowledge of tourism. *: and, ~: negation.

'causal model', 'algorithm', and 'configuration'. For example, the negative impact of tourism appears as an insufficient but necessary factor in predicting the support for tourism development. The fsQCA can provide a deeper insight into the role of each predictor, in combination with other predictors of the outcome (Table 4). The fsQCA outputs include three types of solutions, namely 'complex', 'parsimonious', and 'intermediate', described by Pappas as follows:

> The complex solution presents all the possible combinations of conditions when traditional logical operations are applied. In general, because the number of configurations identified can be very large, the number of complex solutions can be large and these may include configurations with several terms. This makes the interpretation of the solutions difficult and in most cases impractical.... For this reason, they are usually simplified further into parsimonious and intermediate solutions. (Pappas et al., 2019, p. 654)

Unlike Olya and Gavylian's (2017) study, which concentrated on complex solutions, this study focuses on parsimonious and intermediate solutions, not only to integrate the results of current knowledge on support for tourism development but also to improve readings of fsQCA results.

As evident in Table 4, four solutions for both high (S1-S4) and low (L1-L4) levels of support for tourism respectively are calculated. Consistency, which is similar to correlation, represents the proportion of cases consistent with the desired outcome (support for tourism development). 99% of cases follow S1-S4 solutions that explain conditions in which residents support tourism development; 55% of cases do not support tourism development in conditions matched with L1-L4. Coverage, which is analogous to R^2 in a symmetrical analysis, is relatively high (.49 and .87). This means solutions S1-S4 explain 49% of the variation in residents' support for tourism development and solutions L1-L4 explain 87% of the variation in low levels of support for tourism development (Table 4). The unique coverage meanwhile shows the relative importance of solution in predicting the outcome (high or low support). As shown in Table 4, among the solutions for predicting high support, S2 receives unique coverage of .03, while among solutions for low support, L1 has unique coverage of .045. These results indicate that relative to other solutions, S2 and L1 are more important in explaining the conditions for predicting high and low levels of residents' support respectively. Table A2, Supplementary material in appendix as presents different types of community groups (%) correspondence with the solutions.

S1 indicates that in both the lack of trust in the government and the presence of negative impacts of tourism, residents who are satisfied with their quality of life and are knowledgeable about tourism would support tourism development if they were attached to and involved in a community and perceive potential personal benefits. S2 explains a condition in which there is a lack of trust in the government, knowledgeable residents who are satisfied with their quality of life would support tourism development if they are attached to and involved in a community, feel the power to influence tourism, and perceive positive impacts of tourism. According to the third solution (S3), tourism development is supported by knowledgeable residents who are happy with their quality of life as well as the perceived personal benefits and positive impacts from tourism. These residents are attached to and involved in a community and feel the ability to influence tourism, however, perceive the negative impacts of tourism. Solution 4 explains a condition for tourism development support by residents who are not satisfied with their quality of life, feel no power to influence tourism, and do not perceive the negative impacts of tourism. Nonetheless, they have knowledge about tourism and trust the government, and perceive personal benefits and positive impacts of tourism; they are also attached to and involved in a community.

Unlike symmetrical modelling, which considers the models for low outcomes and mirror opposites of models for high outcomes (e.g., low personal benefit → low support level and high personal benefit → high support level), fsQCA explores solutions for low scores of support which are unique and different from the mirror opposites of solutions for high scores (c.f. Table 4). Solutions for low levels of support are important to consider in preventing anti-tourism community groups. According to L1, residents would not support tourism development if they do not trust in the government and are not attached to a community. Alternatively, residents who do not perceive personal benefits and positive impacts from tourism most likely would not support tourism development (L2). According to L3, residents would not support tourism development if they do not perceive personal benefits and perceive the negative impacts of tourism. The responses from residents matched with L2 and L3 are well explained by social exchange theory. Another solution (L4) that explains the conditions of not supporting tourism development represents the view of residents who are not satisfied with their quality of life and are not involved in a community (Table 4).

The results of asymmetrical modelling show that stimulating the residents' behaviours to support tourism is rather complex, as the fsQCA results confirm the 'equifinality' tenet of complexity theory. This tenet postulates that more than one solution explains conditions in which the expected outcome can be achieved. In this study, the fsQCA explores four solutions to high or low support for tourism development, in which each solution represents the voices of different groups of local residents. As Olya et al. (2018) have noted, this is important for including views of residents with different backgrounds and interests, whose voices should be heard in order to achieve sustainable tourism development. Otherwise, anti-tourism community groups opposing tourism development may begin to form (Olya et al., 2019a). Furthermore, the fsQCA results confirm the 'causal asymmetry' tenet of complexity theory, which indicates that solutions for achieving support for tourism development do not merely depend on the presence or absence of a specific predictor. For example, the fsQCA offers a solution for residents who do not trust in the government, are not happy with their quality of life, feel no power to influence tourism, and are not attached to and involved in a local community.

From a theoretical perspective, the fsQCA results reveal that predictors can play both negative and positive roles in stimulating support for tourism development. This helps justify heterogeneous results reported in the literature (Látková & Vogt, 2012; McCool & Martin, 1994; Sharpley, 2014). The fsQCA calculates configurations (solutions) that show the combinational effect of the factors in which the role of one factor may vary, depending on the attributes (absence or presence) of other factors. Practically speaking, residents as individuals may consider the combinational effects of predictors simultaneously, in order to come to a decision or take action (Olya & Al-Ansi, 2018; Olya & Han, 2020). Sharpley (2014) has argued that residents support tourism development as positive impacts may outweigh the negative impacts of tourism. Therefore, as a pragmatic approach, fsQCA helps to explore solutions describing complex conditions in which residents support sustainable tourism development and prevent anti-tourism local community groups from emerging.

4. Discussions and implications

This section includes discussions on the methodological and theoretical advances in modelling residents' behaviours to support sustainable development. It also presents common good practices for conducting symmetrical analytical approaches (e.g., regressions and SEM). To maximise the impacts of tourism research, this section ends with a framework illustrating the transformation of research outputs into social, economic, and ecological impacts.

4.1. SEM: a building with no walls or roof

A review of the discourse on analytical approaches shows that there are different views on the functionality of symmetrical techniques. For example, Sarstedt et al. (2020) highlight the superiority of PLS-SEM over the PROCESS approach proposed by Hayes (2017), emphasising the advantages of PLS-SEM (e.g., the flexibility of testing a complex model that includes formative observed variables and estimating smaller standard errors and functionality using a small sample size). They believe these advantages can address the pitfalls of symmetrical modelling, such as the limitation on generating interaction terms, the need for a large sample size, and the assumption of linearity of association between predictor and outcome variables (Sarstedt et al., 2020). Furthermore, SEM provides flexibility for the model's specifications and options in dealing with missing data (Hayes et al., 2017). However, a number of scholars have questioned the advantages of PLS-SEM. For example, Guide and Ketokivi (2015, p. vii) indicated that "use of PLS is (incorrectly) justified by saying that PLS is suitable for small samples, that it should be used when one has formative indicators in a measurement model, or that it is suitable when the Maximum Likelihood estimator fails to converge to a solution. All are poor excuses for using PLS. Claiming that PLS fixes problems or overcomes shortcomings associated with other estimators is an indirect admission that one does not understand PLS."

This study attempts to analyse the credibility of the aforementioned advantages of SEM from different angles. First, researchers are encouraged to test the structural model with a large sample size as it most likely represents a more accurate picture of a normal population. Furthermore, the most common estimation technique of SEM (i.e., maximum likelihood, or ML) requires a large sample. As Hayes et al. (2017) indicate, using a small sample to test structural models by ML computes standard errors that tend to be biased downwards. Therefore, 'the apparent advantage of SEM evidenced by the smaller standard errors is likely illusory in this case and similar ones. Smaller standard errors are not better when they are wrong' (Hayes et al., 2017, p. 79). There are other estimation techniques (e.g., generalised least squares, asymptotically distribution-free, unweighted least squares, and consistent PLS) in which a large sample size is not the key assumption for running the model. However, selecting the estimation approach does not necessarily rely on the size of the sample; rather, it does depend on the research objective (such as testing a theory or structure or investigating key predictors), access to SEM software (e.g. AMOS or PLS-SEM), and researchers' statistical literacy and skills (Dijkstra, 2014; Hair et al., 2017; Reinartz et al., 2009).

Second, it seems that amidst such contestations on the superiority of applying techniques to work with a small sample size, we almost neglect the importance of large sizes of data as a requirement for satisfying generalisability—one of the key quality criteria for quantitative research. Many researchers of applied symmetrical approaches acknowledge the small sample size as a limitation of their research. Suggestions to test the model using a larger sample is not a solution, as it questions the basis of the research process used. Furthermore, in the case of residents' perceptions, justifying generalisability criteria can be addressed by collecting multi-source data (i.e., regions with various contextual conditions) from residents with different profiles.

Third, 24.14% of research applied PLS-SEM aimed at using this approach as a technique to develop theory (e.g. Ali et al., 2018). A reality check on the claim of PLS-SEM functionality for theory development is hence required. As grounded theory is recognised as a well-established qualitative approach for theory development (Patten & Newhart, 2017), this study calls for a clarification on how PLS-SEM—as a quantitative analytical approach—can contribute to theory development. Indeed, what are the differences between these two approaches for theory development? This study contends that such assertions (i.e. PLS-SEM is a method for theory development) may mislead junior researchers to incautiously develop and test a conceptual model with poor (or no) theoretical underpinnings. As an editor and reviewer, I have handled and reviewed some manuscripts that mainly reference misleading information, and fail to develop a theoretically-sound model. Due to a lack of theory, researchers are either confused about the cause-effect role of factors and the flow of the conceptual model (namely, the sequence—e.g., cognitive stimulus ➔ affective stimulus ➔ behaviours), or end up with many rejected hypotheses. Using irrelevant theory can yield the same outcome. For example, one study that had developed a structural model based on social exchange theory had to ultimately drop 18 hypotheses from the model due to either non-significant results or model fitness issues (Gursoy & Rutherford, 2004). The present study argues that using relevant theories (e.g., the collaboration theory and distributive or restorative justice theory) and analytical approaches which satisfy the study's objective (e.g., the net effect of the predictors on the outcome) can provide more accurate results about residents' behaviours toward tourism development.

The fourth objection relates to the name 'structural equation modelling'. According to the Oxford dictionary, 'structure' is defined as 'the state of being well organized or planned with all the parts linked together; a careful plan', while 'structural' means 'connected with the way in which something is built or organized' (Oxford English Dictionary, 2020). This point may seem pedantic, but a review of tourism and hospitality studies using SEM shows that R^2—an indicator of the predictive power of the model—varies from .07 (Altinay et al., 2019) to .41 (Rasoolimanesh et al., 2015). This means that factors used to craft the structure of a model can explain 7 to 41% of the variation of the model's given outcomes. That percentage could increase if either more factors, or more effective factors, were added into the model. Nonetheless, it is less likely that a model produces R^2 over .9 (90%). Due to the complexity of this social phenomena, it is not realistic that researchers could identify, measure, and include all outcome predictors into a structural model. If they do so, there is no guarantee that software can run such a complex model. To this end, it is not the fault of researchers who make significant efforts to develop implications according to SEM results, but this may mislead practitioners that are supposed to rely on a structure demonstrating incomplete insight into a phenomenon. In other words, the present study argues that with so-called 'structural' equation modelling, we are overselling a building with no roof or walls to the potential beneficiaries. Nonetheless, this study acknowledges that the functionality of SEM in testing models involves mediation factors. By all means, mediation hypotheses aim at investigating a sequence (e.g., X ➔ M ➔ Y) that can be considered a structure and can be investigated using SEM.

To summarise: this study appreciates the merits of SEM as a technique enabling the inclusion of formative and reflective constructs along with observed variables and measurement models into the model, as well as providing a rigorous factor analysis for measurement model testing

(Hair et al., 2013). Furthermore, SEM software (e.g., PLS-SEM and AMOS) are more user-friendly and efficient, especially in the case of complex models (Sarstedt et al., 2020). Nonetheless, the primary objective of a cause-effect analysis is the investigation of the net effect of the independent variable (X) on the dependent variable (Y). The ontological position of the quantitative paradigm is based on the objectivity and generalisability of the research. Therefore, rather being method-driven, positivist researchers should use analytical approaches as a means of addressing the research objectives. It is worth noting that one tourism researcher was keen to use a new version of a statistical package because it was considered a trendy approach; their rationale was '*I should run the model using this technique as now all using this*'. Using the correct analytical approach and appropriate package should be based on reasonable criteria such as 1) meeting the research objective, 2) fitting the data and design of the research, 3) calculating accurate, reliable, and valid outputs, 4) having the knowledge and skills to conduct the analysis and interpretation of the results, and 5) accessing the resources (having availability to the package licence). This study recommends 10 steps for ensuring common good practices when conducting a symmetrical approach, outlined in Table 5.

A recent study by Ryan (2020) provides a series of suggestions for reviewers assigned to review manuscripts using SEM. He recommends that the reviewer should check the source of the scale items and the adaptation process (a copy of the survey should be available), the details of the data and methods (e.g., date of data collection, target groups, sampling techniques, justification of sample size), and the descriptive statistics (i.e., the means, standard deviation, and normality tests).

4.2. Methodological innovations

This study argues that results from a symmetrical analysis are necessary but insufficient to develop policy implications based on a comprehensive view of residents' behaviours towards sustainable tourism development. A series of complementary analytical approaches should be conducted to deepen the understanding of the complex interactions between predictors of the residents' perceptions and behaviours. The fsQCA is a set-theoretic technique that explores recipes for predicting residents' support for sustainable tourism development (Olya & Gavylian, 2017). The symmetrical analysis (e.g., SEM/SRA) and fsQCA show the net effect of predictors and solutions for predicting the residents' support for sustainable tourism development, respectively. Nonetheless, the necessary predictors (those factors whose outcome is less likely to occur in their absence) are unknown. This is the first empirical study that uses an analysis of the necessary conditions to identify the necessary predictors of residents' behaviour toward tourism development.

Residents' behaviours toward sustainable tourism development in each destination or tourist attraction are highly related to their geographical location. For example, residents that are living near to the tourist attraction or heritage site may rate the socio-economic and environmental impacts of tourism development differently. For example, Alipour et al. (2017) have used a geographical information system (GIS) to show the spatial variation of environmental impacts of religious tourism from the residents' perceptions. As the residents' preferences and expectations may change by location and time (Olya et al., 2018), a spatial-temporal analysis of tourism's impacts can help prioritise plans for sustainable tourism development (Olya et al., 2019a). Similarly, Bayesian spatial modelling can demonstrate the variation of residents' viewpoints and predict their behavioural changes based on their vicinity to the attraction or heritage site (Shaddick et al., 2013).

A majority of the quantitative research on residents' attitude toward tourism mainly focuses on self-reported data obtained from surveys that are subject to various types of common method biases (Olya et al., 2020a). Furthermore, model testing results are based on residents' perceptions or intentions—which might be different from their actual behaviours. The intention-

Table 5. Common good practices for symmetrical modelling.

Step	Description
1. Develop hypotheses and/or model based on solid theory and/or theoretical reasoning	Theory helps researchers to understand how social phenomena behave and how influential factors can be related and/or affected to predict an outcome. 'A theory should be generalizable; in other words, it should apply in a range of specified contexts and settings' (Wilkins et al., 2019. p 4). Ideally, researchers can support the proposed model using an overarching theory. In some cases, researchers employ more than one theory to explain different interactions within the conceptual model.
2. Scanning and screening data	Researchers can scan and screen data to identify 'yea-saying or nay-saying' responses, which indicate cases that have rated the questions carelessly and just responded to all questions with either a 1 or 7 (assuming items are gauged using the 7-Likert scale). These cases can be dropped as they are an indication of acquiescence biases. The next step is treating missing data. To calculate an accurate estimation, cases with more than 15% missing values should be dropped from further analysis (Sarstedt et al., 2017). If the percentage of missing values across the factors used to build the model is less than 5%, they can be treated using replacement techniques such as mean substitution, regression imputation, expectation maximisation, tree imputation, and multiple imputation (Mohd Jamil, 2013; Taheri et al., 2020).
3. Normality of data and linearity of the associations between predictors and outcomes	The Kolmogorov-Smirnov (K-S) and Shapiro-Wilk (S-W) tests identify the normality of data (the results of K-S and S-W should not be significant, which means the data is normally distributed). Moreover, skewness and kurtosis are two measures for a normality check, and values for the normal data should fall within the acceptable range of ±3. The linearity of the linkage predictor and outcome should be assessed before conducting the analytical approaches (e.g., SRA), and linearity is an assumption (Olya & Han, 2020).
4. Existence of contrarian cases	This is important in the contexts in which a minority community (e.g., a pressure group or influencers) can play a key role in sustainable tourism development. If contrarian cases represent the views of key actor groups, the study is less likely to rely on the results of symmetrical analysis (e.g., SEM). This is because it may disregard views of contrarian cases and calculate the results based on ratings obtained from the majority of the sample. This can be checked using Cramer's V test (Olya & Nia, 2020).
5. Multi-collinearity issue	In a symmetrical analysis that aims to assess the net effect of independent factors on the dependent factors if two independent factors are highly correlated, the results most likely suffer from a multi-collinearity issue. This can be tested using the variance inflation factor (VIF); the rule of thumb for the VIF value is 5 (Hair et al., 2017) or 10 (Kutner et al., 2005). In interpretations of VIF, O'brien (2007) has suggested that the influence of other predictors that stimulate the stability of the specific estimate of the regression coefficient should be considered in interpretations of multi-collinearity checks. Freund and Wilson (1998) have suggested that the overall fitness of the model (based on R^2) should be taken into account when assessing the multi-collinearity issue.
6. Common method bias	Common method biases result from common rater effects (e.g., acquiescence bias), item characteristic effects (e.g., item social desirability), item context effects (e.g., context-induced mood) and measurement context effect (collecting data for dependent and independent variables from the same time/location/medium). There is a wider range of procedural and statistical remedies to control common method biases recommended by Podsakoff et al. (2003). Collecting data from different sources, times, and standard processes in the designing of surveys are

(*continued*)

Table 5. Continued.

Step	Description
	examples of procedural recommendations for reducing biases caused by measurement context effects. Statistical remedies vary from simple approaches (e.g., Harman's one-factor test) to more advanced techniques (e.g., marker variables) that researchers need to verify, specifically if the data are obtained from a self-reported survey.
7. Correct analysis and software	This is important to achieve accurate and reliable results for hypothesis testing. For example, the types of regression analysis depend on the number of independent variables, the pattern of the regression line, and the type of dependent variable(s). For instance, a linear regression analysis is used when the link of predictor and outcome is linear and the aim is to investigate the effect of one predictor on the model outcome. MRA is used when the association between the predictor and outcome is linear and the model involves more than one predictor. In MRA, the model outcome should be normally distributed. Probit regression and logistic regression is used when the outcome of the model is binary (i.e., dichotomous). Poison regression is used to test a model with count data. For example, poison regression can be used to predict numbers of tourism businesses and services around heritage sites as an outcome of the model or numbers of jobs created by a series of experiential tourism projects. There are other types of regression analyses that researchers may use based on the aforementioned criteria.
8. Assessment of results with theory	Much of tourism research's well-described theory—and its application in the theoretical background and hypothesis development sections of the paper—misses the opportunity to examine the statistical results with the theory. This is important in ensuring that the theory fits with the proposed conceptual model (Olya et al., 2019b). Furthermore, it is important to reflect on the justification of using the appropriate analytical approach to test the research objectives. For example, CB-SEM is suitable for 'theory testing, theory confirmation, or comparison of alternative theories', whereas PLS-SEM is appropriate for 'predicting key target constructs or identifying key "driver" constructs' (Hair et al., 2011, p. 144).
9. Cautious development of implications	Researchers need to develop implications based on the study results. Some studies may focus on generic suggestions related to study subjects. For example, if a study develops and tests a model that involves social-related factors for sustainable tourism development, the implications should be proposed based on social factors—and not economic and environmental factors, which are not measured and involved in the conceptual model. If researchers believe there is a connection between the social factors with other dimensions of sustainability, they should develop the implications with caution and justify their approach in their recommendation of development. Furthermore, the implications should be insightful, practical, cost-effective, and workable.
10. Acknowledgement of the study's limitations with honesty and integrity	Researchers should follow the code of practice and acknowledge the theoretical and methodological drawbacks of their research. This highlights the significance of conducting research responsibly and sustainably. It is also important to conduct impactful research in a way that policymakers and managers can rely on the research outputs while being aware of the limitations of the study.

behaviour gap among residents related to an event (Lee et al., 2014) and pro-environmental practice (Echegaray & Hansstein, 2017) have been identified. One solution to address this gap is to measure the actual behaviours of residents associated with tourism development initiatives, using an experimental research design in which residents have the opportunity to evaluate actual or projected scenarios on tourism development initiatives. The difference in the estimation method is also recommended to measure the actual impacts of implementing an intervention or policy on residents' behaviours toward tourism development (Romero, 2009).

As sustainable development of tourism with the involvement of residents is a complex social phenomenon, soft computing techniques such as fuzzy cognitive maps contribute to tackling the complexity of the residents' behavioural modelling (Olazabal & Pascual, 2016). The application of a fuzzy-set-based model enables the translation of linguistic terms expressed by residents to numerical data—a key step in conducting advanced modelling of residents' behaviours (Olya & Alipour, 2015). Artificial intelligence and neural network analyses are recommended for modelling the behaviours of residents (Olya & Alipour, 2015). Data from real cases can be obtained and used for machine learning to predict future behaviours of residents towards sustainable tourism development. Future research on modelling residents' behaviours towards sustainable tourism could follow innovative mixed- and multi-method approaches (e.g., social design for the visualization of tourism planning outcomes, data mining, eye-tracking technology, LEGO Serious Play)—a recent proposal in a special issue of the International Journal of Contemporary Hospitality Management (Olya et al., 2020c).

4.3. Theoretical advances

Tourism development and residents' perceptions thereof are among the well-studied research areas in the tourism field. Social exchange theory is frequently used to explain the residents' perceptions and responses towards tourism development. However, there is a need for theoretical advances through both applying different theories and pruning theory to tackle the complexity of tourism development from the lens of stakeholders (including residents). In this vein, Olya and Gavilyan (2017) have argued that social exchange theory is unable to explain the connections of residents' perceptions with their behaviours towards tourism development. This is in line with Boley et al. (2014, p. 36), who have referred to McGehee and Andereck (2004) critique characterising social exchange theory "as harbouring two incorrect assumptions: first, that individuals always make decisions with personal gain in mind as a top priority, and second, residents may think they are making the most prudent choice at the time, but will later come to realize that certain choices were not beneficial. Woosnam et al. (2009) critique SET for treating the relationship between residents and tourists as solely economic and not including other factors affecting the relationship." Therefore, similar to this research, the application of complexity theory explains the complex behaviours of residents who consider the negative impacts of tourism along with other factors (e.g., personal benefits, positive impacts of tourism, community involvement, and community attachment) and ultimately support tourism development. An alternative theory to the complexity theory is Bronfenbrenner's (1994) ecological systems theory, which includes both factors at individual levels (e.g., demographics and perceived personal benefit) and a wider range of environmental and situational levels (e.g., community attachment, socio-economic impacts) as predictors of residents' behaviours (Kline et al., 2013; Olya et al., 2020b).

As residents can be categorised in various community groups based on different criteria (e.g., jobs), theories that focus on including key actor groups' perspectives should be used to ensure inclusivity and sustainable tourism development. Social identity theory and collaboration theory are examples of theories that can potentially support models that involve views of different community groups (Jamal & Getz, 1995; Olya et al., 2018). In places where there is a lack of interest among community groups, nudge theory can enable those communities to rethink and revisit their outlooks toward tourism development. Nudge theory focuses on re-designing the individuals' thinking systems and generating conditions in which residents instinctively make decisions towards supporting tourism development (Murakami & Tsubokura, 2017). In other words, according to the nudge theory, residents choose to support tourism development as they have been influenced to both include human beings (not just their personal benefit) in their decision-

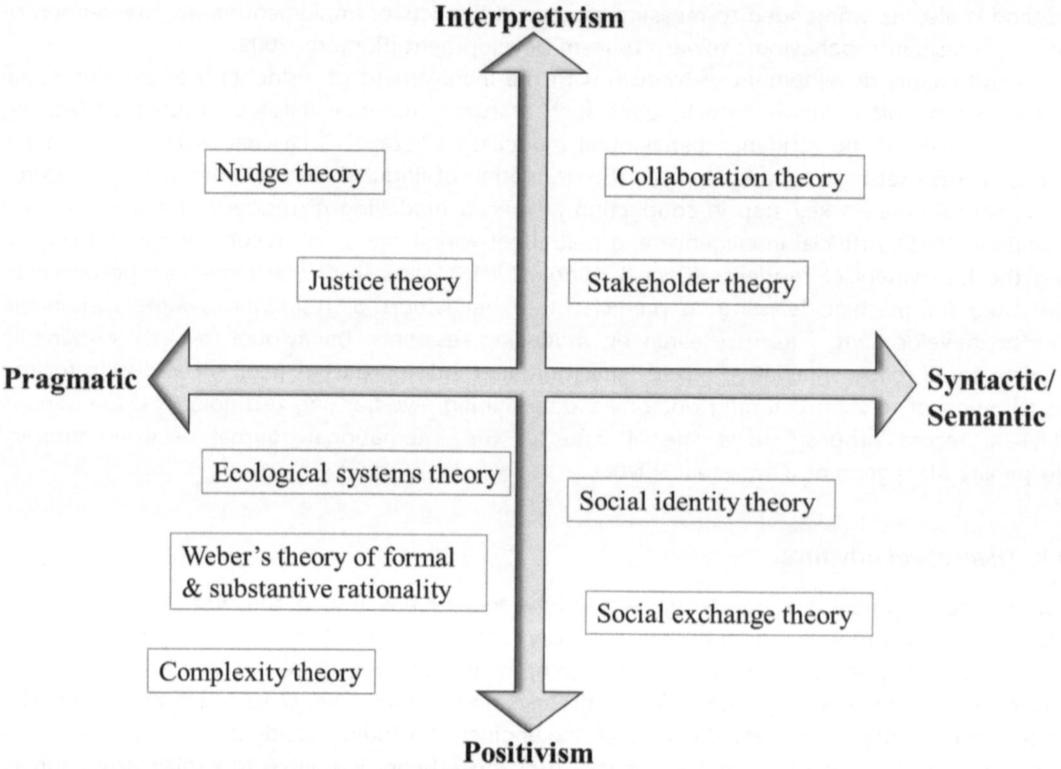

Figure 2. A classification of theories on residents' support for tourism research.

making system and to think about the common interests of broader society (i.e., the sustainability of the social, economic, and ecological impacts of tourism development).

In places where there is a conflict of interest among the community groups, the theory of justice can explain models for mitigating conflict escalation (Rawls, 2009). According to the multi-level distributive justice theory, 'individuals form judgments about the propriety of reward allocations based upon social comparisons across individuals, groups or standards, and that all such comparisons are potential sources for feelings of injustice and justice-restoring behaviors' (Markovsky, 1985, p. 822). Inequalities and injustice cause anti-tourism community groups to emerge. This theory can be used to address socio-economic inequalities—a key barrier against sustainable tourism development.

This study presents a classification of theories on residents' attitude towards tourism development based on two criteria of the level and function of theory (Figure 2). The two levels of theories are pragmatic and syntactic/semantic. In terms of the level of theory, the pragmatic view focuses on observing the attitude and behaviour of residents toward tourism development. By contrast, the syntactic paradigm relies on logical thoughts with a set of assumptions (e.g., number and types of local communities) and semantic viewpoint represents theories that used to explain real-world events (e.g., economic benefits encourage residents to support tourism development). The function of these theories — which connects to the ontological and epistemological assumptions of the research — involves positive and interpretive approaches. In the positivistic approach, reality is objective and can be investigated using deductive methods (e.g., developing and testing hypotheses on social impact has a negative effect on residents' support). In the interpretive approach, reality is to some extent subjective, emphasizing an individual's interpretations of residents' attitudes toward tourism development, and which can be explored through inductive methods.

As illustrated in Figure 2, social exchange theory and social identity theory are recognised as part of the syntactic/semantic and positive class (c.f., Gannon et al., 2020; Olya et al., 2018), while stakeholder theory (Nicholas et al., 2009; Retolaza et al., 2014) and collaboration theory (Jamal & Getz, 1995) represent the syntactic/semantic and interpretive approach. Nudge theory and Justice theory are classified in the pragmatic and interpretive category. To the best of our knowledge, there is no study that uses these two theories to study residents' attitudes toward sustainable tourism development. This study recommends applying these two theories to advance our knowledge on residents' attitudes due to recent changes (e.g., the growth of anti-tourism community groups and social and economic equalities in tourism development) (Alipour et al., 2017; Olya et al., 2019a). The ecological systems theory (Kline et al., 2013), Weber's theory of formal and substantive rationality (Boley et al., 2014; Gannon et al., 2020), and the complexity theory (Olya & Gavilyan, 2017) are three theories that follow the pragmatic and positivistic approach in studies on the attitudes of residents towards tourism development. These theories value the inclusion of multiple factors — ranging from personal (e.g., quality of life and demographics) to external (e.g., social and environmental impacts of tourism) — into the research model. Therefore, the results of model testing provide a comprehensive and accurate insight into residents' attitudes toward tourism development.

To tackle the complexity of resident's support for sustainable tourism development, future research needs to move beyond the syntactic/semantic outlook by using a mixed-method approach to pragmatically study this social complex phenomenon. This study encourages researchers to apply new theories (e.g., the nudge theory and justice theory) to address the socio-economic and environmental challenges of tourism development by involving residents. Future research can contribute to the advancement of theory within the field in three ways: 1. developing new theories using a grounded theory approach; 2. using key theories that match with exploratory research methods (e.g., the complexity theory with fsQCA); and 3. pruning theories and advocating for a rival theory based on the results of empirical studies (i.e., the corroboration/falsification of theory).

4.4. Pathway to impact

Tourism scholars need to go beyond developing theoretical and managerial implications by ensuring the impacts of their own research outputs. As shown in Figure 3, scholars need to ensure the impacts of their research by developing a logical model that illustrates the sequence of achieving the desired impacts (outputs ➔ outcomes ➔ impacts) (Alvarez et al., 2010). Research articles, book chapters, conference proceedings, workshops, seminars, blogs, social media posts, broadcasts, and a media/press presence are all examples of outputs produced by tourism researchers. According to the theory of change, to secure the practicality of the logical model, key actor groups should be identified and involved in the co-design, co-production, and co-delivery of outputs, outcomes (e.g., support of development and policy or intervention in it), and impacts of the research. The theory of change represents a causal framework illustrating why and how a cause (e.g., research outputs) lead to a desired change (e.g., social impacts) in a particular context (e.g., residents' support for sustainable tourism development).

Researchers need to conduct the stakeholder mapping analysis to develop a matrix of internal and external stakeholders (including local residents, businesses and services, DMOs, governments, NGOs, universities, and research centres) (Olya et al., 2019a). The extent of communication, involvement, and collaboration of key actor groups should be prioritised based on their interests and level of influence. According to the results of the stakeholder mapping analysis, more time, effort, and resources should be allocated to key actor groups to engage the latter in co-producing the outcomes and outputs (in the form of policies or interventions), as well as co-implementing the impacts. Apart from the power and interests of key actor groups, gender and

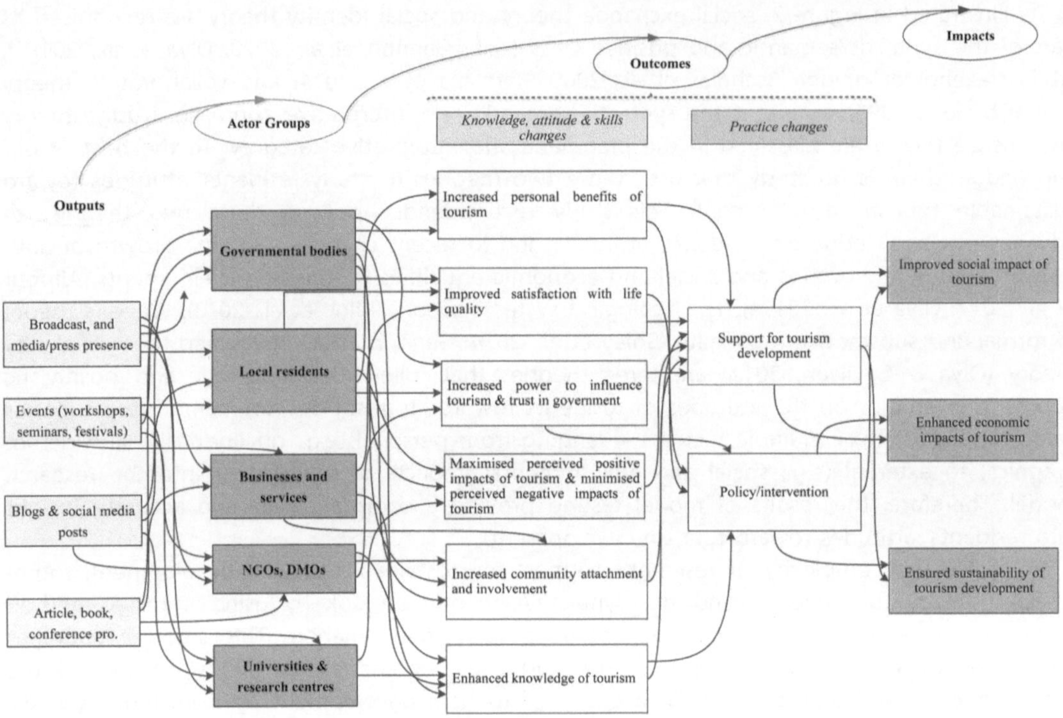

Figure 3. The logical model of the pathway to impact.

social equalities should also be considered criteria for involving the actor groups throughout the research project.

5. Conclusion

This study has contributed to the current knowledge of residents' support for tourism development in several ways. First, this is the first attempt to identify the necessary conditions for residents' support for tourism development. It provides an overview of the most frequent analytical approaches for model testing in the area of residents' attitudes toward sustainable tourism development. Second, four symmetrical analytical approaches have been used to investigate the net effect (sufficiency) of the predictors on residents' support for tourism development. The results of four techniques have been critically assessed. This study raises a red flag with respect to the precarious claims (e.g., a tool for theory development) and critiques (e.g., accuracy of estimator) of PLS-SEM. This study calls for further clarifications on the purpose and functionality of this tool in quantitative research, as it may offer misleading information and inaccurate results. A 10-step guideline for using a symmetrical analysis is presented. In addition to sufficiency and necessity, it has used fsQCA to explore the sufficient combinations of the predictors (i.e., solutions) for high and low scores of residents' support for tourism development. This study argues that each analytical approach (symmetrical and asymmetrical methods) provides different insights into the support for tourism development from residents' viewpoints. In fact, the results of the analytical approaches used in this study are complementary, providing a deeper understanding of sustainable support for tourism development. Third, this study has discussed methodological and theoretical innovations to advance research on residents' support for tourism development. This study proposes a classification of theories on residents' support for tourism development which highlights the importance of a theory-method fit in conducting impactful research. Fourth, drawing upon the theory of change, it develops a logical model demonstrating the pathways to impact. It encourages tourism scholars to focus on the pathways to impact to highlight the significance of research activities in improving tourism's impacts on society.

Acknowledgment

Support of Yaqub Gavylian and Dr. Mostafa Rasoolimanesh in completion of this research is highly appreciated.

Disclosure statement

No potential conflict of interest was reported by the author.

ORCID

Hossein G. T. Olya https://orcid.org/0000-0002-0360-0744

References

Armstrong, J. S., Brodie, R. J., & Parsons, A. G. (2001). Hypotheses in marketing science: Literature review and publication audit. *Marketing Letters*, *12*(2), 171–187. https://doi.org/10.1023/A:1011169104290

Bronfenbrenner, U. (1994). Ecological models of human development. *Readings on the Development of Children*, *2*(1), 37–43.

Markovsky, B. (1985). Toward a multilevel distributive justice theory. *American Sociological Review*, *50*(6), 822–839.

Ali, F., Rasoolimanesh, S. M., Sarstedt, M., Ringle, C. M., & Ryu, K. (2018). An assessment of the use of partial least squares structural equation modeling (PLS-SEM) in hospitality research. *International Journal of Contemporary Hospitality Management*, *30*(1), 514–538. https://doi.org/10.1108/IJCHM-10-2016-0568

Alipour, H., Olya, H. G., & Forouzan, I.. (2017). Environmental impacts of mass religious tourism: From residents' perspectives. *Tourism Analysis*, *22*(2), 167–183. https://doi.org/10.3727/108354217X14888192562285

Altinay, L., Song, H., Madanoglu, M., & Wang, X. L. (2019). The influence of customer-to-customer interactions on elderly consumers' satisfaction and social well-being. *International Journal of Hospitality Management*, *78*, 223–233. https://doi.org/10.1016/j.ijhm.2018.09.005

Alvarez, S., Douthwaite, B., Thiele, G., Mackay, R., Córdoba, D., & Tehelen, K. (2010). Participatory impact pathways analysis: A practical method for project planning and evaluation. *Development in Practice*, *20*(8), 946–958. https://doi.org/10.1080/09614524.2010.513723

Boley, B. B., McGehee, N. G., Perdue, R. R., & Long, P. (2014). Empowerment and resident attitudes toward tourism: Strengthening the theoretical foundation through a Weberian lens. *Annals of Tourism Research*, *49*, 33–50. https://doi.org/10.1016/j.annals.2014.08.005

Dijkstra, T. K. (2014). PLS'Janus face–response to professor Rigdon's 'rethinking partial least squares modeling: In praise of simple methods. *Long Range Planning*, *47*(3), 146–153.

Echegaray, F., & Hansstein, F. V. (2017). Assessing the intention-behavior gap in electronic waste recycling: The case of Brazil. *Journal of Cleaner Production*, *142*, 180–190. https://doi.org/10.1016/j.jclepro.2016.05.064

Feng, K., Altinay, L., & Olya, H. (2019). Social well-being and transformative service research: Evidence from China. *Journal of Services Marketing*, *33*(6), 735–750. https://doi.org/10.1108/JSM-10-2018-0294

Freund, R. J., & Wilson, W. J. (1998). *Regression analysis: Statistical modeling of a response variable*. Academic Press.

Gannon, M., Rasoolimanesh, S. M., & Taheri, B. (2020). Assessing the mediating role of residents' perceptions toward tourism development. *Journal of Travel Research*, https://doi.org/10.1177/0047287519890926.

Guide, V. D. R., Jr., & Ketokivi, M. (2015). Notes from the Editors: Redefining some methodological criteria for the journal. *Journal of Operations Management*, *37*(1), v–viii.

Gursoy, D., & Rutherford, D. G. (2004). Host attitudes toward tourism: An improved structural model. *Annals of Tourism Research*, *31*(3), 495–516. https://doi.org/10.1016/j.annals.2003.08.008

Hair, J. F., Hult, G. T. M., Ringle, C. M., & Sarstedt, M. (2017). *A primer on partial least squares structural equation modeling (PLS-SEM)* (2nd ed.). Sage.

Hair, J. F., Ringle, C. M., & Sarstedt, M. (2011). PLS-SEM: Indeed a silver bullet. *Journal of Marketing Theory and Practice*, *19*(2), 139–152. https://doi.org/10.2753/MTP1069-6679190202

Hair, J. F., Ringle, C. M., & Sarstedt, M. (2013). Partial least squares structural equation modeling: Rigorous applications, better results and higher acceptance. *Long Range Planning*, *46*(1–2), 1–12. https://doi.org/10.1016/j.lrp.2013.01.001

Hayes, A. F. (2017). *Introduction to mediation, moderation, and conditional process analysis: A regression-based approach*. Guilford publications.

Hayes, A. F., Montoya, A. K., & Rockwood, N. J. (2017). The analysis of mechanisms and their contingencies: PROCESS versus structural equation modeling. *Australasian Marketing Journal (Amj))*, *25*(1), 76–81. https://doi.org/10.1016/j.ausmj.2017.02.001

Jamal, T. B., & Getz, D. (1995). Collaboration theory and community tourism planning. *Annals of Tourism Research*, *22*(1), 186–204. https://doi.org/10.1016/0160-7383(94)00067-3

Kayat, K. (2002). Power, social exchanges and tourism in Langkawi: Rethinking resident perceptions. *International Journal of Tourism Research*, *4*(3), 171–191.

Khan, H. U. R., Ali, M., Olya, H. G., Zulqarnain, M., & Khan, Z. R. (2018). Transformational leadership, corporate social responsibility, organizational innovation, and organizational performance: Symmetrical and asymmetrical analytical approaches. *Corporate Social Responsibility and Environmental Management*, *25*(6), 1270–1283. https://doi.org/10.1002/csr.1637

Kline, C., Gard McGehee, N., Paterson, S., & Tsao, J. (2013). Using ecological systems theory and density of acquaintance to explore resident perception of entrepreneurial climate. *Journal of Travel Research*, *52*(3), 294–309. https://doi.org/10.1177/0047287512465962

Kutner, M. H., Nachtsheim, C. J., Neter, J., & Li, W. (2005). *Applied linear statistical models* (Vol. 5). McGraw-Hill Irwin.

Látková, P., & Vogt, C. A. (2012). Residents' attitudes toward existing and future tourism development in rural communities. *Journal of Travel Research*, *51*(1), 50–67. https://doi.org/10.1177/0047287510394193

Lee, T. H. (2013). Influence analysis of community resident support for sustainable tourism development. *Tourism Management*, *34*, 37–46.

Lee, C. K., Mjelde, J. W., Kim, T. K., & Lee, H. M. (2014). Estimating the intention–behavior gap associated with a mega event: The case of the Expo 2012 Yeosu Korea. *Tourism Management*, *41*, 168–177.

Madrigal, R.. (1993). A tale of tourism in two cities. *Annals of Tourism Research*, *20*(2), 336–353. https://doi.org/10.1016/0160-7383(93)90059-C

McCool, S. F., & Martin, S. R. (1994). Community attachment and attitudes toward tourism development. *Journal of Travel Research*, *32*(3), 29–34. https://doi.org/10.1177/004728759403200305

McGehee, N. G., & Andereck, K. L. (2004). Factors predicting rural residents' support of tourism. *Journal of Travel Research*, *43*(2), 131–140. https://doi.org/10.1177/0047287504268234

Mehran, J., Olya, H. G., Han, H., & Kapuscinski, G. (2020). Determinants of canal boat tour participant behaviours: An explanatory mixed-method approach. *Journal of Travel & Tourism Marketing*, *37*(1), 112–127.

Mohd Jamil, J. B. (2013). *Partial least squares structural equation modelling with incomplete data. An investigation of the impact of imputation methods* [Doctoral dissertation]. University of Bradford. http://hdl.handle.net/10454/5728.

Murakami, M., & Tsubokura, M. (2017). Evaluating risk communication after the Fukushima disaster based on nudge theory. *Asia-Pacific Journal of Public Health*, *29*(2_suppl), 193S–200S. https://doi.org/10.1177/1010539517691338

Nicholas, L. N., Thapa, B., & Ko, Y. J. (2009). Residents' perspectives of a world heritage site: The pitons management area, St. Lucia. *Annals of Tourism Research*, *36*(3), 390–412. https://doi.org/10.1016/j.annals.2009.03.005

Nunkoo, R., & So, K. K. F. (2016). Residents' support for tourism: Testing alternative structural models. *Journal of Travel Research*, *55*(7), 847–861. https://doi.org/10.1177/0047287515592972

O'brien, R. M. (2007). A caution regarding rules of thumb for variance inflation factors. *Quality & Quantity*, *41*(5), 673–690.

Olazabal, M., & Pascual, U. (2016). Use of fuzzy cognitive maps to study urban resilience and transformation. *Environmental Innovation and Societal Transitions*, *18*, 18–40. https://doi.org/10.1016/j.eist.2015.06.006

Olya, H. G., & Al-Ansi, A. (2018). Risk assessment of halal products and services: Implication for tourism industry. *Tourism Management*, *65*, 279–291. https://doi.org/10.1016/j.tourman.2017.10.015

Olya, H., & Alipour, H. (2015). Modeling tourism climate indices through fuzzy logic. *Climate Research*, *66*(1), 49–63. https://doi.org/10.3354/cr01327

Olya, H. G., Alipour, H., & Gavilyan, Y. (2018). Different voices from community groups to support sustainable tourism development at Iranian World Heritage Sites: Evidence from Bisotun. *Journal of Sustainable Tourism*, *26*(10), 1728–1748.

Olya, H., Altinay, L., Farmaki, A., Kenebayeva, A., & Gursoy, D. (2020a). Hotels' sustainability practices and guests' familiarity, attitudes and behaviours. *Journal of Sustainable Tourism*, 1–19. Doi: https://doi.org/10.1080/09669582.2020.1775622.

Olya, H., Bagheri, P., & Tümer, M. (2019b). Decoding behavioural responses of green hotel guests: A deeper insight into the application of the theory of planned behaviour. *International Journal of Contemporary Hospitality Management*, *31*(6), 2509–2525.

Olya, H. G., & Gavilyan, Y. (2017). Configurational models to predict residents' support for tourism development. *Journal of Travel Research*, *56*(7), 893–912. https://doi.org/10.1177/0047287516667850

Olya, H. G., & Han, H. (2020). Antecedents of space traveler behavioral intention. *Journal of Travel Research*, *59*(3), 528–544. https://doi.org/10.1177/0047287519841714

Olya, H., Jung, H., Dieck, M. C. T., & Ryu, K. (2020b). Engaging visitors of science festivals using augmented reality: Asymmetrical modelling. *International Journal of Contemporary Hospitality Management*, *32*(2), 769–796.

Olya, H. G., & Mehran, J. (2017). Modelling tourism expenditure using complexity theory. *Journal of Business Research*, *75*, 147–158. https://doi.org/10.1016/j.jbusres.2017.02.015

Olya, H., & Nia, T. H. (2020). The medical tourism index and behavioral responses of medical travelers: A mixed-method study. *Journal of Travel Research*, https://doi.org/10.1177/0047287520915278.

Olya, H. G., Shahmirzdi, E. K., & Alipour, H. (2019a). Pro-tourism and anti-tourism community groups at a world heritage site in Turkey. *Current Issues in Tourism*, *22*(7), 763–785.

Olya, H., Van Niekerk, M., Taheri, B., & Gannon, M. J. (2020c). Guest editorial: Innovative mixed and multi method approaches to hospitality and tourism research. *International Journal of Contemporary Hospitality Management*, *32*(4), 1385–1391.

Oxford English Dictionary. (2020). *Structural*. Retrieved July 4, 2020 from https://www.oxfordlearnersdictionaries.com

Pappas, I. O., Giannakos, M. N., & Sampson, D. G. (2019). Fuzzy set analysis as a means to understand users of 21st-century learning systems: The case of mobile learning and reflections on learning analytics research. *Computers in Human Behavior*, *92*, 646–659.

Patten, M. L., & Newhart, M. (2017). *Understanding research methods: An overview of the essentials*. Routledge.

Podsakoff, P. M., MacKenzie, S. B., Lee, J. Y., & Podsakoff, N. P. (2003). Common method biases in behavioral research: A critical review of the literature and recommended remedies. *The Journal of Applied Psychology*, *88*(5), 879–903. https://doi.org/10.1037/0021-9010.88.5.879

Rasoolimanesh, S. M., Jaafar, M., Kock, N., & Ramayah, T. (2015). A revised framework of social exchange theory to investigate the factors influencing residents' perceptions. *Tourism Management Perspectives*, *16*, 335–345. https://doi.org/10.1016/j.tmp.2015.10.001

Rawls, J. (2009). *A theory of justice*. Harvard university press.

Reinartz, W., Haenlein, M., & Henseler, J. (2009). An empirical comparison of the efficacy of covariance-based and variance-based SEM. *International Journal of Research in Marketing*, *26*(4), 332–344. https://doi.org/10.1016/j.ijresmar.2009.08.001

Retolaza, J. L., San-Jose, L., & Ruiz-Roqueñi, M. (2014). Ontological stakeholder view: An innovative proposition. *Global Business Review*, *15*(1), 25–36. https://doi.org/10.1177/0972150913515590

Romero, R. G. (2009). Estimating the impact of England's area-based intervention 'new deal for communities' on employment. *Regional Science and Urban Economics*, *39*(3), 323–331.

Rönkkö, M., McIntosh, C. N., Antonakis, J., & Edwards, J. R. (2016). Partial least squares path modeling: Time for some serious second thoughts. *Journal of Operations Management*, *47–48*(1), 9–27. https://doi.org/10.1016/j.jom.2016.05.002

Ryan, C. (2020). Refereeing articles including SEM–what should referees look for? *Tourism Critiques: Practice and Theory*, *1*(1), 47–61. https://doi.org/10.1108/TRC-03-2020-0002

Sarstedt, M., Hair, J. F., Jr, Nitzl, C., Ringle, C. M., & Howard, M. C. (2020). Beyond a tandem analysis of SEM and PROCESS: Use of PLS-SEM for mediation analyses! *International Journal of Market Research*, *62*(3), 288–299. https://doi.org/10.1177/1470785320915686

Sarstedt, M., Ringle, C. M., & Hair, J. F. (2017). Treating unobserved heterogeneity in PLS-SEM: A multi-method approach. In: Latan, H., Noonan, R. (eds) *Partial least squares path modeling* (pp. 197–217). Springer. Doi:https://doi.org/10.1007/978-3-319-64069-3_9.

Shaddick, G., Yan, H., Salway, R., Vienneau, D., Kounali, D., & Briggs, D. (2013). Large-scale Bayesian spatial modelling of air pollution for policy support. *Journal of Applied Statistics*, *40*(4), 777–794. https://doi.org/10.1080/02664763.2012.754851

Sharpley, R. (2014). Host perceptions of tourism: A review of the research. *Tourism Management*, *42*, 37–49.

Taheri, B., Olya, H., Ali, F., & Gannon, M. J. (2020). Understanding the influence of airport servicescape on traveler dissatisfaction and misbehavior. *Journal of Travel Research*, *59*(6), 1008–1028. https://doi.org/10.1177/0047287519877257

Westland, J. C. (2015). From paths to networks: The evolving science of networks. In Structural equation models (pp. 161–172). Springer, Cham. https://doi.org/10.1007/978-3-319-16507-3_9.

Wilkins, S., Neri, S., & Lean, J. (2019). The role of theory in the business/management PhD: How students may use theory to make an original contribution to knowledge. *The International Journal of Management Education*, *17*(3), 100316. Doi: 10.1016/j.ijme.2019.100316. https://doi.org/10.1016/j.ijme.2019.100316

Woo, E., Kim, H., & Uysal, M. (2015). Life satisfaction and support for tourism development. *Annals of Tourism Research*, *50*, 84–97. https://doi.org/10.1016/j.annals.2014.11.001

Woodside, A. G. (2013). Moving beyond multiple regression analysis to algorithms: Calling for adoption of a paradigm shift from symmetric to asymmetric thinking in data analysis and crafting theory. *Journal of Business Research*, *66*(4), 463–472.

Woodside, A. G. (2014). Embrace• perform• model: Complexity theory, contrarian case analysis, and multiple realities. *Journal of Business Research*, *67*(12), 2495–2503.

Woosnam, K. M., Norman, W. C., & Ying, T. (2009). Exploring the theoretical framework of emotional solidarity between residents and tourists. *Journal of Travel Research*, *48*(2), 245–258. https://doi.org/10.1177/0047287509332334

The effect of minority residents' attitudes and emotional solidarity on ethnic neighborhood tourism: a multigroup invariance analysis

Naho Ueda Maruyama, Manuel Alector Ribeiro and Kyle Maurice Woosnam

ABSTRACT

In urban areas where ethnic composition is not homogeneous, interethnic relations play a critical role in attitudes tourism representing a particular ethnic culture, or ethnic neighborhood tourism. However, few studies have explored minority residents' perceptions of interethnic relations, solidarity, and attitudes towards tourism. Given members of the dominant and minority ethnic groups differ in social status and power, they may have different orientations towards interethnic relationships and cultural representation. Furthermore, no study has compared multiple diasporic groups whose social relationships with the dominant ethnic group might differ within a country. By applying the common ingroup identity model and the interactive acculturation model, this study aims to explore whether and how interethnic relations with members of a dominant ethnic group influence attitudes about ethnic neighborhood tourism among members of ethnic minority groups whose culture is represented for tourism. This research focuses on Korean and Brazilian immigrants and their descendants. Analysis of 343 completed questionnaires revealed that interethnic relationships within the local communities influence minority residents' attitudes tourism development. However, the comparison of the two diasporic groups revealed have different orientations towards interethnic relations and tourism. This may be because of various acculturation experiences and socioeconomic conditions.

Introduction

Because a positive host-guest interaction is considered critical for tourists' satisfaction (Pizam et al., 2002; Ribeiro et al., 2018), researchers (e.g., Giles et al., 2013; Joo et al., 2018) have predominantly examined whether and how to reduce intergroup bias between locals and tourists. More recently, interethnic relations within a local community and their influence on residents' attitudes regarding tourism have gained some research momentum (Maruyama & Woosnam, 2015; Woosnam et al., 2016a, 2016b). Especially in urban areas where the ethnic composition is not homogeneous (i.e., "ethnic neighborhoods"), interethnic relations may play a critical role to form attitudes towards tourism that represent a particular ethnic culture, or ethnic neighborhood tourism.

Ethnic neighborhoods are areas with a high concentration of members of a particular ethnic group. Such neighborhoods were typically characterized by high rates of unemployment and crime, and thus, considered "no-go" areas. However, more recently, interest in these ethnic neighborhoods has developed among tourists, giving way to 'ethnic neighborhood tourism' (ENT).

ENT has also been linked to interethnic relationships within local communities. More precisely, though the percentage of minority residents is higher than the other areas, ethnic neighborhoods, particularly in Japan, still consist of residents from both the dominant and minority ethnic groups (Maruyama et al., 2017). In these areas, relationships between the two groups may influence local residents' perspectives concerning ENT because the niche form of tourism is associated with issues such as the local identity (Maruyama et al., 2017), access and distribution of resources (Aytar & Rath, 2012), and power (Maruyama et al., 2016).

The relationship between interethnic relations and support for tourism can be explained by the common in-group identity model (i.e., CIIM) which suggests positive effects of inclusive, one-group representation of various social groups (Dovidio et al., 2009). Indeed, previous studies (Maruyama & Woosnam, 2015; Woosnam et al., 2016a, 2016b) reveal that residents' positive attitudes towards their ethnic counterparts develop from a sense of emotional solidarity with counterparts, and ultimately, support for ENT especially among members of the dominant ethnic group. However, previous studies have largely neglected to explore these relationships from the perspectives of minority residents. Given members of the dominant and minority ethnic groups differ in social status, access to resources, and power, they may have different orientations towards interethnic relationships and cultural representation (Dovidio et al., 2009; Tropp & Pettigrew, 2005). Despite this, few studies have concurrently examined minority residents' perceptions of interethnic relations, solidarity, and attitudes concerning ENT.

Furthermore, no studies have compared multiple diasporic groups whose social relationships with the dominant ethnic groups might differ within a particular country. The interactive acculturation model (Bourhis et al., 1997) explains that immigrants' relations with the host society may differ among distinct ethnic groups depending on each group's acculturation experiences. As such, how each minority group values the relationship with the dominant group and individuals' formulated attitudes about tourism may differ.

The aim of this study is to explore whether and how interethnic relations and solidarity with members of a dominant ethnic group influence attitudes about ENT among members of ethnic minority groups whose culture is represented for tourism. This research focuses on members of two diasporic communities in Japan; namely, Korean and Brazilian immigrants and their descendants. To apply the common in-group identity model (Dovidio et al., 2009) and the interactive acculturation model (Bourhis et al., 1997), this study utilizes several constructs. They include ethnic attitudes (Amir & Ben-Ari, 1985; Pizam et al., 2002), emotional solidarity (Woosnam, 2012; Woosnam et al., 2009), and attitudes about ENT (Maruyama & Woosnam, 2015). Specific research questions include:

1. To what extent do ethnic attitudes towards Japanese residents influence minority residents' (i.e., Brazilian and Korean residents) attitudes about ENT?
2. To what extent do ethnic attitudes of minority residents influence perceived levels of emotional solidarity with Japanese residents?
3. To what extent do minority residents' perceived levels of emotional solidarity with Japanese residents influence the formers' attitudes about ENT?
4. Do the influence of ethnic attitudes and emotional solidarity on the attitudes towards ENT differ between Korean and Brazilian residents?

Literature review

Ethnic neighborhood and interethnic contacts within the local community

Ethnic neighborhoods are areas where there is a disproportionate presence of members of a particular ethnic group (Logan et al., 2011). Historically, these neighborhoods were created for the purpose to control the spread of ethnic culture to the wider society, typically characterized by high rates of unemployment, crime, congestion, and isolation from the surrounding area (Santos & Yan, 2008; Trevor & Monder, 2003). However, the deindustrialization and ideological shift of immigrants from an expectation of assimilation to one focused on embracing multi-culturalism has refashioned these disadvantaged areas. Owing to the change of social norms, ethnic neighborhoods that embrace ethnic diversity have begun to be utilized as tourism resources, serving to attract visitors eager to experience "authentic" ethnic foods and entertainment (Santos & Yan, 2008). Research has highlighted the origin of such neighborhoods and tourism from western, multi-ethnic regions, such as the United States (Drew, 2011; Santos & Yan, 2008) and Australia (Mak, 2003), spreading onto Asian countries, including Singapore (Lee et al., 2016), Malaysia (Khan, 2015), and Japan (Maruyama & Woosnam, 2015).

Some studies (see Aytar & Rath, 2012; Santos & Yan, 2008) have identified both merits and problems associated with ENT. In a positive vein, ENT stimulates employment and entrepreneurial opportunities (Santos & Yan, 2008). It also raises awareness of and appreciation for ethnic diversity among members of the dominant group, providing underrepresented groups the opportunity to gain a sense of control of their own culture (Pang & Rath, 2007: Maruyama et al., 2016). Contrary to this, ENT may reinforce neighborhood isolation from the surrounding area given the desire to showcase stereotypical images of the minority ethnic group used to attract visitors (Santos & Yan, 2008). ENT may also "sanitize" the area while fabricating the social reality that the area has been facing, such as ethnic prejudices and social inequalities (Aytar & Rath, 2012; Santos & Yan, 2008).Ethnic neighborhoods also provide suitable contexts to examine the relationships between tourism initiation, interethnic relationships and solidarity within a local community, and residents' attitudes concerning tourism. Historically, interactions have been limited between members of dominant ethnic and diasporic groups within ethnic precincts (Sturgis et al., 2014). However, with the advent of tourism, interaction between the two groups may increase because tourism that utilizes local resources often requires cooperation of different social groups within a community (Okazaki, 2008). Such cooperation has the potential to bring historically divided groups together and foster solidarity, which leads to the support for tourism as a common goal (Henderson, 2003; Maruyama & Woosnam, 2015).

This link between interethnic relations and support for tourism development can be explained by the common in-group identity model (i.e., CIIM) (Dovidio et al., 2009; Gaertner & Dovidio, 2000). The model, derived from the contact hypothesis and social identity theory (Tajfel & Turner, 1979), suggests that if the contact situation is structured in such a way that members of different social groups are re-categorized into and represented as one group, members may develop favorable attitudes towards former out-group members. Then, the favorable attitudes leads to the development of a sense of closeness and solidarity between such individuals. Research (e.g., Beaton et al., 2012; Gaertner et al., 2001) further indicates that this intergroup solidarity motivates individuals to endorse inclusive, one-group representation of the group and cooperative behavior. Indeed, Nier et al. (2001) reveal that when African-American and Caucasian football players interacted as members of the same group, Caucasian players evaluated African-Americans more favorably and presented cooperative attitudes with African-American interviewers' request. Other studies similarly indicate a positive effect of inclusive representation in the classroom (Houlette et al., 2004), banking mergers (Van Knippenber & Van Leeuwen, 2001), and stepfamilies (Gaertner et al., 2001).

Representation of ethnic neighborhoods for tourism can be thought of as a case of one-group representation because ENT often represents an ethnically diverse neighborhood, where members of both ethnic minority and dominant ethnic groups often reside together, as one community (Maruyama et al., 2017). ENT has the potential to allow locals of different ethnic backgrounds to interact and improve the intergroup relationship, develop solidarity with each other, and endorse ENT as a viable means of revitalizing their community. To this point, Maruyama and Woosnam (2015) reveal that residents' positive relationship with ethnic counterparts influence their attitudes regarding ENT. Such findings helped to foster the creation of H_1 and H_2. Furthermore, Woosnam et al. (2016a) as well as Woosnam et al. (2016b) reveal that positive relationships with ethnic counterparts predict levels of solidarity with other ethnic group representatives, and this solidarity predicts cooperative attitudes regarding ENT. Such findings prompted the creation of H_3, H_4, and H_5.

However, while previous studies support the idea that a common in-group identity model can explain locals' attitudes regarding ENT among members of the dominant ethnic group (i.e., Japanese residents), whether the same model can explain minority residents' attitudes towards ENT has yet to be examined. According to Dovidio et al. (2009), while members of the high-status group (i.e., dominant ethnic group) generally endorse a one-group representation, members of a disadvantaged group (i.e., ethnic minority groups) may refuse it. This may be because in a one-group representation, distinctions between subgroups become merged by focusing on commonality, and the merger often favors members of the powerful group. Therefore, this type of representation may particularly be resisted in the context where the disadvantaged groups' distinctive culture is central to members' functioning (Dovidio et al., 2009). Indeed, studies that support a common in-group identity model have predominantly been conducted from the perspectives of dominant ethnic group residents (e.g., Nier et al., 2001; Woosnam et al., 2016a, 2016b).

Therefore, scholars have suggested that for members of lower-status groups (who may refuse a one-group form of representation), a dual identity form of representation may be a viable alternative (Dovidio et al., 2009; Hewstone & Brown 1986). A dual identity form of representation focuses on maintaining the original group boundaries and distinctiveness of each group, rather than focusing on commonality, while keeping a positive connection with members of the other groups. Dovidio et al. (2009) hypothesize that minority group members prefer a dual identity form of representation while members of a dominant group typically prefer an inclusive representation.

In the context of ENT, members of a minority group may stand in a unique position. On one hand, they are members of a "disadvantaged group" who are typically concerned about the distinctiveness of their culture being relinquished in cultural representation (Dovidio et al., 2009). This can be particularly true in an ENT destination because the ethnic group's unique culture is a central resource that helps members improve their economic and social status. Therefore, while the minority residents may appreciate the emerging relationship with members of the majority group, they may not welcome tourism development as a "community project" (Dovidio et al., 2009). On the other hand, members of the minority group in an ENT destination may possibly be in a "high-status" position because they have privileged access to tourism resources (Maruyama et al., 2016), and thus they may see the value, rather than a threat to interethnic solidarity and the inclusive representation of a particular neighborhood. However, the connections between positive relationships with ethnic counterparts, interethnic solidarity, and support for tourism have not yet been explored from the perspective of ethnic minority residents. Therefore, the main goal of this study is to examine such relationships with the utilization of the common in-group identity model.

Measuring interethnic relations and solidarity

In the previous studies mentioned above (Maruyama & Woosnam, 2015; Woosnam et al., 2016a, 2016b), interethnic relationships are often measured through the concept of ethnic attitudes.

Ethnic attitudes can be defined as attitudes held toward a group by non-members of the group, and considered to have strong links to concepts of tolerance, prejudice, and social inclusion/exclusion of ethnic groups (Kalin & Berry, 1982). To examine these ethnic attitudes, the Ethnic Attitude Scale (EAS) developed by Osgood et al. (1957) has been employed and well-accepted within tourism studies (Amir & Ben-Ari, 1985; Nyaupane et al., 2008). The scale includes a series of adjectives in a semantic differential format, with recent studies (Maruyama & Woosnam, 2015) yielded two distinct dimensions, namely *character and intelligence* and *social evaluation.*

Beyond this, to measure intergroup solidarity, the concept of emotional solidarity (Woosnam & Norman, 2010) has often been applied in tourism studies, while several constructs, such as helping behavior (Nadler, 2002) or self-disclosure (Gaertner et al., 1998) have been used in social psychology. Emotional solidarity, having its roots back in the work of the classical sociologist, Emile Durkheim (1995 [1915]), is defined as the affective bonds and closeness experienced and perceived between individuals (Woosnam, 2012). In tourism research, Woosnam et al. (2009) first introduced and operationalized the concept to measure intergroup emotions from multiple perspectives. According to Woosnam et al., emotional solidarity is comprised of three key dimensions: *welcoming nature*, *emotional closeness*, and *sympathetic understanding*. Based on the above discussion, the following hypotheses are advanced.

H_1: Minority residents' level of ethnic attitudes (i.e., *character and intelligence*) of residents in the dominant ethnic group will significantly predict the former's attitudes towards ethnic neighborhood tourism.

H_2: Minority residents' level of ethnic attitudes (i.e., *social evaluation*) of residents in the dominant ethnic group will significantly predict the former's attitudes towards ethnic neighborhood tourism.

$H_{3(a)}$: Minority residents' attitudes towards the level of *character and intelligence* of residents in the dominant ethnic group will significantly predict the former's degree of emotional solidarity (a: *welcoming nature*) with the latter.

$H_{3(b)}$: Minority residents' attitudes towards the level of *character and intelligence* of residents in the dominant ethnic group will significantly predict the former's degree of emotional solidarity (*b: emotional closeness*) with the latter.

$H_{3(c)}$: Minority residents' attitudes towards the level of character and intelligence of residents in the dominant ethnic group will significantly predict the former's degree of emotional solidarity (*c: sympathetic understanding*) with the latter.

$H_{4(a)}$: Minority residents' attitudes towards the level of *social evaluation* of residents in the dominant ethnic group will significantly predict the former's degree of emotional solidarity (a: *welcoming nature*) with the latter

$H_{4(b)}$: Minority residents' attitudes towards the level of *social evaluation* of residents in the dominant ethnic group will significantly predict the former's degree of emotional solidarity (b: *emotional closeness*) with the latter

$H_{4(c)}$: Minority residents' attitudes towards the level of social evaluation of residents in the dominant ethnic group will significantly predict the former's degree of emotional solidarity (c: sympathetic understanding) with the latter

$H_{5(a)}$: Minority residents' degree of emotional solidarity (a: *welcoming nature*) with residents in the dominant ethnic group will predict the former's attitudes towards of ethnic neighborhood tourism.

$H_{5(b)}$: Minority residents' degree of emotional solidarity (b: *emotional closeness*) with residents in the dominant ethnic group will predict the former's attitudes towards of ethnic neighborhood tourism.

$H_{5(c)}$: Minority residents' degree of emotional solidarity (c: *sympathetic understanding*) with residents in the dominant ethnic group will predict the former's attitudes towards of ethnic neighborhood tourism.

Lastly, there is potential for interethnic relations and its influence on attitudes concerning ENT to differ among ethnic minority groups even within a country. Indeed, Woosnam et al. (2019) reveal that two ethnic communities in Japan have different perceptions towards tourism in general, and identify possible reasons for the differences, including the presence of other economic activities (outside of tourism) in the neighborhoods or levels of maturity as a tourism destination of the areas. In addition to these reasons, distinct natures of the acculturation experiences might differentiate various ethnic communities' attitudes towards ENT. Such reasoning led us to formulate H_6, H_7, H_8, H_9, and H_{10}. The interactive acculturation model (Bourhis et al., 1997) argues that immigrants' interethnic relations with the host society are mutually influenced by acculturation expectations of the immigrant and host society. According to Van Oudenhoven et al. (2006), most immigrants prefer "integration" that signifies maintaining both original cultural and positive relations with the host society. On the contrary, the acculturation expectation held by the host society towards the immigrant group may vary by integration, assimilation, or marginalization depending both on the host country's policy towards immigrants and the national origin of the immigrants being assessed. Van Oudenhoven et al. further maintain that if the acculturation expectations by the immigrants and the host society do not match, perceived discrimination and distress would pervade.

In Japan, Korean and Brazilian residents have experienced a different nature of acculturation expectation within society, which may contribute to different levels of distress for the two ethnic minority groups. On the one hand, most of the Korean residents are descendants of the colonial migrants to Japan. Because of the nation's exclusion policy towards the post-colonial foreigners before the 1980s, they were excluded from the welfare system and pressured to assimilate in Japanese society to reinforce the idea of Japan as an ethnically homogeneous nation (Chapman, 2006; Tai, 2009). Owing to such experiences, many Korean residents are keenly aware of their ethnic heritage even after the immigration policy in Japan became more inclusive in the late 1980s. In other words, their acculturation experiences are characterized by the struggle for integration. The international relations between Japan and Korea has also long been characterized with negative sentiments, due to the colonial history and current political conflicts (e.g., disputes over sexual slavery of Korean women during WWII, unresolved territorial issues over Takeshima/Dokdo Islands, and issues regarding the trade arrangement). Such issues make Korean residents the "stigmatic immigrant group" (Bourhis et al., 1997), towards which residents of Japan often hold a segregation and exclusion orientation.

On the other hand, most of the Brazilian residents immigrated to Japan between the late 1980s and early 1990s, when Japan's immigration policies began to change from excluding foreigners to *kyosei* (integration). During this time, a number of local governments that hosted large populations of foreign workers were proactive to integrate these immigrants into the local community and protect their rights (Tai, 2009). Owing to the changes of the social norm towards foreign residents, Japanese residents relatively tolerated the presence of the Brazilian population (Tsuda, 2003). In addition, relations between Japan and Brazil have been friendly, and immigrants were flowing from Japan to Brazil before WWII with an economic motivation. Indeed, current Brazilian residents in Japan are descendants of the Japanese migrants to Brazil, or "return migrants" (Tsuda, 2003). Because of such historical and ethnic connection with Brazil, Japanese residents perceive Brazilians as a "favored" immigrant group. Therefore, Brazilian residents' experiences of acculturation seem less contentious compared to the experiences of Koreans, with the former less likely to perceive of their social status as "disadvantaged."

H_6: Minority residents' perceived level of ethnic attitudes (i.e., *character and intelligence*) of residents in the dominant ethnic group will significantly predict the former's attitudes towards ethnic neighborhood tourism and this relationship will be stronger for Brazilian residents.

H_7: Minority residents' perceived level of ethnic attitudes (i.e., *social evaluation*) of residents in the dominant ethnic group will significantly predict the former's attitudes towards ethnic neighborhood tourism and this relationship will be stronger for Brazilian residents.

$H_{8(a)}$: Minority residents' perceived level of *character and intelligence* of residents in the dominant ethnic group will significantly predict the former's degree of emotional solidarity (a: *welcoming nature*) and this relationship will be stronger for Brazilian residents.

$H_{8(b)}$: Minority residents' perceived level of character and intelligence of residents in the dominant ethnic group will significantly predict the former's degree of emotional solidarity (b: emotional closeness) and this relationship will be stronger for Brazilian residents.

$H_{8(c)}$: Minority residents' perceived level of character and intelligence of residents in the dominant ethnic group will significantly predict the former's degree of emotional solidarity (c: sympathetic understanding) and this relationship will be stronger for Brazilian residents.

$H_{9(a)}$: Minority residents' perceived level of *social evaluation* of residents in the dominant ethnic group will significantly predict the former's degree of emotional solidarity (a: *welcoming nature*) with the latter, and this relationship will be stronger for Brazilian residents.

$H_{9(b)}$: Minority residents' perceived level of *social evaluation* of residents in the dominant ethnic group will significantly predict the former's degree of emotional solidarity (b: *emotional closeness*) with the latter, and this relationship will be stronger for Brazilian residents.

$H_{9(c)}$: Minority residents' perceived level of *social evaluation* of residents in the dominant ethnic group will significantly predict the former's degree of emotional solidarity (c: *sympathetic understanding*) with the latter, and this relationship will be stronger for Brazilian residents.

$H_{10(a)}$: Minority residents' degree of emotional solidarity (a: *welcoming nature*) with residents in the dominant ethnic group will predict the former's attitudes towards ethnic neighborhood tourism and this relationship will be stronger for Brazilian residents.

$H_{10(b)}$: Minority residents' degree of emotional solidarity (b: *emotional closeness*) with residents in the dominant ethnic group will predict the former's attitudes towards ethnic neighborhood tourism and this relationship will be stronger for Brazilian residents.

$H_{10(c)}$: Minority residents' degree of emotional solidarity (c: *sympathetic understanding*) with residents in the dominant ethnic group will predict the former's attitudes towards ethnic neighborhood tourism and this relationship will be stronger for Brazilian residents.

Methods

Study sites and context

The survey in this study was conducted in two ethnic neighborhoods in Japan; namely, the Brazilian neighborhood in Oizumi town, Gunma prefecture, and the Korean neighborhood in Ikuno ward, Osaka prefecture. These two sites were chosen given geographical- and social-accessibility to the researchers.

Oizumi town, a well-known manufacturing town, in Gunma prefecture is located approximately 110 km northwest of Tokyo. According to the 2019 census, 41,841 people, including 7,686 foreign residents, reside in Oizumi (Oizumi Town, 2019). Among the foreign residents, 4,367 people (i.e., 57% of the foreign residents) are Brazilians—the largest concentration of Brazilian residents in Japan.

The Brazilian neighborhood in Oizumi town emerged in the early 1990s, when Japan was experiencing a serious labor deficiency in the midst of a booming economy. To respond to the urgent need to address the labor shortage, the Japanese government amended a new immigration act that granted residence visas to Japanese emigrants, their descendants, and family members (Tsuda, 2003). Following the policy change, the mayor of Oizumi town actively recruited Brazilian workers, which resulted in a substantial influx of Brazilians to the town. Most of the Brazilian residents in Oizumi town have racial connections to Japan because they are second or third generations of Japanese migrants to Brazil. Yet, they are generally considered "foreigners" because of their distinct culture, including language, food, custom, and ways of communication

(Tsuda, 2003). The increased concentration of Brazilian residents stimulated the development of a couple of ethnic business districts.

Ethnic neighborhood tourism focusing on Brazilian culture started in 2007 as the local government attempted to revitalize the town's declining economy due to the closing of the largest factory in Oizumi town. To attract tourists, several events centered on Brazilian heritage and culture were organized by the tourism bureau, including "Gourmet Yokocho," –a street food festival held every fourth Sunday and the "Oizumi Samba Festival" –a major, annual event to celebrate Brazilian culture. Other tourism resources in town include two Brazilian supermarkets, several small snack stores, and restaurants, all of which were predominantly used by local Brazilian residents before the introduction of tourism.

The second study site for this research is the Korean town in Ikuno ward, located near the central area of Osaka. Tsuruhashi station, the closest train station to the town, is approximately 15 minutes away from the Osaka Station. According to most recent data, 127,783 people live in the Ikuno ward, and 23,499 claim to be Korean nationals (Ikuno Ward, 2019). In addition, the number of "Japanese residents" includes those of Korean descent who have been naturalized as Japanese citizens. This makes Ikuno ward the largest Korean community in Japan. Japanese and Korean populations not only share common physical characteristics (i.e., color of skin and hair) but also similar cultural traits in food, customs, and language (Minauskaitė, 2016). Yet, the Korean neighborhood in Ikuno ward has long been isolated from the surrounding area because of the historical and political conflicts between Korean and Japan.

The Korean neighborhood in Ikuno ward emerged at the beginning of the 20th Century when many colonial migrants from Korea emigrated to Japan to work on maintenance projects surrounding the Hirano river (which began in 1919) (Maher & Kawanishi, 2012). Owing to this history of colonization, Korean residents have faced harsh discrimination on a daily basis. Many were forced to assimilate into Japanese society by changing their names and language (Murphy-Shigematsu, 2000). By the same token, the Korean residential area had long been considered by the dominant society as a place to avoid. However, the "Korean Wave," also known as *Hallyu* (the South Korean pop culture) swept through Japan in the early 2000s, providing a significant turning point for the area. The wave not only changed the social climate to encourage Korean residents to express their ethnic heritage but also re-fashioned the Korean town as a "sacred place" for *Hallyu* fans. At the heart of the current Korean town is a central shopping district with approximately 120 restaurants, grocery stores, clothing stores, souvenir shops, and other Korean businesses.

Questionnaire design

While previous research has undertaken factor analysis on the Ethnic Attitudes Scale to identify salient factors of the measure, the number of factors identified varies from two (Amir & Ben-Ari, 1985) to three (Pizam et al., 1991). This study followed a two-factor structure (i.e., *character and intelligence* and *social evaluation*) in keeping with recent studies yielding the same factors (Maruyama & Woosnam, 2015; Woosnam et al., 2016b).

To examine differences in emotional solidarity between residents of ethnic minority and dominant ethnic groups, the authors utilized the 10-item Emotional Solidarity Scale (see Table 2) as developed by Woosnam and Norman (2010). The 10 items comprise three factors (i.e., *welcoming nature*, *emotional closeness*, and *sympathetic understanding*).

To measure individuals' concerns about ethnic neighborhood tourism, an eight-item scale developed by Maruyama and Woosnam (2015) was used. Among the eight, four items concern the employment and social connection with their co-ethnics and ethnic-counterparts, and the other four items pertain to cultural representation. This scale is treated as unidimensional. The Ethnic Attitude Scale used a 7-point semantic differential scale, where each item included bipolar adjectives describing Japanese residents (i.e., "1" indicates negative perceptions, whereas "7" indicates positive perceptions). For the other two scales, respondents were asked to indicate their level of agreement by using a 7-point scale, where 1 = strongly disagree and 7 = strongly agree. Finally, residents were asked standard questions concerning socio-demographic information (e.g., gender, age, marital status, education, annual household income, and length of residency).

Data collection and sample characteristics

The target populations for this study were Brazilian residents (in Oizumi) and Korean residents (in Ikuno) living in the two Japanese study sites. Heads of households or their spouses in both areas were sampled on weekends from November, 2013 to June, 2015. To collect data, a multistage cluster sampling method was utilized (Babbie, 2014). Oizumi town was reduced to 30 administrative areas designated by the town office of Oizumi, and 28 of them were randomly selected to visit. Similarly, Ikuno ward was reduced to 19 administrative areas designated by the town office, 15 of them were randomly selected. There are more administrative areas in Oizumi town than Ikuno ward because Oizumi town is larger than Ikuno. In the selected areas, the research teams visited every second household between the hours of 11:00am and 4:00pm on weekends, starting in randomly-selected locations within each area.

Student assistants from numerous local universities comprised the data collection research teams. They described the nature of the study and asked each head of household or their spouse to participate. If the resident agreed, a questionnaire was distributed and collected later that day (Woosnam, 2012). If no one answered the residence, assistants visited the next house, and the second-house sequence was restarted. Questionnaires were first prepared in English, and translated into Japanese, Portuguese, and Korean languages. For all languages, back translation was performed following Brislin's (1970) recommendation.

In Oizumi town, 5,566 households were visited in total. Among them, no one answered the door at 4,022 households, with 1,544 individuals contacted. Of those, 700 individuals agreed to participate (an acceptance rate of 45%) while 854 declined. In total, 662 surveys were completed (a completion rate of 94%). This resulted in an overall response rate of 42%. Of the 662 surveys completed, 183 survey instruments completed by Brazilian residents were used for this study. The rest includes 12 survey instruments less than 50% completed and 467 were completed by Japanese residents, neither of which was included in this study.

Similarly, in Ikuno ward, 5,930 households were visited, and 2,489 residents answered the door. From those that were contacted, 733 surveys were distributed (an acceptance rate of 29%). In total, 640 surveys were returned (overall response rate of 26% and a completion rate of 89%). Of the 640 surveys returned, 160 usable instruments completed by Korean residents were used in this study. The rest included 14 less-than-half completed surveys and 466 surveys completed by Japanese residents, neither of which was included in this study. The response rate of this study was relatively low. This may be because both Oizumi town and Ikuno ward consists of more single family houses than apartment complexes. Miwa (2008) pointed out that surveying single family houses resulted in a higher refusal rate than visiting apartment complexes in Japan possibly for a security reason.

Arguably, the data may seem dated. Yet, as Zimmerman (2008) states, researchers can expand our knowledge and better identify the problems by asking new questions based on data collected previously. Particularly, exploring the relationship between members of the ethnic minority and dominant ethnic groups from the perspective of tourism is an emerging field of research in Japan. Therefore, looking at previously collected data with newly acquired knowledge of literature, social movements, and political and economic conflicts helps deepen our knowledge and fill research gaps.

Table 1. Residents characteristics by ethnic group.

	Brazilian Residents		Korean residents	
Variables	*n*	%	*n*	%
Gender *(n = 330)*				
Male	79	45.4	70	44.9
Female	95	54.6	86	55.1
Age *(n = 312)*				
18-35 years	63	37.7	20	13.8
36-50 years	61	36.5	39	26.9
51-65 years	38	22.8	44	30.3
66 and over	5	3.0	42	29.0
Marital status *(n = 322)*				
Single	45	26.2	33	22.0
Married	123	71.5	115	76.7
Divorced or Separated	2	1.1	1	0.7
Widowed	2	1.1	1	0.7
Education *(n = 333)*				
Junior high or high school graduate	122	66.2	83	53.9
Technical school or junior college	26	15.4	39	25.3
Four year college or graduate school	31	18.4	32	20.8
Annual household income *(n = 287)*				
Less than ¥2,000,000	60	37.3	33	26.2
¥2,000,000-3,999,999	81	50.3	34	27.0
¥4,000,000-5,999,999	18	11.2	31	24.6
¥6,000,000-7,999,999	1	0.5	11	8.7
¥8,000,000 or more	1	0.5	17	13.5
Length of Residence *(n = 326)*				
1–10 years	113	66.1	17	11.0
11–20 years	49	28.7	20	12.9
21–30 years	7	4.1	21	13.5
31–40 years	2	1.2	22	12.4
40 + years	0	0.0	75	48.4
Interaction with Japanese residents *(n = 325)*				
No interaction at all	68	39.3	33	21.7
Once or twice per year	28	16.2	16	10.5
Once or twice per month	38	22.0	16	10.5
Once or twice per week	16	9.2	19	12.5
More than twice per week	23	13.3	68	44.7

Regarding the respondents' gender, 55% were female ($n = 180$), 45% were male ($n = 147$). The respondents' average age was 37 years. Concerning the educational level, 61.56% of the respondents indicated they were either junior high school or high school graduate, and only 10.3% revealed that they completed a graduate degree. Regarding income level, approximately 12.31% revealed that their annual household income was between ¥2,000,000 and 3,999,999 (40.5%), followed by those who earned less than ¥2,000,000 annually (31.3%), and between ¥4,000,000 and 5,999,999. For a detailed description of the sample profile, see Table 1.

Data analysis and preliminary statistical verification

Preliminary statistical verification for the six composite variables were conducted in SPSS 25. Mplus 8.2 (Muthén & Muthén, 1998–2017) was used to test the model fit of the multigroup confirmatory factor analysis (MGCFA) and the proposed structural invariance of the ethnic neighborhood tourism model across Brazilian and Korean residents (Figure 1). Specifically, the maximum likelihood (ML) estimation in combination with the two-stage approach put forth by Anderson and Gerbing (1988) was used to analyse data. To begin, a confirmatory factor analysis (CFA) was conducted to establish structural validity and reliability of the six constructs in the pooled sample.

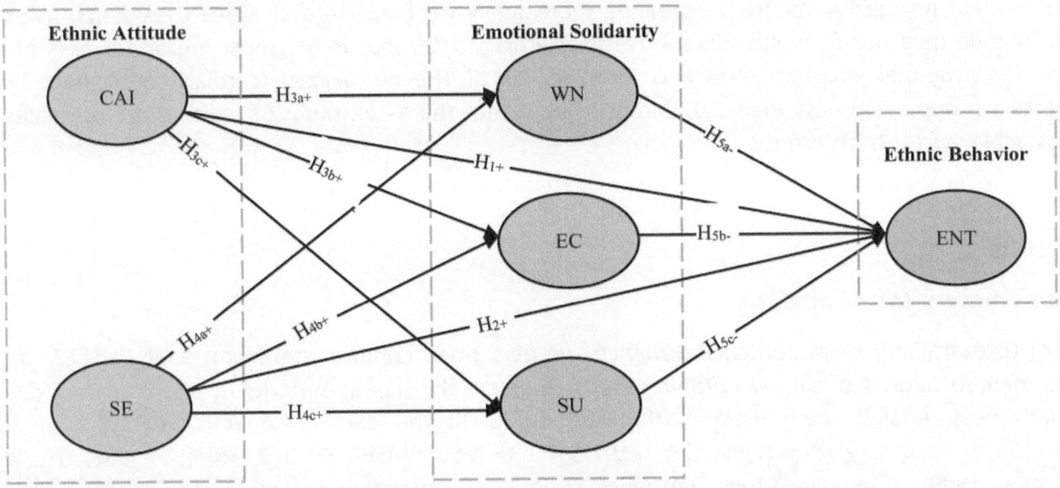

Figure 1. Hypothesized conceptual framework.
Note: CAI = Character and Intelligence; SE = Social Evaluation; WN = Welcoming Nature; EC = Emotional Closeness; SU = Sympathetic Understanding; ENT = Ethnic Neighborhood Tourism.

Since χ^2 is sensitive to sample size (Brown, 2015; Kline, 2011), other metrics should be used when assessing model fit (Hu & Bentler, 1999). The χ^2 statistics, comparative fit index (CFI), Tucker-Lewis index (TLI), root mean square error of approximation (RMSEA) along with its associated 90% confidence interval (CI), and the standardized root mean square residual (SRMR) were assessed for model fit. Following the standards used in preceding literature (Browne & Cudeck, 1989; Hu & Bentler, 1999), an acceptable model fit is considered with CFI and TLI figures higher than .90, and RMSEA and SRMR figures below .08. Measurement invariance (MI) was assessed across ethnic groups (i.e., Brazilians and Koreans) by using multigroup confirmatory factor analysis (MGCFA) (Jöreskog, 1971), which is considered the most commonly used method to test MI. Assessing MI involves the use of progressively restrictive equality constraints across groups to assess model fit (Muthén & Asparouhov, 2018; Vandenberg & Lance, 2000).

To assess the MGCFA, we initially assessed our model fit considering the three different degrees of measurement invariance (Meade et al., 2008; Millsap, 2001; Vandenberg & Lance, 2000). Firstly, configural invariance stipulates that the same factor loadings and measurement intercepts are invariant across groups; afterwards we assessed the metric invariance (in which factor loadings were constrained across groups) and this model was compared to the configural invariance (Vandenberg & Lance, 2000). Lastly, we compared the metric invariance model fit to the scalar invariance model fit, where factor loadings and item intercepts were fixed to be invariant across groups.

The chi-square difference test ($\Delta\chi^2$) was used to compare the model fit. It is recommended that multiple indicators be used to assess variation in model fit (Meade et al., 2008; Sass et al., 2014; Vandenberg & Lance, 2000); then, we also examined differences in RMSEA (ΔRMSEA), CFI (ΔCFI) and TLI (ΔTLI). Cheung and Rensvold (2002) suggested thresholds of ΔCFI > .010 and ΔRMSEA > .015 as suggestive of non-invariance. Since there are some difficulties to evaluate variation in model fit for ordinal data (Fredrick et al., 2019), we decided to use multiple metrics ($\Delta\chi^2$, ΔCFI, ΔTLI, ΔRMSEA) to evaluate the measurement invariance. As recommended, if metrics surpassed standards for non-invariance, partial invariance would be assessed.

This study used a maximum likelihood (ML) estimation approach that is known to produce biased outcomes when the normality assumption is not met (Curran et al., 1996). The normality of each construct was assessed considering their kurtosis and skewness values. Following recommendations of severe nonmorality (i.e., skew > 2; kurtosis > 7) recommended by West et al. (1995), the normality assumption of all the six factors was below the recommended threshold.

Multivariate normality was then examined individually for Brazilian and Korean residents, using the Mardia measure of multivariate kurtosis (Mardia, 1970), that is the most important assumption for structural equation modeling (Browne, 1982). The normalized Mardia's coefficients for the two groups were less than 2.0, demonstrating that the assumption of multivariate normality was achieved for both groups.

Results

Confirmatory factor analysis

Prior to examining measurement invariance, we performed confirmatory factor analysis (CFA) on the merged dataset of the two ethnic groups to assess the baseline model fit and construct distinctiveness. Analysis demonstrated that the model fit the data reasonably well ($\chi^2_{(343)}$ = 548.194, $p < .001$, $\chi^2/df = 1.598$, CFI = .971, TLI = .966, RMSEA = .042; SRMR = .038) (Hu & Bentler, 1999). Afterwards, the CFA were conducted separately for each ethnic group (i.e., Brazilian and Korean). The models demonstrated a reasonable fit for Brazilian ($\chi^2_{(350)}$ = 537.50, $p < .001$, $\chi^2/df = 1.536$, CFI = .949, TLI = .941, RMSEA = .054; SRMR = .054) and Korean residents ($\chi^2_{(350)}$ = 533.29, $p < .001$, $\chi^2/df = 1.523$, CFI = .950, TLI = .942, RMSEA = .057; SRMR = .050). All factor loadings for three models (i.e., pooled sample, Brazilian and Korean data) were significant and in the expected direction, ranging from .63 to .92 (see Table 2), suggesting convergent validity of the three measurement models. Likewise, all AVEs for each dimension were higher than 0.5, demonstrating that indicators were able to represent each construct respectively (Anderson & Gerbing, 1988). Furthermore, discriminant validity was met, as all AVEs that ranged from 0.51 to 0.76, were higher than squared correlations between the constructs (Fornell & Larcker, 1981). Likewise, Cronbach's alpha values were all higher than 0.7, which confirmed that the measurement models in three samples were both valid and reliable (see Tables 2 and 3).

Measurement invariance across ethnic groups

Multigroup analyses were performed to test measurement invariance across the ethnic groups (i.e., Brazilian and Korean). This involved comparing a baseline (unconstrained) model (which specified the same factor structure for each group (i.e., configural model)) to a model that constrained factor loadings (i.e., metric model) and thresholds (i.e., scalar model) across groups (Vandenberg & Lance, 2000). The configural model revealed reasonable fit, $\chi^2_{(686)}$ 1046.02, $p < .001$, CFI = .951, TLI = .942, SRMR = .050, and RMSEA = .055, with 90% CI [.049-.062]. All factor loadings across the two ethnic groups were significantly different from zero ($p < .001$) and ranged from .59 to .88 for Brazilian and .63 to .92 for Koreans. The metric model was then assessed by constraining the factor loadings across groups, $\chi^2_{(709)}$ = 1064.30, $p < .001$, CFI = .951, TLI = .944, SRMR = .057, and RMSEA = .054, with 90% CI [.047-.061] and this model was compared to the configural model; the chi-square difference test was nonsignificant, $\chi^2_{(23)}$ = 18.26 ($p < .74$). Additional fit indices also provided evidence for metric invariance across ethnic groups (ΔRMSEA = .001, ΔCFI = .000, ΔTLI = 002). Then, scalar invariance was assessed by constraining both the factor loadings and thresholds to be equal across groups, $\chi^2_{(732)}$ = 1082.86, $p < .001$, CFI = .952, TLI = .947, SRMR = .058, and RMSEA = .053, with 90% CI [.046-.069]. The scalar model was then compared to the metric model; the chi-square difference test was nonsignificant, $\chi^2_{(23)}$ = 18.56 (p = .72), and the values for additional fit indices were below the recommended thresholds (ΔRMSEA = .001, ΔCFI = .001, ΔTLI = .003), demonstrating that scalar invariance across ethnic group was also achieved (see Table 4).

Table 2. Results of the pooled and multi-group measurement model: reliability and convergent validity.

Pooled Sample (n = 343)			Brazilian (*n = 181*)		Korean (*n = 162*)	
Constructs and Indicators	*β*	*T-Values*	*β*	*T-Values*	*β*	*T-Values*
Character and Intelligence[a]	α = 0.93		α = 0.91		α = 0.91	
Rigid – flexible	0.75	14.86***	0.74	13.77***	0.75	9.01***
Stupid – intelligent	0.87	16.57***	0.88	12.16***	0.85	11.21***
Illiterate – educated	0.83	15.84***	0.84	15.56***	0.81	10.68***
Dirty – clean	0.88	16.83***	0.88	12.10***	0.88	11.66***
Not at all like me – like myself	0.78	14.80***	0.80	10.93***	0.77	9.92***
Abrupt – polite	0.78	14.20***	0.77	10.11***	0.80	10.10***
Immoral – moral	0.80	14.88***	0.80	10.75***	0.80	10.36***
Social Evaluation[a]	α = 0.87		α = 0.81		α = 0.89	
Cold-hearted – warm-hearted	0.82	20.45***	0.83	14.41***	0.83	12.59***
Bad-good	0.87	22.20***	0.86	15.16***	0.89	16.39***
Cruel-kind	0.88	19.21***	0.85	13.06***	0.91	14.22***
Old-fashioned – modern	0.91	19.41***	0.88	13.51***	0.92	13.87***
Discriminate against women – egalitarian	0.83	18.00***	0.80	12.27***	0.85	13.11***
Tense-relaxed	0.74	15.32***	0.73	10.65***	0.76	11.05***
Weak – powerful	0.74	15.35***	0.74	11.11***	0.73	10.64***
Welcoming Nature[b]	α = 0.84		α = 0.70		α = 0.76	
I appreciate Japan residents for the contribution they make to the local economy	0.73	12.60***	0.72	11.91***	0.79	10.03***
I treat Japan residents fairly.	0.76	13.91***	0.70	8.41***	0.82	11.66***
I am proud to have Japan residents in Oizumi/Osaka	0.91	16.82***	0.91	10.44***	0.91	13.47***
I feel the community benefits from having Japanese residents on Oizumi/Osaka	0.92	17.03***	0.92	10.49***	0.92	13.71***
Emotional Closeness[b]	α = 0.86		α = 0.88		α = 0.86	
I have made friends with some Japanese residents in Oizumi/Osaka	0.84	11.30***	0.82	10.41***	0.83	7.20***
I feel close to some Japanese residents I have met in Oizumi/Osaka	0.84	11.10***	0.81	8.60***	0.91	7.63***
Sympathetic Understanding[b]	α = 0.85		α = 0.83		α = 0.83	
I understand Japanese residents	0.70	10.59***	0.67	7.98***	0.81	11.16***
I feel affection towards Japanese residents in Oizumi/Osaka	0.76	11.22***	0.72	7.77***	0.66	8.00***
I identify with Japanese residents in Oizumi/Osaka	0.70	12.02***	0.79	8.19***	0.73	8.85***
I have a lot in common with Japanese residents in Oizumi/Osaka	0.73	11.28***	0.68	7.45***	0.74	9.01***
Ethnic Neighborhood Tourism[b]	α = 0.75		α = 0.76		α = 0.70	
Tourism in Oizumi/Osaka Town should NOT focus only on Brazilian/Korean culture but also other foreign cultures.	0.70	9.58***	0.68	9.65***	0.72	9.92***
Tourism that focuses on Brazilian/Korean culture can be a medium to get to know my Japanese neighbors.	0.63	10.48***	0.61	7.26***	0.65	7.59***
Oizumi/Ikuno has other things to present to tourists than Brazilian/Korean culture.	0.81	12.93***	0.81	9.15***	0.80	9.16***
Japanese residents will feel out of place in Oizumi/Ikuno if tourism development focuses on Japanese culture only.	0.71	14.20***	0.73	10.25***	0.69	9.86***
Local Brazilian/Korean should be more involved in tourism development in Oizumi/Ikuno.	0.85	13.90***	0.78	10.04***	0.72	9.54***

Note. ***$p < 0.001$ level (Two-tailed).
a. 7-point semantic differential scale, where each item included bipolar adjectives describing Japanese residents (i.e. 1 indicates negative perception, whereas 7 indicates positive perception);.
b. 1 = Strongly disagree, 7 = Strongly agree.

Structural models

After establishing validity and full invariance of the measurement model, a structural model was then performed to test Hypotheses 1 to 5. The results of the structural model using the pooled

Table 3. Descriptive statistics and discriminant validity for pooled sample, Brazilian and Koreans residents in Japan.

Pooled sample ($n = 343$)	CAI	SE	WN	EC	SU	ENT	CR	AVE
Character and Intelligence (CAI)	**0.81**[a]						0.94	0.68
Social Evaluation (SE)	0.48	**0.83**					0.93	0.67
Welcoming Nature (WN)	0.38	0.56	**0.83**				0.90	0.70
Emotional Closeness (EC)	0.50	0.32	0.25	**0.84**			0.83	0.71
Sympathetic Understanding (SU)	0.34	0.30	0.17	0.39	**0.72**		0.81	0.52
Ethnic Neighborhood Tourism (ENT)	0.52	0.64	0.75	0.40	0.31	**0.76**	0.84	0.58
Mean	4.60	4.66	4.95	5.11	4.98	4.86		
SD	1.54	1.50	1.54	2.24	1.49	1.54		
Skewness	−0.31	−0.35	−0.66	−0.11	0.04	−0.1		
Kurtosis	0.24	0.29	1.32	−0.65	0.46	1,84		
Brazilian ($n = 181$)	CAI	SE	WN	EC	SU	ENT	CR	AVE
Character and Intelligence (CAI)	**0.82**						0.93	0.67
Social Evaluation (SE)	0.45	**0.81**					0.93	0.66
Welcoming Nature (WN)	0.37	0.49	**0.81**				0.88	0.65
Emotional Closeness (EC)	0.55	0.37	0.33	**0.82**			0.80	0.67
Sympathetic Understanding (SU)	0.31	0.25	0.10	0.41	**0.71**		0.81	0.51
Ethnic Neighborhood Tourism (ENT)	0.51	0.64	0.63	0.56	0.32	**0.72**	0.85	0.52
Mean	5.25	5.03	5.67	4.74	5.37	6.61		
SD	1.50	1.51	1.44	1.89	1.42	1.35		
Skewness	−0.86	−0.67	−1.45	−0.63	−1.01	−1.14		
Kurtosis	0.53	0.20	0.28	−0.64	1.02	1.33		
Korean ($n = 163$)	CAI	SE	WN	EC	SU	ENT	CR	AVE
Character and Intelligence (CAI)	**0.81**						0.93	0.66
Social Evaluation (SE)	0.51	**0.84**					0.95	0.71
Welcoming Nature (WN)	0.41	0.63	**0.86**				0.92	0.75
Emotional Closeness (EC)	0.44	0.35	0.16	**0.87**			0.86	0.76
Sympathetic Understanding (SU)	0.34	0.26	0.25	0.38	**0.74**		0.82	0.54
Ethnic Neighborhood Tourism (ENT)	0.57	0.66	0.67	0.21	0.33	**0.72**	0.84	0.51
Mean	3.95	4.40	4.55	5.51	4.54	4.17		
SD	1.26	1.33	1.42	2.54	1.43	1.30		
Skewness	0.05	−0.05	−0.42	−1.19	−0.27	−0.08		
Kurtosis	0.52	0.09	0.34	−0.51	0.13	0.18		

[a]The bold diagonal elements in the matrix are the square root of the average variance extracted; interconstruct correlations are shown off-diagonal.

CR = Composite reliability; AVE = Average variance extracted.

Note: All correlations are significant at the $p < 0.001$ level (Two-tailed).

sample demonstrated that the proposed model fit the data reasonably well ($\chi^2_{(346)} = 563.81$, $p < .001$, $\chi^2/df = 1.629$, CFI = .969, TLI = .964, RMSEA = .043; SRMR = .042). As presented in Figure 2, ten of the 11 hypotheses were supported. The model explained 39% of variance in welcoming nature, 26% in emotional closeness, 15% in sympathetic understanding and 70% in ethnic neighbourhood tourism. Consequently, the proposed model can be assumed to adequately predict minority residents' emotional solidarity with residents of the dominant group and ethnic neighbourhood tourism.

A significant positive relationship was identified between residents' level of *character and intelligence* and *social evaluation* significantly predicting residents ethnic neighborhood tourism ($\beta = .167$, $p < .05$ and $\beta = .239$, $p < .001$), supporting H_1 and H_2. Minority residents' level of *character and intelligence* regarding residents of the dominant group significantly predicted their emotional solidarity with such group ($H_{3a\_WN:}$ $\beta = .147$, $p < .05$; $H_{3b\_EC:}$ $\beta = .454$, $p < .001$; $H_{3c\_SU:}$ $\beta = .268$, $p < .001$), lending support to support to Hypotheses 3.

Moreover, the proposed effect of minority residents' levels of *social evaluation* of residents of the dominant group significantly predicted three dimensions of emotional solidarity with the latter group ($H_{4a\_WN:}$ $\beta = .487$, $p < .001$, $H_{4b\_EC:}$ $\beta = .118$, $p < 0.05$, and $H_{4cSU:}$ $\beta = .168$, $p < 0.05$), supporting hypotheses H_{4a}, H_{4b} and H_{4c}.

Lastly, H_5 postulated an effect of minority residents' degree of emotional solidarity ($H_{5a\_WN:}$ $\beta = -.468$, $p < .001$; $H_{5b\_EC:}$ $\beta = -.089$, $p < .05$; and $H_{5c\_SU:}$ $\beta = -.066$, $p > .05$) with residents in the dominant ethnic group and minority residents of ethnic neighborhood tourism. Results

Table 4. Test of six-factor Measurement Invariance (MI) across ethnicity groups.

Model	χ^2	*df*	$\Delta\chi^2$	Δdf	RMSEA[CI 90%]	ΔRMSEA	SRMR	CFI	ΔCFI	TLI	ΔTLI
Configural	1046.02*	686	–	–	.055[.049 – .062]	–	.050	.951	–	.942	–
Metric	1064.30*	709	18.28	23	.054[.047 – .061]	.001	.057	.951	.000	.944	.002
Scalar	1082.86*	732	18.56	23	.053[.046 – .059]	.001	.058	.952	.001	.947	.003

Note. *$p < .001$; $N = 343$ (Korean = 162; Brazilian = 181); cut-off values for measurement invariance are: Bonferroni-Holm corrected $p < .01$ for 2 groups, RMSEA $\leq .015$, CFI $\leq .010$, and TLI $\leq .010$ (Cheung and Rensvold, 2000; Meade et al., 2008).

showed that H_{5a} and H_{5b} were significant, and therefore supported by the data. H_{5c} was rejected given the effect was not significant.

Structural invariance model assessment

Because we were interested in significant differences in paths in our model, we further assessed whether structural invariance might exist across the two ethnic groups (Muthén & Asparouhov, 2018; Vandenberg & Lance, 2000). A metric invariance test (using a structural invariance test) was utilized to measure the effect of minority residents' level of ethnic attitudes and degree of emotional solidarity with residents of the dominant ethnic group and determine whether they would predict minority residents' attitudes towards ethnic neighborhood tourism and whether these were different across the two across ethnic groups. Initially, a baseline (unconstrained) model with a maximum likelihood approach was tested. Results revealed that the model fit the data reasonably well ($\chi^2_{(700)} = 1160.899$; $p < .001$; $\chi^2/df = 1.58$; RMSEA = 0.044; SRMR = .066; TLI = 0.926; CFI = 0.937). Table 5 summarises the results of the baseline model. Following this, we compared the baseline model in which the paths were allowed to vary freely (i.e., Brazilian and Korean) to a constrained model in which the paths were constrained equally across the ethnic groups. As we predicted, results of the χ^2 difference test showed that the two ethnic groups vary at the model level ($\Delta\chi^2_{(10)} = 55.96$, $p < .001$), demonstrating that there are differences in the path relationships between Brazilian and Korean residents living in these two Japanese cities.

To understand which path estimates vary between the two ethnic groups, we calculated statistical difference analyses by each path. The results depicted in Table 5 demonstrated that minority residents' perceived level of *character and intelligence* ($\Delta\chi^2_{(1)} = 3.917$, $p < .05$) of residents in the dominant ethnic group on the former's attitudes toward ethnic neighborhood tourism was only significant for Korean residents, rejecting H_6. Findings showed that minority residents' perceived level of *social evaluation* ($\Delta\chi^2_{(1)} = 3.758$, $p < .05$) of residents in the dominant ethnic group on the former's attitudes toward ethnic neighborhood tourism was only significant for Brazilian residents, giving support to H_7. The effect of minority residents' perceived level of *character and intelligence* of residents in the dominant ethnic on the former's degree of *welcoming nature* ($\Delta\chi^2_{(1)} = 4.018$, $p < .05$) was significant only for Brazilian residents, supporting H_{8a}. The same effect on *emotional closeness* ($\Delta\chi^2_{(1)} = 2.158$, $p < .05$) and *sympathetic understanding* ($\Delta\ \chi^2 = 3.039$, $p < .05$) were significantly different between groups but stronger for Brazilian residents. Consequently, H_{8b} and H_{8c} were also supported. Moreover, when assessing the effects of minority residents' perceived level of *social evaluation* of residents in the dominant ethnic group on the former's degree of *welcoming nature* ($\Delta\chi^2_{(1)} = 0.758$, $p > .05$) was not significantly different between Brazilian and Korean residents. However, the same effect on *emotional closeness* ($\Delta\chi^2_{(1)} = 2.859$, $p < .05$) was found to be different between groups and stronger for Brazilians. The same effect on *sympathetic understanding* ($\Delta\chi^2_{(1)} = 0.758$, $p > .05$) was found to be significant only for Koreans. As such, H_{9b} was supported, while H_{9a} and H_{9c} were rejected.

Lastly, the effect of minority residents' degree of *welcoming nature* ($\Delta\chi^2_{(1)} = 0.512$, $p > .05$) of residents in the dominant ethnic group on the former's attitudes toward ethnic neighborhood tourism was not significantly different between the two groups, however, the relationship was stronger for Korean residents. Conversely, the same effect of *emotional closeness* ($\Delta\chi^2_{(1)} = 7.723$,

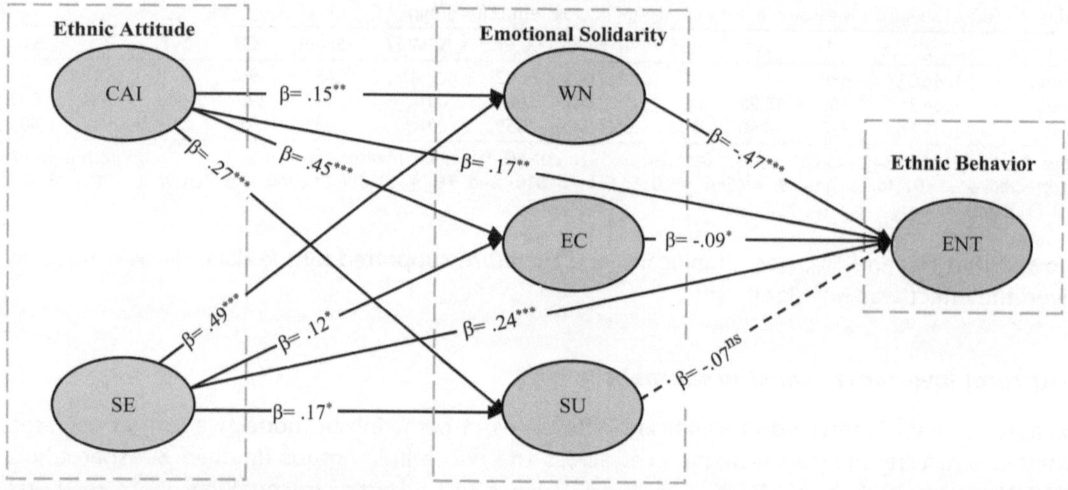

Figure 2. Result of structural equation modeling.
Note₁: Model fit: $\chi^2_{(346)} = 563.81$, $p = < .001$, $\chi^2/df = 1.63$, CF = .97, TLI = .96, RMSEA=.043, SRMR=.042.
Note₂: CAI = *Character and Intelligence*; SE = *Social Evaluation; WN = Welcoming Nature; EC = Emotional Closeness; SU = Sympathetic Understanding; ENT = Ethnic Neighborhood Tourism.*
$^*p < 0.05$; $^{**}p < 0.01$; $^{***}p < 0.001$ (one-tailed test). The dot line indicates an insignificant path.

Table 5. Invariance test of hypothesized paths for Brazilian and Korean residents in Japan.

Paths	Brazilian ($n = 181$)		Korean ($n = 162$)		Baseline Model (Freely estimated)	Nested Model (constrained to be equal)	Chi-Square test
	β	*t-value*	*β*	*t-value*			
H6: CAI → ENT	.052	.668	.320***	3.780	$\chi^2_{(700)} = 1160.866$	$\chi^2_{(701)} = 1164.783$	$\Delta\chi^2$ (1) = 3.917, p < .05
H7: SE → ENT	.320***	4.233	.137	1.586	$\chi^2_{(700)} = 1160.866$	$\chi^2_{(701)} = 1164.624$	$\Delta\chi^2$ (1) = 3.758, p < .05
H8a: CAI → WN	.186*	2.242	.119	1.473	$\chi^2_{(700)} = 1160.866$	$\chi^2_{(701)} = 1164.884$	$\Delta\chi^2$ (1) = 4.018, p < .05
H8b: CAI → EC	.487***	5.022	.442***	3.875	$\chi^2_{(700)} = 1160.866$	$\chi^2_{(701)} = 1163.024$	$\Delta\chi^2$ (1) = 2.158, p < .05
H8c: CAI → SU	.259**	2.649	.233*	2.306	$\chi^2_{(700)} = 1160.866$	$\chi^2_{(701)} = 1163.905$	$\Delta\chi^2$ (1) = 3.039, p < .05
H9a: SE → WN	.408***	4.497	.562***	6.131	$\chi^2_{(700)} = 1160.866$	$\chi^2_{(701)} = 1161.624$	$\Delta\chi^2$ (1) = 0.758, p > .05
H9b: SE → EC	.156*	1.800	.044	.475	$\chi^2_{(700)} = 1160.866$	$\chi^2_{(701)} = 1163.725$	$\Delta\chi^2$ (1) = 2.859, p < .05
H9c: SE→ SU	.130	1.370	.234*	2.344	$\chi^2_{(700)} = 1160.866$	$\chi^2_{(701)} = 1161.624$	$\Delta\chi^2$ (1) = 0.758, p > .05
H10a: WN → ENT	.417***	5.219	.505***	5.661	$\chi^2_{(700)} = 1160.866$	$\chi^2_{(701)} = 1161.378$	$\Delta\chi^2$ (1) = 0.512, p > .05
H10b: EC → ENT	.207**	2.608	.048	.697	$\chi^2_{(700)} = 1160.866$	$\chi^2_{(701)} = 1168.589$	$\Delta\chi^2$ (1) = 7.723, p < .05
H10c: SU → ENT	.118*	1.825	.041	.589	$\chi^2_{(700)} = 1160.866$	$\chi^2_{(701)} = 1166.185$	$\Delta\chi^2$ (1) = 5.319, p < .05

Note: CAI = Character and Intelligence; SE = Social Evaluation; WN = Welcoming Nature; EC = Emotional Closeness; SU = Sympathetic Understanding; ENT = Ethnic Neighborhood Tourism.
Goodness-of-fit index for the baseline model: $\chi^2_{(700)} = 1160.899$; p < .001; $\chi^2/df = 1.58$; RMSEA = 0.044; SRMR = .066; TLI = 0.926; CFI = 0.937;.
*p < 0.05; **p < 0.01; ***p < 0.001 (one-tailed test).
Explained variance:.
R^2_{WN}: Brazilian = 0.27; Korean = .40.
R^2_{EC}: Brazilian = 0.33; Korean = .20.
R^2_{SU}: Brazilian = 0.11; Korean = .16.
R^2_{ENT}: Brazilian = 0.71; Korean = .69.

$p < .05$) to residents in the dominant ethnic group on the minority group's attitudes toward ethnic neighborhood tourism was significantly different between groups and stronger for Brazilian residents. Yet, the effect of *sympathetic understanding* ($\Delta\chi^2_{(1)} = 5.319$, $p < .05$) to residents in the dominant ethnic group on the minority group's attitudes towards ethnic neighborhood tourism was significant only for Brazilian residents. Thus, while H_{10a} was rejected, H_{10b} and H_{10c} were therefore supported.

Discussion and implications

The primary goal of this study was to explore how Brazilian and Korean residents' (as minority ethnic groups) interethnic relationships with Japanese residents (as the dominant ethnic group) influence the former's attitudes towards ethnic neighborhood tourism representing each of the minority groups by applying the common in-group identity model (CIIM). To our knowledge, this is the first study that explores such relationships from the perspectives of minority residents.

The first research question explored the extent to which minority residents' ethnic attitudes towards Japanese residents influence the formers' attitudes regarding ENT focused on each of the ethnic minority groups. The analysis of the pooled sample indicated that both factors of EAS significantly predict the level of concern about ENT development. These findings run parallel to previous studies that indicate the link between residents' ethnic attitudes and their attitude towards tourism among members of the dominant ethnic groups (Luo et al., 2015; Maruyama & Woosnam, 2015). Perhaps, this study suggests that positive ethnic attitudes lead to foster positive attitudes towards tourism not only among residents of the dominant ethnic group but also those residents of the ethnic minority group.

The second research question concerned how minority residents' ethnic attitudes towards Japanese residents factor into the former's emotional solidarity with the latter. The analysis of the pooled sample indicated that all six relationships between EAS and ESS were significant, in a positive direction. This is consistent with previous studies (Woosnam et al., 2016a) conducted from the perspective of the ethnic dominant group. Arguably, by representing members of both the dominant ethnic and ethnic minority group as members of "a local community" and fostering positive attitudes towards their ethnic counterparts, ENT has the potential to forge solidarity among members of different ethnic groups.

The third research question regarded how emotional solidarity explained attitudes regarding ENT. Contrary to the second research question, two out of the three relationships were negatively significant. This result indicates that, while the ethnic minority residents appreciate the newly created solidarity with their Japanese counterparts, they may be resistant to focusing on commonality with Japanese residents regarding tourism representation, and are willing to keep the duality. As Dovidio et al. (2009) argue, the minority residents may be concerned that the distinctiveness of their ethnic culture might be relinquished in the representation that focuses on commonality and cooperation with their Japanese counterparts. However, the analysis indicated that the link among EAS, ESS, and ENT constructs differ between Korean and Brazilian residents (research question 4). More precisely, the analysis of the data from the Korean sample indicated that the first factor of EAS, namely, *character and intelligence* (which pertains to general characteristics and perceived levels of intelligence of others), had a significant relationship with the attitudes towards ENT. On the contrary, the analysis indicated that among Brazilian residents, the second factor of EAS regarding how people relate to others (i.e., *social evaluation*) had a significant relationship with the attitudes towards ENT. Regarding the relationships between EAS and ESS, two relationships (H_{8a}, H_{9b}, H_{10b}, and H_{10c}) were significant for Brazilians while only one relationship (H_{9c}) was significant for Koreans. Furthermore, among four relationships (H_{8b}, H_{8c}, H_{9a}, and H_{10a}) were significant for both groups, with one relationship (H_{8c}) considered stronger for the Brazilian residents. These differences may be explained by the interactive acculturation model that proposes interethnic relations between immigrants and their host society are mutually influenced by the acculturation expectations of the both groups (Bourhis et al., 1997).

More precisely, Korean residents who have been marginalized in Japan had only limited social contacts with Japanese residents until recently. This may contribute to a more abstract evaluation of Japanese residents by Korean residents (Lammers et al., 2012). In addition, the mismatch of acculturation expectation between Koreans' eagerness to be integrated into Japanese society and Japanese residents' exclusion orientation may have caused strong distress among Korean residents (Van Oudenhoven et al., 2006). Thus, while the Korean residents appreciate the newly created relationship with their Japanese neighbors, Korean residents may fear that an overemphasis on solidarity with Japanese residents lead to greater conflicts, and concern that the distinctiveness of their culture may be relinquished. In contrast, Brazilian residents, most of whom have been granted a long-term working visa and are employed by a Japanese-owned company, may evaluate Japanese counterparts in terms of how they treat others. In addition, Brazil and Japan have historically had good relations with one another, and indeed, many of Brazilian immigrants to Japan are the descendants of Japanese migrants to Brazil. In addition to the ethnic connection, Brazilian residents migrated to Japan when the society's view of foreign residents was becoming more accommodating (Takaya, 2007). In other words, the acculturation expectation mismatch that Brazilian residents experienced with Japanese residents may be relatively minor compared to Korean residents' experiences, and thus Brazilian residents feel less distress. As a result, Brazilian residents may feel less threatened that the distinctiveness of their culture will be controlled by the dominant culture through tourism. Rather, they perceive the advantage of cooperating with the dominant ethnic group.

These findings may also indicate that Brazilian residents are more inclined to an inclusive representation, while Korean residents are more resistant to it and rather willing to keep the duality as it relates to tourism representation. (Dovidio et al., 2009). This may partly be because Korean and Japanese culture have much commonality, and thus Korean residents are more acutely aware that their cultural distinctiveness might be easily surrendered in an inclusive representation (Tai, 2009). On the contrary, Brazilian residents whose cultural traits differ from Japanese might be less concerned with losing their cultural uniqueness in such a representation. Furthermore, this finding may indicate the disparity in social status and power between the two diasporic groups within Japan. More precisely, Dovidio et al. (2009) hypothesize that members of a high-status group typically prefer a one-group form of representation while members of disadvantaged groups tend to refuse it. The findings of this study indicate that, in an ENT destination where members of ethnic minority groups have a privileged access to the resources, Korean residents still consider themselves disadvantaged while Brazilian residents feel empowered. In other words, the degree to which residents resist a form of an inclusive representation and prefer a dual form representation may differ even among minority groups depending on social conditions.

Further implications

Findings from the current study have implications for both theory and practice. In regards to theory, with over several decades of research examining the mechanism and implication of interethnic relations in the tourism literature (e.g., Giles et al., 2013; Yu & Lee, 2014) and general literature (e.g. Coenders et al., 2008; Savelkoul et al., 2011), what is lacking is a discussion of how interethnic relations within a community influence residents' attitudes towards tourism particularly among ethnic minority residents in ENT destinations. As levels of ethnic diversity increases in many cities, and the potential of ENT for the local revitalization is recognized, filling this gap is crucial. Arguably, results of this study indicated that overall, interethnic relations influence the attitudes towards tourism among minorities. At the same time, the findings suggest that preference for collaboration may be different depending on acculturation experiences of a particular ethnic minority group. This indicates that by applying the interactive acculturation model (Bourhis et al., 1997), researchers may better understand ways of representation that may most accurately reflect various ethnic groups with different acculturation experiences.

The findings of this study indicated that minority residents' ethnic attitudes towards Japanese residents influence the formers' attitudes regarding ENT. To foster residents' support for tourism and make it more sustainable, tourism practitioners need to create opportunities for residents to interact with their counterparts in a friendly manner (i.e., a community event and leisure activities) to foster intercultural understanding. At the same time, the results from the pooled sample indicated that minority residents may prefer a dual identity form of representation over an inclusive, one-group representation. Therefore, the tourism practitioners may need to acutely aware of the distinctive culture that minority residents want to sustain.

At the same time, tourism practitioners also need to be aware that each ethnic group may have different preferences regarding the nature of collaboration with the members of dominant ethnic group when initiating ENT. As Yang et al. (2013) suggest, often the government considers a social group as homogeneous and thereby overlooks the different needs and demands of different segments. Indeed, these findings indicate that Brazilian residents are more inclined to an inclusive representation, while Korean residents are more resistant to it. Such differences may be particularly important to understand in fostering support for ENT among local residents. Therefore, tourism practitioners, including people in the local governments, need to understand that each ethnic group has unique histories of immigration, experiences of discrimination, and preferences regarding what and how to represent their ethnic culture to tourists.

Study limitations and directions for future research

This study focused on two diasporic groups who have historical connections to Japan. More precisely, Korean residents in Ikuno ward are descendants of those who immigrated to Japan around 1920, and a large number of them have been naturalized to Japan (Lim, 2009). In terms of Brazilian residents, a large number of them share the lineage with Japanese residents because they are the descendants of Japanese immigrants who migrated to Brazil before WWII (Tsuda, 2003). These historical relationships within Japanese society may have particular influence on their ethnic relations with the Japanese counterparts. Therefore, future studies need to explore other ethnic groups whose relationship with Japan is relatively new (i.e., Indian or Vietnamese residents). In terms of scales used in this study, the ENT scale includes a mixture of positively- and negatively-worded statements. Therefore, for the purpose of this study, the authors of this study employed reverse coding for particular items within constructs. In the future, revising the wording of the items may make the scale stronger. Yet, another limitation of the study was the sample size gathered from both groups. While this did not represent a threat in the data analysis, larger samples from each ethnic minority group may contribute to more powerful findings.

This study indicated that interethnic relationships within a local community influence locals' attitudes towards tourism development particularly from the minority residents' perspective. The findings of this study also indicated that how interethnic relationships influence the residents' attitudes towards tourism differ between two diasporic groups. This may indicate that, while ENT has potential to provide better access to resource minority residents and empower them (Drew, 2011; Maruyama et al., 2016), in reality, social status and power differ among the diasporic groups in ENT destinations. To maximize the potential of ENT as a strategy of regional development, research needs to explore the roles of interethnic relationship to form residents' attitudes towards ENT among residents of distinctively different social status backgrounds.

Disclosure statement

No potential conflict of interest was reported by the authors.

Funding

JSPS KAKENHIThis work was supported by JSPS KAKENHI Grant Number 24611010.

ORCID

Manuel Alector Ribeiro http://orcid.org/0000-0003-4484-1082

References

Amir, Y., & Ben-Ari, R. (1985). International tourism, ethnic contact, and attitude change. *Journal of Social Issues, 41*(3), 105–115. https://doi.org/10.1111/j.1540-4560.1985.tb01131.x

Anastasopoulos, P. G. (1992). Tourism and attitude change: Greek tourists visiting Turkey. *Annals of Tourism Research, 19*(4), 629–642.

Anderson, J. C., & Gerbing, D. W. (1988). Structural equation modeling in practice: A review and recommended two-step approach. *Psychological Bulletin, 103*(3), 411–423. https://doi.org/10.1037/0033-2909.103.3.411

Aytar, V., & Rath, J. (2012). *Selling ethnic neighborhoods: The rise of neighborhoods as places of leisure and consumption.* : Routledge.

Babbie, E. R. (2014). *The basics of social research* (6th ed.). Wadsworth.

Beaton, A. M., Monger, T., Leblanc, D., Bourque, J., Levi, Y., Joseph, D. J., Richard, J., Bourque, P., & Chouinard, O. (2012). Crossing the divide: The common in-group identity model and intergroup affinity. *International Journal of Intercultural Relations, 36*(3), 365–376.

Bourhis, R. Y., Moise, L. C., Perreault, S., & Senecal, S. (1997). Towards an interactive acculturation model: A social psychological approach. *International Journal of Psychology, 32*(6), 369–386. https://doi.org/10.1080/002075997400629

Brislin, R. W. (1970). Back-translation for cross-cultural research. *Journal of Cross-Cultural Psychology, 1*(3), 185–216.

Brown, T. A. (2015). Confirmatory factor analysis for applied research (2nd ed). Guilford Press.

Browne, M. W. (1982). Covariance structures. In D. M. Hawkins (Ed.), *Topics in applied multivariate analysis* (pp. 72–141). Cambridge University Press.

Browne, M. W., & Cudeck, R. (1989). Single sample cross-validation indices for covariance structures. *Multivariate Behavioral Research, 24*(4), 445–455. https://doi.org/10.1207/s15327906mbr2404_4

Chapman, D. (2006). Discourses of multicultural coexistence (tabunka kyōsei) and the 'old-comer'Korean residents of Japan. *Asian Ethnicity, 7*(1), 89–102.

Cheung, G. W., & Rensvold, R. B. (2002). Evaluating goodness-of-fit indexes for testing measurement invariance. *Structural Equation Modeling: A Multidisciplinary Journal, 9*(2), 233–255. https://doi.org/10.1207/S15328007SEM0902_5

Coenders, M., Lubbers, M., Scheepers, P., & Verkuyten, M. (2008). More than two decades of changing ethnic attitudes in the Netherlands. *Journal of Social Issues, 64*(2), 269–285. https://doi.org/10.1111/j.1540-4560.2008.00561.x

Cote, R. R., & Erickson, B. H. (2009). Untangling the roots of tolerance: How forms of social capital shape attitudes toward ethnic minorities and immigrants. *American Behavioral Scientist, 52*(12), 1664–1689.

Curran, P. J., West, S. G., & Finch, J. F. (1996). The robustness of test statistics to nonnormality and specification error in confirmatory factor analysis. *Psychological Methods, 1*(1), 16–29. [Database] https://doi.org/10.1037/1082-989X.1.1.16

Dovidio, J. F., Gaertner, S. L., & Saguy, T. (2009). Commonality and the complexity of "we": Social attitudes and social change. *Personality and Social Psychology Review, 13*(1), 3–20. https://doi.org/10.1177/1088868308326751

Drew, E. M. (2011). Strategies for antiracist representation: Ethnic tourism guides in Chicago. *Journal of Tourism and Cultural Change, 9*(2), 55–69.

Durkheim, E. (1995[1915]). *The elementary forms of the religious life.* Free Press.

Fornell, C., & Larcker, D. F. (1981). Evaluating structural equation models with unobservable variables and measurement error. *Journal of Marketing Research, 18*(1), 39–50. https://doi.org/10.1177/002224378101800104

Fredrick, S. S., Drevon, D. D., & Jervinsky, M. (2019). Measurement invariance of the Student Risk Screening Scale across time and gender. *School Psychology (Washington, D.C.).)*, *34*(2), 159–167. https://doi.org/10.1037/spq0000278

Gaertner, S. L., Bachman, B. A., Dovidio, J. D., & Banker, B. S. (2001). Corporate mergers and stepfamily marriages: Identity, harmony, and commitment. In M. A. Hogg & D. Terry (Eds.), Social identity in organizations (pp. 265-288). Blackwell.

Gaertner, S. L., & Dovidio, J. F. (2000). *Reducing intergroup bias: The common ingroup identity model*. Psychology Press.

Gaertner, L., & Schopler, J. (1998). Perceived ingroup entitativity and intergroup bias: An interconnection of self and others. *European Journal of Social Psychology*, *28*(6), 963–980. https://doi.org/10.1002/(SICI)1099-0992(1998110)28:6<963::AID-EJSP905>3.0.CO;2-S

Giles, H., Ota, H., & Foley, M. (2013). Tourism: An intergroup communication model with Russian inflections. *Russian Journal of Communication*, *5*(3), 229–243.

Henderson, J. (2003). Ethnic heritage as a tourist attraction: The Peranakans of Singapore. *International Journal of Heritage Studies*, *9*(1), 27–44.

Hewstone, M., & Brown, R. (1986). Contact is not enough: An intergroup perspective on the'contact hypothesis.' In M. Hewstone & R. Brown (Eds.), *Social psychology and society. Contact and conflict in intergroup encounters* (p. 1–44). Basil Blackwell: Hoboken, New Jersey.

Hewstone, M., Cairns, E., Voci, A., Hamberger, J., & Niens, U. (2006). Intergroup contact, forgiveness, and experience of "The Troubles" in Northern Ireland. *Journal of Social Issues*, *62*(1), 99–120. https://doi.org/10.1111/j.1540-4560.2006.00441.x

Houlette, M. A., Gaertner, S. L., Johnson, K. M., Banker, B. S., Riek, B. M., & Dovidio, J. F. (2004). Developing a more inclusive social identity: An elementary school intervention. *Journal of Social Issues*, *60*(1), 35–55. https://doi.org/10.1111/j.0022-4537.2004.00098.x

Hu, L. T., & Bentler, P. M. (1999). Cutoff criteria for fit indexes in covariance structure analysis: Conventional criteria versus new alternatives. *Structural Equation Modeling: A Multidisciplinary Journal*, *6*(1), 1–55. [Database] https://doi.org/10.1080/10705519909540118

Joo, D., Tasci, A. D., Woosnam, K. M., Maruyama, N. U., Hollas, C. R., & Aleshinloye, K. D. (2018). Residents' attitude towards domestic tourists explained by contact, emotional solidarity and social distance. *Tourism Management*, *64*, 245–257. https://doi.org/10.1016/j.tourman.2017.08.012

Jöreskog, K. G. (1971). Statistical analysis of sets of congeneric tests. Psychometrika, *36*(2), 109–133.

Kalin, R., & Berry, J. W. (1982). The social ecology of ethnic attitudes in Canada. *Canadian Journal of Behavioural Science / Revue Canadienne Des Sciences du Comportement*, *14*(2), 97–109. https://doi.org/10.1037/h0081246

Khan, S. (2015). Otherness" of ethnic enclave attractions in multicultural cities: A study of Chinatown and Little India. *Asia-Pacific Journal of Innovation in Hospitality and Tourism*, *4*(1), 63–76.

Kline, R. B. (2011). *Principles and practice of structural equation modeling*. Guilford press.

Lammers, J., Galinsky, A. D., Gordijn, E. H., & Otten, S. (2012). Power increases social distance. *Social Psychological and Personality Science*, *3*(3), 282–290.

Lee, S., Phau, I., Hughes, M., Li, Y. F., & Quintal, V. (2016). Heritage tourism in Singapore Chinatown: A perceived value approach to authenticity and satisfaction. *Journal of Travel & Tourism Marketing*, *33*(7), 981–998.

Lim, T. (2009). Who is Korean? Migration, immigration, and the challenge of multiculturalism in homogeneous societies. *Asia-Pacific Journal*, *30*(1).1–21.

Logan, J. R., Spielman, S., Xu, H., & Klein, P. N. (2011). Identifying and bounding ethnic neighborhoods. *Urban Geography*, *32*(3), 334–359. https://doi.org/10.2747/0272-3638.32.3.334

Luo, X., Brown, G., & Huang, S. S. (2015). Host perceptions of backpackers: Examining the influence of intergroup contact. *Tourism Management*, *50*, 292–305. https://doi.org/10.1016/j.tourman.2015.03.009

Maher, J. C., & Kawanishi, Y. (2012). Ikuno-Ku, Osaka: Centre of hope and struggle. In J. C. Maher & G. M. Macdonald (Eds.), *Diversity in Japanese culture* (pp. 160–177). Routledge.

Mak, A. L. (2003). Negotiating identity: Ethnicity, tourism and Chinatown. *Journal of Australian Studies*, *27*(77), 93–100.

Mardia, K. V. (1970). Measures of multivariate skewness and kurtosis with applications. Biometrika, *57*(3), 519–530.

Maruyama, N., & Woosnam, K. M. (2015). Residents' ethnic attitudes and support for ethnic neighborhood tourism: The case of a Brazilian town in Japan. *Tourism Management*, *50*, 225–237. https://doi.org/10.1016/j.tourman.2015.01.030

Maruyama, N. U., Woosnam, K. M., & Boley, B. B. (2016). Comparing levels of resident empowerment among two culturally diverse resident populations in Oizumi, Gunma, Japan. *Journal of Sustainable Tourism*, 24(10),1442-1460.

Maruyama, U. N., Woosnam, M. K., & Boley, B. B. (2017). Who is ethnic neighborhood tourism for anyway? Considering perspectives of the dominant cultural group. *International Journal of Tourism Research*, *19*(6), 727–735. https://doi.org/10.1002/jtr.2144

Meade, A. W., Johnson, E. C., & Braddy, P. W. (2008). Power and sensitivity of alternative fit indices in tests of measurement invariance. *The Journal of Applied Psychology, 93*(3), 568–592. https://doi.org/10.1037/0021-9010.93.3.568

Millsap, R. E. (2001). When trivial constraints are not trivial: The choice of uniqueness constraints in confirmatory factor analysis. *Structural Equation Modeling, 8*(1), 1–17.

Minauskaitė. (2016).Japan and South Korea: The Differences and Similaritie. The Voice Space. Retrieved October 10, 2020, from https://thevoice.space/2016/09/japan-and-south-korea-the-differences-and-similarities-by-julija-minaus-kaite-8b/

Miwa, S. (2008). Hatarakikatato lifestyle no henkanikansruzenkokuchousa 2007 niokeru hyouhontokuseito keppyou-nituiteno kisobunseki [Sample bias and unit nonresponse cases for JLPS-2007]. *University of Tokyo Institute of Social Science Panel Survey Discussion Paper Series, 10*, 1–17.

Murphy-Shigematsu, S. (2000). Identities of multiethnic people in Japan. In Douglass, M., & Roberts, G. S. (Eds.), *Japan and global migration: Foreign workers and the advent of a multicultural society* (pp. 196–216). University of Hawaii Press.

Muthén, B., & Asparouhov, T. (2018). Recent methods for the study of measurement invariance with many groups. *Sociological Methods & Research, 47*(4), 637–664.

Muthén, L. K., & Muthén, B. O. (2017). *Mplus user's guide* (8th ed.). Muthén & Muthén.

Nadler, A. (2002). Inter-group helping relations as power relations: Maintaining or challenging social dominance between groups through helping. *Journal of Social Issues, 58*(3), 487–502.

Nier, J. A., Gaertner, S. L., Dovidio, J. F., Banker, B. S., Ward, C. M., & Rust, M. C. (2001). Changing interracial evaluations and behavior: The effects of a common group identity. *Group Processes & Intergroup Relations, 4*(4), 299–316.

Nyaupane, G. P., Teye, V., & Paris, C. (2008). Innocents abroad: Attitude change toward hosts. *Annals of Tourism Research, 35*(3), 650–667. https://doi.org/10.1016/j.annals.2008.03.002

Oizumi Town. (2019). *Oizumi's population*. [cited May 21 2019]. https://www.town.oizumi.gunma.jp/01soshiki/06jyu-min/01jyumin/data/3104gaikokujinkou.pdf.

Okazaki, E. (2008). A community-based tourism model: Its conception and use. *Journal of Sustainable Tourism, 16*(5), 511–529.

Osgood, S., Suci, G. J., & Tannenbaum, P. H. (1957). *The measurement of meaning*. University of Illinois Press.

Pang, C. L., & Rath, J. (2007). The force of regulation in the land of the free: The persistence of Chinatown, Washington DC as a symbolic ethnic enclave. *Research in the Sociology of Organizations, 25*, 191–216.

Pizam, A., Fleischer, A., & Mansfeld, Y. (2002). Tourism and social change: The case of Israeli ecotourists visiting Jordan. *Journal of Travel Research, 41*(2), 177–184. https://doi.org/10.1177/004728702237423

Pizam, A., Jafari, J., & Milman, A. (1991). Influence of tourism on attitudes: US students visiting USSR. *Tourism Management, 12*(1), 47–54. https://doi.org/10.1016/0261-5177(91)90028-R

Ribeiro, M. A., Woosnam, K. M., Pinto, P., & Silva, J. A. (2018). Tourists' destination loyalty through emotional solidarity with residents: An integrative moderated mediation model. *Journal of Travel Research, 57*(3), 279–295. https://doi.org/10.1177/0047287517699089

Santos, C. A., & Yan, G. (2008). Representational politics in Chinatown: The ethnic other. *Annals of Tourism Research, 35*(4), 879–899. https://doi.org/10.1016/j.annals.2008.06.006

Sass, D. A., Schmitt, T. A., & Marsh, H. W. (2014). Evaluating model fit with ordered categorical data within a measurement invariance framework: A comparison of estimators. *Structural Equation Modeling: A Multidisciplinary Journal, 21*(2), 167–180. https://doi.org/10.1080/10705511.2014.882658

Savelkoul, M., Scheepers, P., Tolsma, J., & Hagendoorn, L. (2011). Anti-Muslim attitudes in The Netherlands: Tests of contradictory hypotheses derived from ethnic competition theory and intergroup contact theory. *European Sociological Review, 27*(6), 741–758. https://doi.org/10.1093/esr/jcq035

Sturgis, P., Brunton-Smith, I., Kuha, J., & Jackson, J. (2014). Ethnic diversity, segregation and the social cohesion of neighbourhoods in London. *Ethnic and Racial Studies, 37*(8), 1286–1309. https://doi.org/10.1080/01419870.2013.831932

Tai, E. (2009). Between assimilation and transnationalism: The debate on nationality acquisition among. *Koreans in Japan. Social Identities, 15*(5), 609–629.

Tajfel, H., & Turner, J. C. (1979). An integrative theory of intergroup conflict. *The Social Psychology of Intergroup Relations, 33*, 47.

Takaya, S. (2007). Multiculturalism in Japan: A victory over assimilationism or subjection to neo-liberalism? *Ritsumeikan Gengobunka-Kenkyu, 18*(3), 45–57.

Trevor, J., & Monder, R. (2003). South Asian businesses in retreat? The case of the UK. *Journal of Ethnic and Migration Studies, 29*(3), 485–500.

Tropp, L. R., & Pettigrew, T. F. (2005). Relationships between intergroup contact and prejudice among minority and majority status groups. *Psychological Science, 16*(12), 951–957. https://doi.org/10.1111/j.1467-9280.2005.01643.x

Tsuda, T. (2003). *Strangers in the ethnic homeland: Japanese Brazilian return migration in transnational perspective*. Columbia University Press.

Van Knippenberg, D., & Van Leeuwen, E. (2001). Organizational identity after a merger. In M.A. Hogg(Ed.), *Social identity processes in organizational contexts* (pp. 249–264). Routledge.

Van Oudenhoven, J. P., Ward, C., & Masgoret, A.-M. (2006). Patterns of relations between immigrants and host societies. *International Journal of Intercultural Relations, 30*(6), 637–651. https://doi.org/10.1016/j.ijintrel.2006.09.001

Vandenberg, R. J., & Lance, C. E. (2000). A review and synthesis of the measurement invariance literature: Suggestions, practices, and recommendations for organizational research. *Organizational Research Methods, 3*(1), 4–70. https://doi.org/10.1177/109442810031002

Ward, I. (2019). *Statistics in Ikuno 2016* [cited May 21 2019]. https://www.city.osaka.lg.jp/ikuno/page/0000450484.html.

West, S. G., Finch, J. F., & Curran, P. J. (1995). *Structural equation models with nonnormal variables: Problems and remedies.* In R. H. Hoyle (Ed.), *Structural equation modeling: Concepts, issues, and applications* (pp. 56–75). Sage Publications, Inc.

Woosnam, K. M. (2012). Using emotional solidarity to explain residents' attitudes about tourism and tourism development. *Journal of Travel Research, 51*(3), 315–327.

Woosnam, K. M., Maruyama, N., & Boley, B. B. (2016a). Perceptions of the 'other'residents: Implications for attitudes of tourism development focused on the minority ethnic group. *Journal of Travel & Tourism Marketing, 33*(5), 567–580.

Woosnam, K. M., Maruyama, N., Boley, B. B., & Erul, E. (2016b). Stereotypes and perceived solidarity in ethnic enclave tourism. *Journal of Tourism and Cultural Change, 16*(2), 138–117.

Woosnam, K. M., Maruyama, N. U., Ribeiro, M. A., & Joo, D. (2019). Explaining minority residents' attitudes of ethnic enclave tourism from general perceptions of tourism impacts. *Journal of Tourism and Cultural Change, 17*(4), 467–484. https://doi.org/10.1080/14766825.2019.1601407

Woosnam, K. M., & Norman, W. C. (2010). Measuring residents' emotional solidarity with tourists: Scale development of Durkheim's theoretical constructs. *Journal of Travel Research, 49*(3), 365–380. https://doi.org/10.1177/0047287509346858

Woosnam, K. M., Norman, W. C., & Ying, T. (2009). Exploring the theoretical framework of emotional solidarity between residents and tourists. *Journal of Travel Research, 48*(2), 245–258.

Yang, J., Ryan, C., & Zhang, L. (2013). Social conflict in communities impacted by tourism. *Tourism Management, 35*, 82–93. https://doi.org/10.1016/j.tourman.2012.06.002

Yu, J., & Lee, T. J. (2014). Impact of tourists' intercultural interactions. *Journal of Travel Research, 53*(2), 225–238.

Zimmerman, A. S. (2008). New knowledge from old data: The role of standards in the sharing and reuse of ecological data. *Science, Technology, & Human Values, 33*(5), 631–652.

Assessing explicit and implicit stereotypes in tourism: self-reports and implicit association test

Wai Tsz Serene Tse and Vincent Wing Sun Tung

ABSTRACT
Stereotyping is a dichotomy system of cognitive information processes reflecting explicit and implicit biases. While existing studies have studied tourist stereotypes with a wide range of approaches, ranging from face-to-face interviews to numerical scale ratings via self-reports, these studies measured only one type of stereotypes: explicit stereotypes, while neglecting the other: implicit stereotypes. The nature of explicit stereotypes allows individuals to access their cognitions, which could increase the chance of socially desirable responses and thus offer incomplete knowledge in tourist stereotyping. Drawing from existing literature on explicit and implicit stereotypes, this study seeks to identify implicit tourist stereotypes via implicit association test (IAT) and measure them against explicit biases from residents of four popular destinations, covering Hong Kong, Malaysia, Singapore, and Thailand. This study contributes by presenting a detailed development of the IAT and demonstrates the applicability of this method for future cognitive-related studies in the tourism literature. Overall, this study provides a timely and impactful methodology for measuring tourist stereotypes, an important cognitive component that could shape residents' attitudes toward sustainable tourism development, and harmonious host-guest relationships.

Introduction

There is an increasing focus in the recent tourism literature on the positive and negative stereotypes that residents hold toward tourists (Hsu & Chen, 2019; Monterrubio, 2018; Shen et al., 2019; Tung et al., 2020). Stereotypes, like other cognitive information processes, consist of two distinct types that mainly differ between conscious and unconscious (Kihlstrom, 1990) that reflects explicit and implicit biases, respectively. The nature of explicit stereotypes allows individuals to access their cognitions, which could increase the chance of socially desirable responses during the research process when individuals are asked to evaluate their biases. Consequently, research findings that focus solely on the explicit stereotype via self-reports would fail to account for underlying (i.e. implicit) biases, and thus offer incomplete insights for policymakers in managing intergroup stereotyping and host-guest relationships.

Different from explicit stereotypes, implicit stereotypes are unconscious attributes that could be measured using an indirect approach, such as the implicit association test (IAT) (Fazio & Olson, 2003). IAT is a computerized program that involves automatic associations between the target (i.e. tourist) and the attributes (i.e. stereotypes) held by an individual (i.e. resident)

(Greenwald & Banaji, 1995). While previous studies in psychology have demonstrated the success of IAT in capturing unconscious evaluations from individuals, it has not been employed in the tourism field to identify residents' stereotypical views toward tourists. Drawing from the literature on explicit and implicit stereotypes and IAT measurement, the objective of this study is to examine the differences in residents' implicit and explicit tourist stereotypes. For instance, would residents' implicit tourist stereotypes be consistent or different from their explicit tourist stereotypes?

To answer the proposed research question, this study aims to measure the implicit and explicit tourist stereotypes as perceived by residents from four different destinations. More specifically, this study identifies the implicit tourist stereotypes through the use of IAT while explicit tourist stereotypes are captured using a self-report scale rating. Furthermore, this study investigates the degree of association between implicit and explicit tourist stereotypes. In this study, the examined target is the Mainland Chinese tourist, an influential tourist market in the international travel, while the respondents are residents from four Asian destinations, Hong Kong, Malaysia, Singapore, and Thailand. The findings of this study will reveal residents' perceptions toward the Mainland Chinese tourist that are theoretically and practically valuable for the area of host-guest relations.

This study contributes by uncovering the dual-information processes in tourist stereotyping research, which would not be possible by measuring only one of the stereotypes. The sole reliance on either explicit or implicit measurements could provide a bias in understanding tourist stereotypes, uncovering the incomplete insights shown in the existing literature. Furthermore, this study presents a detailed development of the IAT and demonstrates the applicability of this method for future cognitive-related studies in the tourism literature. A comprehensive demonstration of the IAT in this study could serve as a template for subsequent studies on implicit cognitive measurements. Overall, this study provides a timely and impactful methodology for measuring tourist stereotypes, an important cognitive component that could shape residents' attitudes toward sustainable tourism development, and harmonious host-guest relations.

Practically, this paper provides insights for destination management organizations (DMOs) and tourism officials, focusing on host-guest relations to evaluate residents' implicit and explicit stereotypes toward tourists. For instance, Woosnam et al. (2018) found that ethnic stereotypes could affect the emotional solidarity between the majority group (i.e. Japanese residents) and minority groups (i.e. Brazilian resident) in terms of welcoming nature, emotional closeness, and sympathetic understanding. As such, there is a need for destination managers to address implicit and explicit stereotypes perceived by the majority on the minority to increase host-guest interactions and thus improve their emotional solidarities. In this regard, this research also contributes to knowledge focusing on stereotypes and host-guest relations. DMOs could manage both implicit and explicit stereotypes as existing studies suggests conscious and unconscious activation of information categorization to produce associated attitudes accordingly (Dovidio et al., 2001). The convergence or divergence between implicit and explicit stereotypes could allow policymakers to implement internal marketing communication to address the representation of stereotypes among residents.

Literature review

Stereotypes and tourist stereotypes

Stereotypes refer to the cognitive function of information processing about the characteristics of individuals belonging to social groups, allowing the categorization of individuals into members of ingroups and outgroups (Taylor et al., 1996). Pieces of information are formulated based on the attributes, characteristics, and conducts accumulated from personal interactions and unsubstantiated gossip from other ingroup members or media reports that may be true, false, or even mixed that are homogenized across all individuals within the same category (Pickering, 2001). Furthermore, stereotypes are associated with discriminatory values (good or bad) and destroy the actual image of individuals (Pickering, 2001), resulting in de-individualization.

More recently, with increasing attention on stereotyping, tourist stereotypes are regarded as a resident's preconceptions of tourists within a destination setting (Tung et al., 2020) that could elicit discrimination and harassment of tourists (van Veelen et al., 2016). The examination of stereotypes is an essential concept in tourism studies where an individual's cognitive evaluation could lead to subsequent emotions and behaviors that are important for sustainable tourism development (Anderson et al., 2015; Rozin & Royzman, 2001). Within tourism studies, the examination of stereotypes has focused on the tourists' and locals' perspectives in various areas such as destination image (Chen et al., 2016), education (Tung & King, 2016), prosocial behaviors (Tung, 2019), product branding (Diamantopoulos et al., 2017), and toward each other (Hsu & Chen, 2019).

Tourist stereotypes refer to the local residents' perceived cognitive evaluation of a particular tourists origin in a tourism setting (Tung et al., 2020), affecting locals behaviors toward the tourist, thereby affecting host-guest relations and destination image. Early tourism studies describe tourist stereotypes following the ethnic characteristics of tourists (Boissevain & Inglott, 1979; Pi-Sunyer, 1977; Sheldon & Var, 1984). Subsequent studies have suggested that tourist stereotyping extends beyond ethnic stereotypes; tourist stereotypes include attributes from physical, temperamental, attitudinal, and behaviors (Hsu & Chen, 2019).

Previous studies have revealed tourist stereotype attributes using different examination pairs of locals and tourists, such as Israeli resident-Egyptian tourists (Milman et al., 1990) and Jordanian tourists (Pizam et al., 2002); Dutch resident-German and East Asian tourists (Moufakkir, 2011); Hong Kong resident-Mainland Chinese tourists (Chen et al., 2018; Shen et al., 2017). Furthermore, these stereotype attributes are categorized along with the valence of positive and negative biases (Chen et al., 2018; Tung et al., 2020).

While existing studies have revealed tourist stereotype attributes, more attention should be given to investigate the activation of tourist stereotypes. Although psychology studies suggested that stereotypes are activated along two dichotomy streams of explicit and implicit, tourism studies often only focus on the explicit activation of tourist stereotypes and conclude the results as the overall tourist stereotype held by the locals.

Dichotomy system of stereotypes

Over the past few decades, stereotyping has been studied as a process of categorization that affects an individual's perception, evaluation as well as subsequent emotions and behaviors (Allport, 1954; Tajfel, 1969). Previous research has emphasized dichotomy in stereotypical categorization between the intended and unintended (Uleman & Bargh, 1989), uncontrolled versus controlled (Devine, 1989), unconsciousness versus consciousness (Kihlstrom, 1990), and mindful versus mindless (Langer, 1989). Recently studies showed that information processes are different between these two stereotypes (Plaza et al., 2017; Rubinstein et al., 2018), which affect the influences of external stimuli acting on them (Clément-Guillotin et al., 2018). They are dichotomous and separated into "explicit" and "implicit" stereotypes (Greenwald & Banaji, 2017). Explicit stereotypes are formed by the conscious awareness where individuals self-report their preconceptions and beliefs of other social groups (Brown & Gaertner, 2008). Conversely, implicit stereotypes are activated unconsciously and indirectly, and thus individuals may not be conscious of the existence of these perceptions (Greenwald & Banaji, 1995). The existence of a dichotomous system of stereotypes was identified in various contexts, ranging from gendered oriented stereotypes (Franceschini et al., 2014; Nowicki & Lopata, 2017), product-origin (Diamantopoulos et al., 2017), to ethnic majority-minority (Glock et al., 2019).

Explicit stereotypes

An explicit stereotype is the result of conscious and controlled evaluation toward individuals who are considered as outgroup members of a social group. Existing tourism research has primarily focused on explicit stereotypes using self-reports; for instance, Ming (2018) suggested that Mainland Chinese tourists are categorized into the quadrants of high competence but low warmth in the Stereotype Content Model. Similarly, Prendergast et al. (2016) adopted the same methodological approach and concluded negative stereotypes held by Hong Kong residents toward Mainland Chinese tourists. A similar conclusion was supported in research by Shen et al. (2017).

There are several challenges, however, when relying solely on self-reports of explicit stereotypes. When individuals are explicitly asked about their biases, they are consciously aware of their stereotypical content, attitudes, and beliefs, which could bias their responses given the presence of opportunities (e.g. time allowances) and extrinsic motivations (Greenwald & Banaji, 1995). Additionally, when respondents are mindful of their decision-making processes, they can become cautious when they answer questions, which could affect their expressions, underlying intentions, or evaluations due to social norms (Nosek et al., 2005).

Previous studies have also noted issues with using numerical scales to evaluate one's views (e.g. the extent of warmth and competence toward outgroup members: Cuddy et al., 2007, 2009; Fiske et al., 2002) due to the presence of self-presentational effects as well as social desirability bias (Fazio & Olson, 2003; Greenwald & Banaji, 1995; Herz & Diamantopoulos, 2013). Individuals may over-state or under-evaluate the target to enhance or reduce discriminations (Dunham et al., 2014; Rohmer & Louvet, 2018). For instance, individuals may evaluate women as high in warmth to compensate for lower ratings in competence (Kervyn et al., 2010).

Relying on just explicit measurements of stereotypes may fail to capture the automatic perceptions of an individual about a social group. Social perceptions are often unconscious or automatic (Bargh et al., 1996), especially on sensitive issues such as stereotyping. The presence of stereotypes toward an outgroup may be done subconsciously without awareness and intentions (Bargh, 1989; Greenwald & Banaji, 1995); hence, it is crucial to measure both explicit and implicit views in social cognition research.

Implicit stereotypes

Implicit stereotypes refer to the unconscious beliefs that an individual holds toward members of outgroups (Greenwald & Banaji, 1995). Unlike explicit stereotypes, implicit stereotypes operate without conscious and controlled intentions (Devine, 1989; Kihlstrom, 1990). The concept of implicit stereotypes is based on two critical theoretical understandings: the associative network in semantic memories and automatic activation.

Semantic memories assumed items are linked with each other in terms of their associative network where related items post stronger links than unrelated items (Collins & Loftus, 1975). These memories are long-term general knowledge of ideas and concepts (Tulving, 2002). For example, "tourists" are more closely associated with "guides" than to dissociate items such as "flowers" or "newspapers". A local network is formed when related concepts are clustered together (Payne & Cameron, 2013), such as tourists, guides, airlines, hotels, and attractions. The activation of one concept leads to the connection to other concepts within the same local network, and the degree of association among concepts could be measured by an individual's reaction time. Reaction time is shorter when the concepts are highly connected (Neely, 1977).

The association of one concept to another is seen as automatic or happening without conscious processing of information (Schneider & Shiffrin, 1977). Information processing happens outside of one's attention and does not require motivation (Cunningham et al., 2001; Greenwald & Banaji, 1995; Hinton, 2017). In terms of stereotyping, the rate of automatic association could be dependent on the frequency of exposure as well as the extent of biased associations with the target; the higher the exposure and strength of subjective linkages, the easier it is for an individual to activate a stereotype (Devine, 1989; Lepore & Brown, 1997). Once this association is learned, it is extremely difficult to unlearn.

As implicit stereotypes reflect the unconscious beliefs of an individual, they could be measured using an indirect approach (Greenwald & Banaji, 1995; Greenwald et al., 1998). There are some approaches to capture implicit stereotypes, such as affective priming task (Fazio et al., 1986), the Go/No Go association task (Nosek & Banaji, 2001), the Sorting Paired Featured task (Bar-Anan et al., 2009) and the IAT (Carpenter et al., 2018; Fazio & Olson, 2003). Among them, the IAT is preferred to capture the implicit tourist stereotypes. It is a computerized programme where individuals have to classify stimuli (e.g. stereotypes) into categories rapidly.

Implicit association test (IAT)

The IAT is an indirect measure to examine an individual's implicit social cognitions (Greenwald et al., 1998). IAT reflects the strength of automatic associations of concepts held by an individual. The main objective of IAT is to evaluate the association between the targets (e.g. tourist) and attribute dimensions (e.g. stereotypes). It requires a respondent to pair the target to an attribute rapidly, and the reaction time is a reflection of the respondent's implicit view. IAT helps to minimize social desirability bias that may arise from self-reported responses (Greenwald et al., 1998; King & Bruner, 2000).

The use of implicit measurement to study the automatic activation of stereotypes started in the 1980s when sequential priming tasks developed from cognitive psychology (Fazio et al., 1986; Gaertner & McLaughlin, 1983). There are many types, among them, Evaluative Priming tasks, and Semantic Priming tasks are the more frequently used measurements. Evaluative Priming task assesses the evaluative responses where respondents are first briefly presented with a prime target and a positive or negative attribute (Fazio et al., 1986). Then, respondents are asked to determine if the attributes belong to the positive or the negative by pressing one of the two responses critical as fast as possible – a faster response to positive words is associated with positive stereotypes and vice versa. For Semantic Priming task, it is analogous to Evaluative Priming, except that respondents are presented with a set of meaningful or meaningless letter strings as attributes. It is testing the association of the prime target with the semantic meaning of the attributes. Although sequential priming tasks have provided significant findings in understanding the cognitive-behavior relationship (Fazio, 2007), they are vulnerable to measurement errors, such as distraction, which lower its reliabilities as demonstrated by Cronbach's Alpha values rarely exceeding 0.50 (Fazio, 2007).

Building on the foundation of the cognitive priming task, the Implicit Association Test (IAT) was developed by Greenwald et al. (1998), and it has been one of the most widely adopted instruments in studying cognitive psychology in recent decades (Gawronski & De Houwer, 2014). IAT consists of congruent and incongruent tasks where the duration for associating target and attributes are recorded. Given the advancement of information technology, online data collection has been the platform for large and high powered samples (Buhrmester et al., 2018; Paolacci & Chandler, 2014). IAT has an interface with online surveys and allows for the merging of data. Given the above discussion, IAT is adopted in this study as the implicit measurement of tourist stereotypes.

The IAT has been employed in various studies, such as gender preferences (Koranyi et al., 2017), political attitudes (Ryan, 2017), consumer preferences (Maison et al., 2004), as well as in the area of intergroup conflicts (Greenwald et al., 1998, 2009). Given its popularity and credibility, continuous improvements have been made to the IAT through various studies. For example, McFarland and Crouch (2002) suggested that an artificially strong association between target and attributes could be achieved by longer reaction time from the disassociated links, thus affecting the calculated scores. To address this confounding effect, a new scoring algorithm was introduced by Greenwald et al. (2003). This scoring algorithm employs a *D* measure to account for contamination from response speed differences. Furthermore, attempts to fake an IAT score is difficult as it could be identifiable by such analysis.

IAT has been used in previous research to examine implicit views of destination image (Chen et al., 2016; Choi et al., 2015; Kim & Chen, 2010; Kim et al., 2011; Yang et al., 2012) and restaurant brands (Lee & Kim, 2013). However, it has not been used to examine tourist stereotypes, an important cognitive component that could affect intergroup relations and shape residents' attitudes toward tourism development. In this regard, this study seeks to provide a detailed development of IAT specifically tailored for tourism stereotype research; demonstrate the applicability of the IAT for assessing implicit stereotypes; and compare the implicit results with explicit stereotype measures via self-reports against Mainland Chinese visitors from residents in Hong Kong, Malaysia, Singapore, and Thailand.

Relationship between mainland China and Hong Kong, Malaysia, Singapore, and Thailand

With the unprecedented growth of China's economy and the associated relaxation of outbound travel restrictions, international travel has become a significant activity of Mainland Chinese residents. A double-digit percentage annual growth of Mainland Chinese outbound travel was recorded in many international destinations, such as Thailand, Singapore, Malaysia, and the Special Administrative Regions of Hong Kong and Macau (Yan, 2018). Apart from the significant economic contributions, the sudden increase of Mainland Chinese tourists have sparked social tensions with the local residents (Piuchan et al., 2018; Tse & Qiu, 2016).

Given its close proximity and ethnic affiliations, Hong Kong has become one of the most visited destinations by Mainland Chinese tourists. From the implementation of the Visiting Friends and Relatives (VFR) scheme in 1983 to the Individual Visit Scheme (IVS) in 2003, a series of policy relaxations has fostered a multifold increase in visitation to Hong Kong (Tourism Commission, 2019). However, this has intensified tensions with increasing conflicts with Hong Kong residents (Rowen, 2016), constructing the residents' stereotypes of Mainland Chinese tourists (Chen et al., 2018; Shen et al., 2017). Despite the vast investigations on Hong Kong residents' perceptions toward the Mainland Chinese tourists, they adopted explicit measurements that encourage the manipulation of stereotypes as a result of self-effect or social desirability. As such, the results may not reflect the underlying stereotypes of Hong Kong residents and thus provide an incomplete understanding. The examination of the implicit stereotypes can potentially aid tourism officials in formulating strategies for the host-guest relationship.

The conflicts between Mainland Chinese tourists and residents have spread to Malaysia, Singapore, and Thailand. Mainland Chinese tourists comprise the largest source of international tourist arrivals and tourism expenditures within the Association of Southeast Asian Nations (ASEAN) region (ASEAN Business Staff, 2019a, 2019b). Disputes between residents have induced negative images of Mainland Chinese tourists, affecting residents' stereotypes. Examining these three destinations may assist tourism officials in understanding residents' stereotypes that contribute to the management of host-guest relationships (Gong et al., 2019). Also, Malaysia, Singapore, and Thailand are the first three foreign countries that were stamped with the Approved Destination Status (ADS); a bilateral tourism agreement with China that allows Chinese citizens to travel in organized tour groups with approved travel agencies in China on visitor visas (Arita et al., 2011). Their long partnerships with China have served as references for other ASEAN countries in formulating their tourism strategies. With a predicted 45–166% growth in Mainland Chinese tourists visiting ASEAN countries (China Tourism Academy, 2016), investigation on both

Table 1. Dimensions and attributes of tourist stereotype model.

Dimension	Attributes
Approachable	Friendly, sincere, good
Competent	Intelligent, industrious, competent
Boastful	Materialistic, loud
Rude	Unreasonable, immoral, rude, uncivilized

Source: Tung et al. (2020).

implicit and explicit stereotypes will allow tourism officials to review their tourism marketing strategies whilst preparing other ASEAN countries for hosting a single large source of the tourism market.

Methodology

This research aims to examine the implicit and explicit tourist stereotypes of Mainland Chinese tourists as perceived by residents from four different Asian destinations. In this section, the components of examination for this research are first discussed. This includes identification of the stereotype target (i.e. Mainland Chinese tourists) and stereotype attributes (i.e. the contents by which Mainland Chinese tourists are labelled) as well as a detailed presentation of the measurement approaches for collecting the explicit and implicit tourist stereotypes of Mainland Chinese tourists. The second section presents the online survey instrument. Finally, the sample and procedure for data collection are discussed.

Components of examination

Stereotype target

Stereotype target, also known as the target of stereotyping, is important in measuring an individual's stereotypes as it ensures the correct identification and association of beliefs or expectations (Dunn & Spellman, 2003; Vescio & Biernat, 1999). The target could be a person, object, or place that allows for the evaluations of its attributes, characteristics as well as conduct. In this study, the stereotype target for measurement is Mainland Chinese tourists. This is an influential and emerging tourist market that receives numerous discussions or even criticism within the news and social media (Hsu & Chen, 2019; Qiu Zhang et al., 2017; Shen et al., 2017).

Stereotype attributes

Stereotype attributes refer to the contents associated with the stereotype target. Despite a variety of attributes available to describe individuals from Mainland China, little effort has been applied in examining the specific stereotype attributes that label Mainland Chinese tourists. Until recently, Tung et al. (2020) developed a tourist stereotype model (TSM) to assess the stereotypical attributes of Mainland Chinese tourists. Using an exploratory study on the stereotyping of Mainland Chinese tourists through a free response tasked by Hong Kong residents, 12 stereotype attributes were identified across four dimensions. Approachable – friendly, sincere, and good; Competent – intelligent, industrious, and competent; Boastful – materialistic and loud; Rude – unreasonable, immoral, rude, and uncivilized. Approachable and Competent were regarded as positive stereotypes, while Boastful and Rude were considered negative stereotypes. The use of adjectives to describe tourists has also been used in previous research, such as Maruyama and Woosnam (2015) and Pizam et al. (2002), in their Ethic Attitude Scale. In the scales, words such as Immoral-Moral, Boastful-Modest, and Stupid-Intelligent were used to capture residents' perceptions of social acceptance or exclusions toward the tourists, reflecting the inter-relations between the two major tourism social groups (Table 1).

Explicit measure of tourist stereotype

The attributes from the TSM were employed to measured explicit stereotypes. Consistent with existing studies on explicit measurements (Cuddy et al., 2009; Fiske et al., 2002; Tung et al., 2020), a 7-point Likert-scale ranging from 1 = Strong Disagree to 7 = Strong Agree, was adopted in this study.

Implicit measure of tourist stereotypes

IAT has been used in previous research. For example, Choi et al. (2015) assessed destination-related top-of-mind awareness by examining respondent reaction time to specific destination icons. This is based on the proposition that shorter reaction time implies a stronger implicit preference. Using self-report surveys and computer-based implicit association tests (IAT) for 87 college students, the authors found that participants' responses vary depending on the two measures. Through a data fuzzification method, the study demonstrated that the IAT could enhance familiarity issues of tourist top-of-mind awareness. Lee and Kim (2013) examined response patterns in explicit and implicit measures toward two known fast food restaurant brands through a self-report survey and IAT, respectively. Furthermore, Lee and Kim (2017) measured the image attributes of destinations using both explicit (i.e. self-report surveys) and implicit cognitions (i.e. reaction times) to understand the nature of destination images. The authors used Single-Target Implicit Association Test (ST-IAT) to measure individuals' explicit and implicit cognitions of image attributes across three countries: China, England, and France.

In this research, respondents completed a full version IAT which adopts two target categories. The first target represents Mainland Chinese tourists, while the second target reflects non-Mainland Chinese tourists. These targets serve to differentiate between the focal and non-focal categories. The IAT consists of seven blocks of target and attributes sorting trials. The target (e.g. *Mainland Chinese tourist* versus *non-Mainland Chinese tourists*) is represented in an image while the attribute (e.g. *positive* versus *negative stereotype content*) is represented in word. In each trial, an image or word appears on the computer screen representing a target or attribute. The respondents sort the target or attributes by pressing a computer key with the assigned hand (e.g. left hand, *Mainland Chinese tourist* or *positive stereotype content*; right hand, *non-Mainland Chinese tourist* or *negative stereotype contents*). In this study, the computer keys of "E" and "I" are employed for the left and right hand, respectively. During the sorting, target trials and attributes are alternating across the seven blocks.

The first and second blocks assigned are the practice blocks with 20 trials each. The purpose of the practice blocks is to allow respondents to become familiar with the computer system. The first block displays only targets (left hand, *Mainland Chinese tourist*, right hand, *non-Mainland Chinese tourists*) while the second block presents only attributes (left hand, *positive stereotype contents*, right hand, *negative stereotype contents*). Next, the third and fourth blocks are a combined block of using both targets and attributes (left hand, *Mainland Chinese tourist* or *positive stereotype content*; right hand, *non-Mainland Chinese tourist* or *negative stereotype contents*) where the computer keys are the same as the initial assignments in the previous two blocks. The third block consists of 20 practice trials, while the fourth block consists of 40 critical trials.

The fifth block presents only attributes with the side switched (left hand, *negative stereotype contents*, right hand, *positive stereotype contents*). This block consists of 40 practice trials in order to eliminate the left/right association learned in the previous blocks (Nosek et al., 2005). Finally, the sixth and seventh blocks are combined blocks with the attributes switched (left hand, *Mainland Chinese tourist* or *negative stereotype content*; right hand, *non-Mainland Chinese tourist* or *positive stereotype contents)*. Similarly, the sixth block consists of 20 practice trials, while the seventh block consists of 40 critical trials. Table 2 summarized the sorting trials of the full version IAT used in this study.

Table 2. Procedure of implicit association test (IAT).

Block	Task description	Assigned to LEFT key	Assigned to RIGHT key
1	Target Practice	Mainland Chinese tourists	Non-Mainland Chinese Tourists
2	Attribute **Congruent** Practice	Positive tourist stereotypes	Negative tourist stereotypes
3	Target-Attribute **Congruent** Practice	Mainland Chinese tourists or Positive tourist stereotypes	Non-Mainland Chinese tourists or Negative tourist stereotypes
4	Target-Attribute **Congruen**t Test	Same as Block 3	
5	Attribute ***Incongruent*** Practice	Negative tourist stereotypes	Positive tourist stereotypes
6	Target-Attribute ***Incongruent*** Practice	Mainland Chinese tourist or Negative tourist stereotypes	Non-Mainland Chinese tourists or Positive tourist stereotypes
7	Target-Attribute ***Incongruent*** Practice	Same as Block 6	

Source: Carpenter et al. (2018).

Table 3. Stereotype contents and target photos used in the implicit association test (IAT).

Dimensions	Attributes
Stereotypes contents	
Positive	Friendly, sincere, good, intelligent, industrious, competent
Negative	Materialistic, loud, unreasonable, rude, immoral, uncivilized
Target	
Mainland Chinese Tourist	
Non-Mainland Chinese Tourist	

In the case of an incorrect association, for instance, respondents pressed the wrong computer key from the assigned key, a red cross "X" is shown. It is mandatory for respondents to correct the errors by pressing the assigned key before proceeding with the IAT test (Table 3 and Figure 1).

The strength of stereotype association is based on the value of the *D* score (Greenwald et al., 2003; Haider et al., 2011) that is depending on the differences in reaction times between the combine blocks (third versus sixth; fourth versus seventh). Hence, the data in the combined, non-practice blocks are recorded for the analysis. The value *D* score is calculated for each respondent indicating in which condition (*Mainland Chinese tourist* or *positive stereotype contents* versus *Mainland Chinese tourist* or *negative stereotype contents)* they reacted faster. It is achieved in three steps. First, the reaction times of the sixth block are subtracted from the third block and then divided by the inclusive standard deviation of them, which is named as D_1. Second, the computation for seventh and fourth blocks is repeated, and is named as D_2. Finally, the value of *D* score is achieved by averaging D_1 and D_2.

To ensure the validity of the sorting, data cleaning procedures were performed to reveal unusual respondents of extremely fast and extremely slow trials. Based on the recommendation

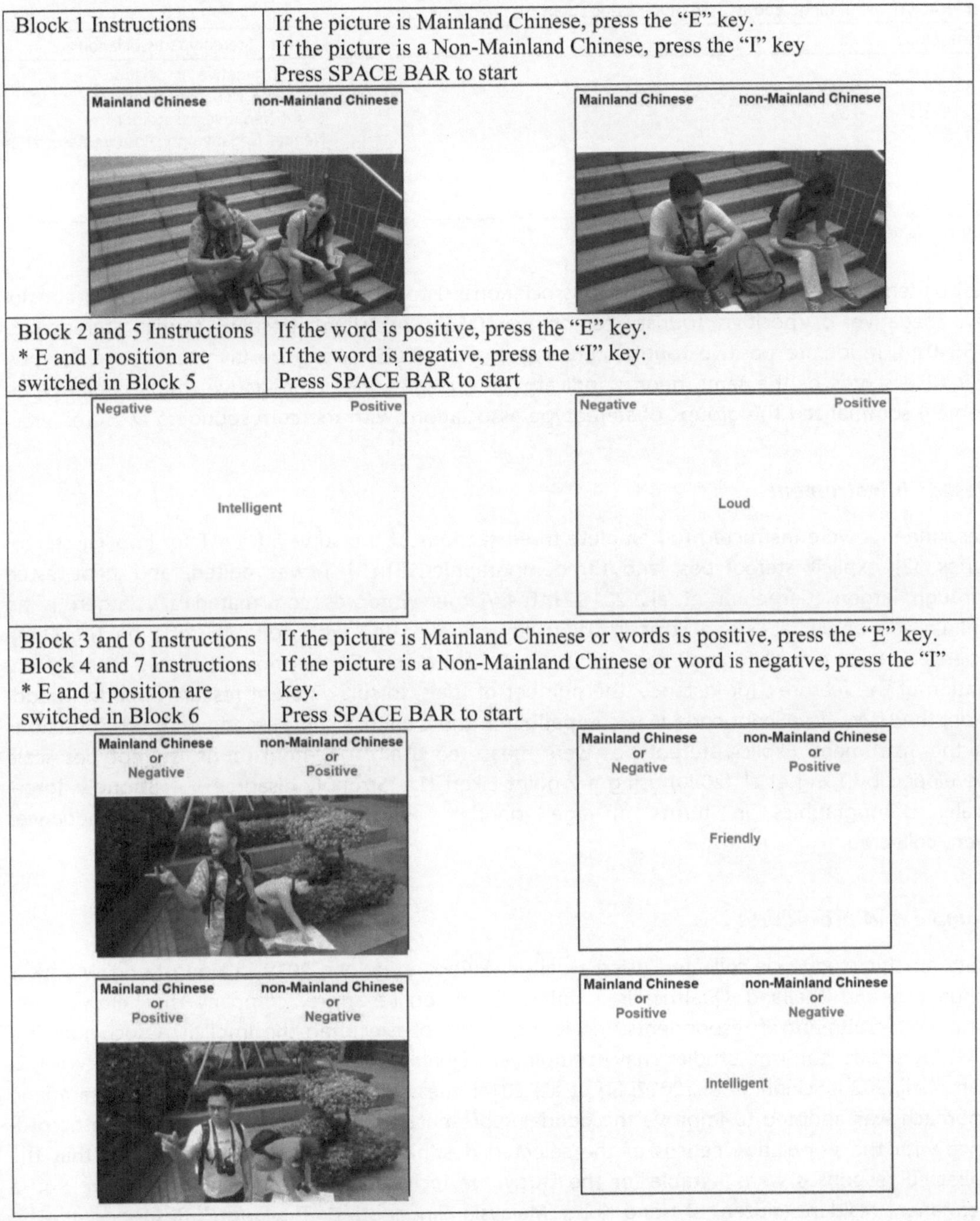

Figure 1. Visual illustration of implicit association test (IAT).

by Greenwald et al. (2003), respondents who employed more than 10,000 ms in any trials and any respondents with more than 10% of trials responded in less than 300 ms were excluded from the calculation.

The calculated *D* score should be within the possible range of −2 to +2, where negative denotes Mainland Chinese tourists with negative stereotype contents, while positive denotes Mainland Chinese tourists with positive stereotype contents (Chevance et al., 2017; Klein, 2020). Based on the psychological conventions for effect size, the value of *D* scores categorized

Table 4. IAT break range and its stereotype associations.

Break range	Stereotype associations
$-0.65 \le x \le -2$	Strong Negative Association
$-0.35 \le x < -0.65$	Moderate Negative Association
$-0.15 \le x < -0.35$	Slight Negative Association
$-0.15 < x < 0.15$	Neither Negative nor Positive Association
$0.15 \le x < 0.35$	Slight Positive Association
$0.35 \le x < 0.65$	Moderate Positive Association
$0.65 \le x \le 2$	Strong Positive Association

Source: Haider et al. (2011).

respondents into groups of stereotype associations: 0 to 0.14 indicating the absence of conclusive negative or positive tourist stereotypes; 0.15–0.34, slight positive tourist stereotypes; 0.35–0.64, moderate positive tourist stereotypes; and >0.65, strong positive tourist stereotypes. Negative scores of the same degree indicate similar categories of negative tourist stereotypes. Table 4 summarized the groups of stereotype associations with the corresponding *D* scores.

Research instrument

Respondents were instructed to complete three sections of the survey: (1) IAT for implicit stereotypes, (2) explicit stereotypes, and (3) demographics. The IAT was edited, and pilot-tested through Iatgen (Carpenter et al., 2018; https://iatgen.wordpress.com/materials/). Iatgen is an online web applet assists in structuralizing survey-based IAT with editable code on the Open Science Framework (OSF) page for various kinds of projects. Furthermore, it allows the customization of the features, for instance, the number of trials, to suit different research needs. In addition, the HTML/JavaScript code was compatible in the Qualtrics survey along with other sections of the instrument. Explicit stereotypes were measured using a 12-item tourist stereotypes scale developed by Tung et al. (2020) using a 7-point Likert (1 – Strongly disagree; 7 – Strongly agree) scale. Demographics in terms of age, gender, education level, working experiences were collected.

Sample and procedures

Data for this study was collected using Qualtrics software in June 2019 in Hong Kong, Malaysia, Singapore, and Thailand. Qualtrics is a United States online survey company that allows international recruitment of respondents and its capability of executing the Implicit Association Test (IAT). Previous tourism studies have employed Qualtrics for data collection (Kubickova & Campbell, 2020; Sciarini et al., 2012; So et al., 2014; Suess et al., 2020). A gender quota sampling approach was adopted to improve the gender representation of the collected sample in accordance with the population census of the selected destinations. Furthermore, to ensure that the collected responses were suitable for the study, we included a screening question, "Are you a permanent resident (citizen) of Hong Kong (Malaysia, Singapore or Thailand)?". Respondents who disagreed with the question were terminated immediately from the main survey. Only those who agreed with the question were invited to complete the survey. After seven days of distribution, 1040 completed responses were collected, 260 from each destination. Since no new responses were received over the next two days, the survey link was deactivated. Forty respondents were removed due to their invalid *D* score from the IAT. They were either too fast or too slow in sorting the trials; thus, calculating the *D* score was impossible. In addition, 10 outliers were detected and removed. Finally, a total of 990 (Hong Kong – 247; Malaysia – 249; Singapore – 246; Thailand – 248) valid responses were used for analysis. The response rate was 95.19%.

Table 5. Demographics of the respondents across the four destinations.

Variable	Hong Kong ($n = 247$) Frequency (%)	Malaysia ($n = 249$) Frequency (%)	Singapore ($n = 246$) Frequency (%)	Thailand ($n = 248$) Frequency (%)
Gender				
Female	133 (53.85)	120 (48.19)	124 (50.41)	126 (50.81)
Male	114 (46.15)	129 (51.80)	122 (49.59)	122 (49.19)
Age				
18 – 24	55 (19.86)	57 (22.89)	48 (19.51)	37 (14.92)
25 – 29	36 (13.00)	49 (19.68)	51 (20.73)	48 (19.36)
30 – 34	50 (18.05)	43 (17.27)	43 (17.48)	47 (18.95)
35 – 39	51 (18.41)	49 (19.68)	38 (15.45)	48 (19.36)
40 – 44	59 (21.30)	23 (9.24)	28 (11.38)	31 (12.50)
45 – 49	9 (3.25)	16 (6.43)	18 (7.32)	16 (6.45)
50 – 54	8 (2.89)	5 (2.00)	7 (2.85)	10 (4.03)
55 – 59	8 (2.89)	2 (0.80)	8 (3.25)	8 (3.23)
60 and above	1 (0.36)	5 (2.00)	5 (2.03)	3 (1.21)
Education				
Up to secondary	27 (10.93)	21 (5.43)	20 (8.13)	11 (4.64)
Post-secondary	52 (21.05)	68 (27.31)	69 (28.05)	27 (11.39)
Bachelor degree	131 (53.03)	99 (39.76)	115 (46.75)	183 (77.22)
Master degree	34 (13.77)	51 (20.48)	40 (16.26)	26 (10.97)
Doctorate degree	3 (0.12)	10 (4.02)	2 (0.81)	1 (0.42)
Working experience in hospitality and tourism				
Yes	48 (19.43)	75 (30.1)	53 (21.55)	80 (32.26)
No	199 (80.57)	174 (69.88)	193 (78.45)	168 (67.74)

Results

Respondent profile

Of the 990 respondents, 247 were from Hong Kong; 249 from Malaysia; 246 from Singapore; 248 from Thailand. The percentage of male respondents was 46.15%, 51.81%, 49.59% and 49.19% in Hong Kong, Malaysia, Singapore, and Thailand respectively, which corresponds closely to the official census provided by the respective governments; in 2018, the proportion of male was 45.77% in Hong Kong, 51.69% in Malaysia, 48.97% in Singapore, and 49.13% in Thailand. The samples covered a broad range of age groups and education levels, with 68.02% of Hong Kong residents, 64.26% of Malaysians, 63.82% of Singaporeans, and 88.61% of Thais holding at least a bachelor degree. Furthermore, 80.57% of the Hong Kong residents, 69.88% of the Malaysians, 78.45% of the Singaporeans, and 67.74% of the Thais have not worked in the Hospitality and Tourism industry. This is important because the results reflect the majority views of residents outside of the industry who would have relatively lower economic dependence on Mainland Chinese tourists (Table 5).

Table 6 shows the Cronbach's alphas of the four stereotype dimensions: Approachable, Competence, Boastful, and Rude were within the range of 0.74 to 0.92 across Hong Kong, Malaysia, Singapore, and Thailand. All the values were greater than the threshold value of 0.7, which represented good internal consistency of the items in each dimension of the tourist stereotype model (Nunnally, 1978).

Implicit tourist stereotypes

Using IAT, the speed of association between targets and attributes were recorded and transformed into the IAT effect score, *D* score. This represented the relative preference for the congruent and incongruent associations. Based on the value of the *D* score, respondents were categorized into six groups of stereotype effects toward the target: Neither Negative nor Positive, Slightly Positive, Medium Positive, Strong Positive, Slightly Negative, Medium Negative, and Strong Negative. The result indicated that a larger proportion of Hong Kong respondents (47.4%) and Singaporeans

Table 6. Internal consistency of stereotype dimensions across four destinations.

Variables	Hong Kong	Malaysia	Singapore	Thailand
Approachable	0.76	0.83	0.86	0.89
Competent	0.80	0.79	0.81	0.82
Boastful	0.82	0.82	0.74	0.84
Rude	0.87	0.92	0.90	0.90

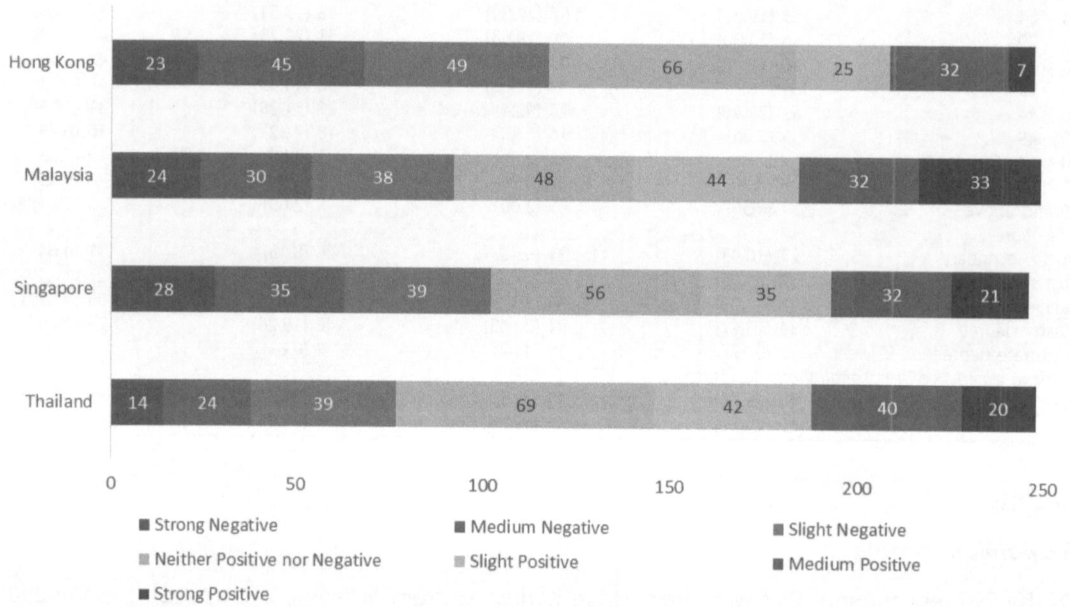

Figure 2. Distribution of implicit tourist stereotype across the four destinations.

Table 7. Means and standard deviations of explicit tourist stereotypes.

Stereotype	Hong Kong	Malaysia	Singapore	Thailand
Approachable	3.83 (1.21)	4.59 (1.13)	4.36 (1.19)	4.91 (1.27)
Competence	4.02 (1.21)	4.68 (1.05)	4.88 (1.04)	5.15 (1.07)
Boastful	5.42 (1.31)	4.61 (1.51)	5.30 (1.27)	4.41 (1.83)
Rude	4.89 (1.26)	4.23 (1.44)	4.68 (1.26)	3.70 (1.48)

Note: Numbers in brackets refer to standard deviation.

(41.5%) reported a negative rather than positive association. In contrast, a larger proportion of Malaysians (43.8%) and Thais (41.1%) reported a positive skew over a negative association toward Mainland Chinese tourists. These results suggest that only a small proportion of respondents displayed neither positive nor negative stereotype effects (see Figure 2).

Explicit tourist stereotypes

The composite means and standard deviations are the averages of all items within each of the four dimensions of explicit tourist stereotypes (see Table 7). Across the four dimensions, the results showed that Mainland Chinese tourists were more strongly viewed as Boastful in Hong Kong ($M = 5.42$, $SD = 1.31$) and Singapore ($M = 5.30$, $SD = 1.27$), as well as the least Approachable ($M_{Hong\ Kong} = 3.83$, $SD_{Hong\ Kong} = 1.21$; $M_{Singapore} = 4.36$, $SD_{Singapore} = 1.19$). On the contrary, Thais rated Mainland Chinese tourists higher on Approachable ($M = 4.91$, $SD = 1.27$) and Competence ($M = 5.15$, $SD = 1.07$). Malaysians also rated Chinese tourists as generally Approachable ($M = 4.59$, $SD = 1.13$) and Competent as well ($M = 4.68$, $SD = 1.05$).

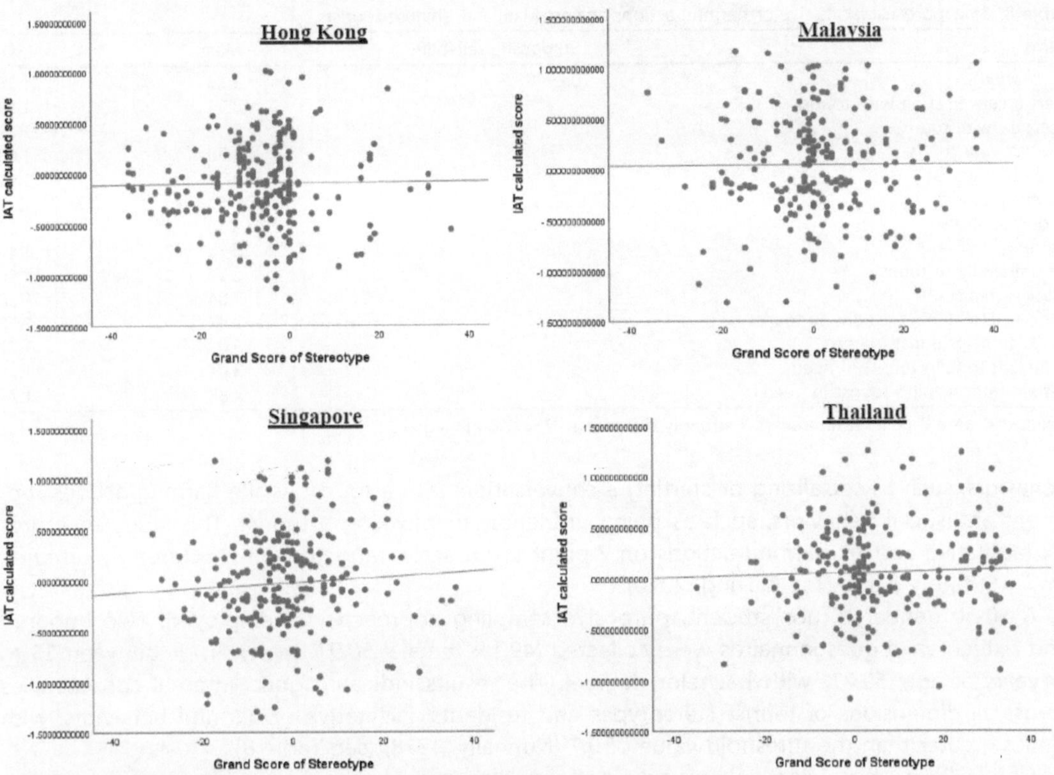

Figure 3. Scatterplots of implicit-explicit tourist stereotype across four destinations.

Correlation between implicit and explicit tourist stereotypes

Pearson's correlation analysis was performed to determine the correlation between implicit and explicit tourist stereotypes. The IAT score was the calculated value of *D*. The explicit stereotype score was the difference between positive (Approachable and Competence) and negative dimensions (Boastful and Rude). A respondent with a positive score suggested positive associations with Mainland Chinese tourists and vice versa. For example, the calculated explicit stereotype score for a respondent who reported Approachable = 5, Competent = 4, Boastful = 6, Rude = 6, would be $[(5+4) - (6+6)] = -3$, indicating a negative association.

Across the four destinations, the correlations were positive, yet weak and nonsignificant [$r_{\text{Hong Kong}} = 0.01$, $n=247$, $p=0.84$; $r_{\text{Malaysia}} = 0.00$, $n=249$, $p=0.94$; $r_{\text{Singapore}} = 0.07$, $n=246$, $p=0.31$; $r_{\text{Thailand}} = 0.04$, $n=248$, $p=0.57$]. The scatterplots of the implicit-explicit tourist stereotype correlations for Hong Kong, Malaysia, Singapore and Thailand are shown in Figure 3. Overall, the results suggest nonsignificant relationships between explicit and implicit stereotypes for all four destinations.

Predictor of residents' actions toward tourists

A follow-up study was conducted with a new sample of Singaporean residents through an online questionnaire in April 2020 to measure implicit and explicit tourist stereotypes as predictors of residents' facilitative or harmful actions toward Mainland Chinese tourists. Facilitative or harmful behaviors reflect dimensions in the Behaviour from Intergroup Affect and Stereotypes (BIAS) Map, which has been used to assess behavioral actions against outgroups members (Cuddy et al., 2007). In the context of tourism, facilitative actions refer to a resident's pro-social

Table 8. Singaporeans' facilitative or harmful actions against Mainland Chinese tourists.

Items[a]	Composite reliability	Mean	SD
Facilitation	0.89	3.33	1.35
Start a conversation with tourist		3.14	1.70
Socialize with tourist		3.12	1.69
Interact with tourist		3.38	1.68
Tolerate a tourist		3.38	1.66
Accept a tourist		3.56	1.71
Endure a tourist		3.38	1.66
Harm	0.92	2.76	1.40
Be unfriendly to tourist		2.27	1.56
Mock a tourist		2.84	1.68
Perform threatening actions to tourist		2.42	1.56
Resist to help tourist in need		3.09	1.71
Reluctant to help tourist in need		3.04	1.75
Refrain from helping tourist in need		2.88	1.75

[a]Measured on a 7-point scale from 1 = strongly disagree to 7 = strongly agree.

behaviors, such as socializing or starting a conversation with a tourist, while harmful actions represent anti-social behaviors, such as being unfriendly or mocking a tourist. This study measures six facilitative and six harmful actions on 7-point Likert scale, ranging from 1 = Strongly disagree to 7 = Strongly agree (Tse & Tung, 2020).

A 50-50 gendered, non-student, purposive sampling approach was employed. Two hundred and sixteen valid questionnaires were collected (49.1% female, 50.9% male; 41.7% between 25 to 34 years of age; 59.7% with bachelor degree). The results indicated good internal consistencies across all dimensions of tourist stereotypes and residents' facilitative or harmful behaviors, with values greater than the threshold value of 0.7 (Nunnally, 1978) (see Table 8).

Simple linear regression analysis was performed to model the relationship between implicit and explicit tourist stereotypes, and residents' facilitative or harmful actions toward Mainland Chinese tourists. The results indicated that explicit tourist stereotypes were more predictive than implicit stereotypes with regards to residents' actions. More specifically, explicit stereotypes better predicted facilitative than harmful behaviors. The results also showed that facilitative actions were induced by positive stereotypes of approachable and competence while reduced by negative stereotypes of boastful and uncivilized. On the other hand, harmful actions were elicited by negative stereotypes of uncivilized but unaffected by the others (see Table 9).

Discussion

Drawing from the social psychology and tourism literature on stereotypes, this study investigated implicit biases against Mainland Chinese tourists and measured them against explicit stereotypes from residents across four major Asian destinations: Hong Kong, Malaysia, Singapore, and Thailand. The study investigated the dual-information processes in tourist stereotyping and presented a detailed methodological development of an IAT tailored for implicit tourism stereotype research that could be useful for future cognitive-related studies in this area.

The findings of IAT indicated that neutral stereotype association about the Mainland Chinese tourist is the highest across the four destinations. The results of this IAT suggest that one's semantic memory of Mainland Chinese tourists is neither negative nor positive because such memories do not correspond to any event, personal experience, or time (Fazio, 2007; Tulving, 1972). The results suggest that harmful views between residents and Mainland Chinese tourists due to social conflicts, for example, may not be necessarily as negative as expected from an implicit perspective.

The results of explicit and implicit tourist stereotypes toward Mainland Chinese tourists were disassociated across the four destinations. These results are similar to the existing intergroup relation literature on measured explicit–implicit stereotypes, especially on highly sensitive topics,

Table 9. Linear regression of tourist stereotypes and residents' actions (follow-up study).

	Facilitative actions					Harmful actions				
	Unstandardized coefficients		Standardized coefficients			Unstandardized coefficients		Standardized coefficients		
	B	Std. error	Beta	T	Sign	B	Std. error	Beta	T	Sign
Implicit tourist stereotypes										
IAT score	0.12	0.18	0.04	0.64	0.53	−0.09	0.19	−0.03	−0.48	0.63
Explicit tourists stereotypes										
Approachable	0.72	0.06	0.64	12.10	$p < 0.001$	−0.00	0.08	−0.00	−0.05	0.96
Competence	0.58	0.07	0.48	8.00	$p < 0.001$	−0.12	0.09	−0.09	−1.36	0.18
Boastful	−0.33	0.06	−0.35	−5.48	$p < 0.001$	0.08	0.07	0.09	1.27	0.21
Uncivilized	−0.36	0.06	−0.39	−6.10	$p < 0.001$	0.25	0.06	0.26	4.00	$p < 0.001$

where the differences are influenced by one's conscious intention to evaluate (Dovidio et al., 2001; Greenwald et al., 2009; Olson & Dunham, 2010). The domain of stereotyping is a socially sensitive topic as it is closely related to one's identity as well as group polarization, which could arouse socially desirable results when measured explicitly due to perceived social consensus in facilitating the preconceptions of a particular social group. In the current study, negative impressions of Mainland Chinese tourists, for example, may have been shared amongst residents in Singapore through social media and online forums; hence, explicit self-reports of negative stereotypes may be considered as socially desirable (Jacobs, 2012; Siswandio, 2019). Implicitly, however, residents in Singapore were more positive. Their automatic IAT responses were unlikely influenced by tendencies to conform, contemplations of social beliefs and values, or motivations to achieve approval in the eyes of the public (Hu et al., 2017).

Theoretical contributions

The research provides significant insights into the knowledge of tourist stereotypes. From an explicit stereotype measurement perspective, this study provided large-scale and cross-cultural support for the tourist stereotype model (Tung et al., 2020). Previous studies have shown the applicability of the model in examining tourist stereotypes in Hong Kong. The present research validates the model in Malaysia, Singapore, and Thailand, holding different cultural and historical backgrounds than Hong Kong. The results indicate that stereotyping Mainland Chinese tourists is not a sole phenomenon in Hong Kong. Still, it could be a worldwide situation, especially with the increasing waves of Mainland Chinese tourists.

This research elevates the understanding of tourist stereotypes through the integration of the Implicit Association Test (IAT) to extract the implicit tourist stereotypes of residents. While previously applied, IAT in other tourist areas such as destination image and restaurant brands (Jang, 2016; Lee & Kim, 2013; Yang et al., 2012), is nevertheless, scarcely applied in tourism research related to social and intergroup relations. The adoption of IAT enhances tourism literature focusing on tourist stereotypes by comprehending the information process of the residents toward tourists. It reveals the residents' unconscious perceptions toward tourists, offering new knowledge about intergroup stereotypes between residents and tourists in a tourism setting. The finding of this study indicates the disassociation between implicit and explicit stereotypes, signifying the existence of the dichotomy system of stereotypes that should be captured separately. Previous studies offer incomplete understandings of tourist stereotypes due to the sole reliance on explicit measurements, failing to capture the unconscious biases of residents toward tourists. As such, this study signifies the needs to incorporate both implicit and explicit measurements that could facilitate the knowledge development of tourist stereotyping.

This research also demonstrated the development of IAT, which can be useful in the tourism literature examining tourist stereotypes. This research has proposed a promising implicit measurement to capture implicit tourist stereotypes of residents, human-human evaluation, which still warrants more attention in tourism studies. Previous studies employed IAT for the association with objects (e.g. destination image) but not in the area of intergroup relations, which is essential for host-guest relations and the sustainable development of tourism. This study provided a detailed explanation of the IAT, including the selection criteria of targets and attributes, descriptions of each block task, choices of analyzed blocks, procedures of data cleaning, calculations of stereotype effects, and associations of stereotype category. The study depicted each step of the IAT to allow for its replication in future studies related to individuals' perception and cognition investigation, serving as an alternative measurement in assessing comparative information processes toward two targets (e.g. two tourist markets).

Additionally, this research also adds value to the literature by demonstrating the predictive association of the tourist stereotype model (Tung et al., 2020) on residents' facilitative or harmful actions. Interestingly, the findings show that tourist stereotypes may be stronger predictors of facilitative actions than harmful actions. Furthermore, explicit stereotypes were found to be better associated with residents' actions than implicit measures, which differs from existing socio-psychology literature. Previous literature suggested that implicit stereotypes could predict social actions and interactions among members of social groups, such as duration of conversation, and sitting distances between individuals (Amodio & Devine, 2006; Kurdi et al., 2019). The predictive association of IAT on residents' actions, in contrast compared to explicit measurements, remains strong in particular for socially sensitive topics (Greenwald et al., 2009). It is possible that tourist stereotyping is no longer a sensitive topic amidst COVID-19, given the unfortunate number of reports of harassment against tourists. As a result, it is important to note that although this study did not find significant predictions of IAT on residents' actions from one Singaporean sample, it should not be definitively concluded that such association does not exist.

Practical implications

As social impacts from tourist concern increases, DMOs and public policymakers should understand their residents' implicit tourist stereotypes. Managing intergroup relations between residents and tourists is one of the most critical challenges faced by destination managers as they need to balance the economic benefits and social conflicts from tourism. The understanding of residents' attitudes helps to formulate responsive strategies that are important for building a harmonious host-guest relation for sustainable tourism development. This research demonstrated the disassociation between implicit and explicit measurements, suggesting the mental discrepancies between implicit and explicit tourist stereotypes. While these differences do not represent the actual one over the other (Kim et al., 2011), it proposes the importance of managing both stereotypes in order to formulate accurate strategies.

Furthermore, the thoughtful mapping of IAT can be efficiently and readily employed as a practical application for destination managers to capture the implicit tourist stereotypes of their residents. This IAT is an online mental sorting test that can be served as a "virtual laboratory" for residents to participate and understand their implicit tourist stereotypes. For instance, Project Implicit initiated by a group of social scientists to measure implicit social cognition on various social issues (Project Implicit, 2020). Destination managers could adopt a similar approach, such as cooperating with experts in host-guest relations, to replicate an IAT that is suitable for residents to access and measure their own implicit tourist stereotypes. Beyond the capturing of implicit tourist stereotypes, destination managers could initiate subsequent courses to educate residents on their implicit tourist stereotypes. The IAT facilitates residents to face their implicit tourist stereotypes, hence creating self-awareness of the external stimuli that affect their explicit views.

The "virtual laboratory" should not be a one-time pop-up platform for residents. It should be a long-term assessment of tourist stereotypes where the results could be stored as longitudinal data. This could allow destination managers to track the changes in residents' implicit stereotypes against tourists over time. For example, Singapore has experienced an unprecedented influx of Mainland Chinese tourists since July 2019 and is predicted to grow in the next few years (China Tourism Academy, 2016; Sim, 2019). A one-time implicit assessment would only provide cross-sectional evaluations and hence, limited information for DMOs and decision-makers for policy refinement. Instead, implicit evaluations over time could map changes in negativity (or positivity) amongst Singaporeans toward Mainland Chinese as this market segment continues to contribute to Singapore's tourism development in the near future. The long-term assessment also allows destination managers to review their internal marketing communication and revise them according to the residents' implicit tourist stereotypes.

Limitations and directions for future research

There are limitations in this study and opportunities for future research. Firstly, this study collected data from Hong Kong, Malaysia, Singapore, and Thailand, which addressed the limitation of a single location study (Tung et al., 2020). While this study provided empirical support for the tourist stereotype model by examining it with other Southeast Asian destinations, the examined target was still focusing on Mainland Chinese tourists. Additionally, different destinations may have different stereotypes toward Mainland Chinese tourists, and the tourist stereotype model may not cover the full range of stereotypes on an implicit-explicit dichotomy. It would have been best to assess stereotypes for each destination, derive a set of explicit and implicit measurements, and then tailor the IAT for each destination specifically. Nevertheless, doing so was beyond the scope of this study. Future studies should examine the model beyond the Mainland Chinese market, such as emerging markets of Indians and Indonesians, which could determine the applicability of the identified stereotype attributes beyond a specific market. Examining tourist stereotypes in different contexts may provide additional insights and stereotype attributes.

Future studies could also examine the circle of friends that respondents often interact with. As noted from the contact hypothesis, it is stated that the increasing positive interaction could affect the stereotype held by an individual. It is suggested by Maruyama and Woosnam (2015) that minority will be accepted, acknowledged, and acquainted if the majority has positive attitudes toward them. Hence, future research could investigate the residents' interpersonal connections and attachment through the number of Mainland Chinese friends they have, how often they meet and socialize with one another, and the perceived closeness. The results could suggest potential connectivity between actual interactions as a possible antecedent on implicit and explicit tourist stereotyping.

Future researchers may further assess the influence of implicit and explicit stereotypes on alternative outcomes, such as emotions. Future studies might examine residents' welcoming nature, emotional closeness, and sympathetic understanding (Woosnam & Norman, 2010) as one of the possible outcomes that shape their emotional solidarity toward the tourists, indicating the residents' attitude and support for tourism development (Woosnam et al., 2018). Besides that, future research could examine the mediating effect of emotions on stereotypes and behaviors. The examination of stereotype-emotion-behavior is a significant topic for the understanding and managing of host-guest relations.

Disclosure statement

No potential conflict of interest was reported by the author(s).

Funding

The work described in this article was fully supported by a grant from the Research Grants Council of the Hong Kong Special Administrative Region, China (Project No. PolyU255017/16B).

ORCID

Wai Tsz Serene Tse http://orcid.org/0000-0001-8508-0465
Vincent Wing Sun Tung http://orcid.org/0000-0001-9560-8761

References

Allport, G. (1954). *The nature of prejudice*. Beacon Press.

Amodio, D. M., & Devine, P. G. (2006). Stereotyping and evaluation in implicit race bias: Evidence for independent constructs and unique effects on behavior. *Journal of Personality and Social Psychology*, *91*(4), 652–661. https://doi.org/10.1037/0022-3514.91.4.652

Anderson, C., Hildreth, J. A. D., & Howland, L. (2015). Is the desire for status a fundamental human motive? A review of the empirical literature. *Psychological Bulletin*, *141*(3), 574–601. https://doi.org/10.1037/a0038781

Arita, S., Edmonds, C., Croix, S. L., & Mak, J. (2011). Impact of Approved Destination Status on Chinese travel abroad: An econometric analysis. *Tourism Economics*, *17*(5), 983–996. https://doi.org/10.5367/te.2011.0076

ASEAN Business Staff. (2019a). China tourism to ASEAN slowing: Report. *The Business Times*. https://www.businesstimes.com.sg/asean-business/china-tourism-to-asean-slowing-report

ASEAN Business Staff. (2019b). South-east Asia scoops up Golden Week tourists from China. *The Business Times*. https://www.businesstimes.com.sg/asean-business/south-east-asia-scoops-up-golden-week-tourists-from-china

Bar-Anan, Y., Nosek, B. A., & Vianello, M. (2009). The sorting paired features task: A measure of association strengths. *Experimental Psychology*, *56*(5), 329–343. https://doi.org/10.1027/1618-3169.56.5.329

Bargh, J. A. (1989). Conditional automaticity: Varieties of automatic influence in social perception and cognition. In J. S. Uleman & J. A. Bargh (Eds.), *Unintended thought*. (pp. 3–51). The Guilford Press.

Bargh, J. A., Chen, M., & Burrows, L. (1996). Automaticity of social behavior: Direct effects of trait construct and stereotype-activation on action. *Journal of Personality and Social Psychology*, *71*(2), 230–244. https://doi.org/10.1037//0022-3514.71.2.230

Boissevain, J., & Inglott, P. S. (1979). Tourism: Passport to development? Perspectives on the social and cultural effects of tourism in developing countries. In E. de Kadt (Ed.), *Tourism in Malta* (pp. 265–284). Oxford University Press for World Bank and UNESCO.

Brown, R., & Gaertner, S. L. (2008). *Blackwell handbook of social psychology: Intergroup processes*. Blackwell Publishers Ltd.

Buhrmester, M. D., Talaifar, S., & Gosling, S. D. (2018). An evaluation of Amazon's Mechanical Turk, its rapid rise, and its effective use. *Perspectives on Psychological Science*, *13*(2), 149–154. https://doi.org/10.1177/1745691617706516

Carpenter, T., Pogacar, R., Pullig, C., Kouril, M., Aguilar, S. J., LaBouff, J. P., Isenberg, N., & Chakroff, A. (2018, September 17). *Survey-software implicit association tests: A methodological and empirical analysis*. Retrieved from https://doi.org/10.31234/osf.io/hgy3z

Chen, C. C., Lai, Y. H. R., Petrick, J. F., & Lin, Y. H. (2016). Tourism between divided nations: An examination of stereotyping on destination image. *Tourism Management*, *55*, 25–36. https://doi.org/10.1016/j.tourman.2016.01.012

Chen, N., Hsu, C. H., & Li, X. R. (2018). Feeling superior or deprived? Attitudes and underlying mentalities of residents towards Mainland Chinese tourists. *Tourism Management*, *66*, 94–107. https://doi.org/10.1016/j.tourman.2017.11.007

Chevance, G., Héraud, N., Guerrieri, A., Rebar, A., & Boiché, J. (2017). Measuring implicit attitudes toward physical activity and sedentary behaviors: Test-retest reliability of three scoring algorithms of the Implicit Association

Test and Single Category-Implicit Association Test. *Psychology of Sport and Exercise*, *31*, 70–78. https://doi.org/10.1016/j.psychsport.2017.04.007

China Tourism Academy. (2016). *Annual report of China outbound tourism development 2016*. Tourism Education Press.

Choi, S., Liu, L., & Kim, D. Y. (2015). Accessing tourists' unconscious associations about international destinations: Data fuzzification of reaction times in the implicit association test. *Journal of Travel & Tourism Marketing*, *32*(5), 578–594.

Clément-Guillotin, C., Rohmer, O., Forestier, C., Guillotin, P., Deshayes, M., & d'Arripe-Longueville, F. (2018). Implicit and explicit stereotype content associated with people with physical disability: Does sport change anything? *Psychology of Sport and Exercise*, *38*, 192–201. https://doi.org/10.1016/j.psychsport.2018.06.014

Collins, A. M., & Loftus, E. F. (1975). A spreading-activation theory of semantic processing. *Psychological Review*, *82*(6), 407–428. https://doi.org/10.1037/0033-295X.82.6.407

Cuddy, A. J., Fiske, S. T., & Glick, P. (2007). The BIAS map: Behaviors from intergroup affect and stereotypes. *Journal of Personality and Social Psychology*, *92*(4), 631–648. https://doi.org/10.1037/0022-3514.92.4.631

Cuddy, A. J. C., Fiske, S. T., Kwan, V. S. Y., Glick, P., Demoulin, S., Leyens, J.-P., Bond, M. H., Croizet, J.-C., Ellemers, N., Sleebos, E., Htun, T. T., Kim, H.-J., Maio, G., Perry, J., Petkova, K., Todorov, V., Rodríguez-Bailón, R., Morales, E., Moya, M., … Ziegler, R. (2009). Stereotype content model across cultures: Towards universal similarities and some differences. *The British Journal of Social Psychology*, *48*(Pt 1), 1–33. https://doi.org/10.1348/014466608X314935

Cunningham, W. A., Preacher, K. J., & Banaji, M. R. (2001). Implicit attitude measures: Consistency, stability, and convergent validity. *Psychological Science*, *12*(2), 163–170. https://doi.org/10.1111/1467-9280.00328

Devine, P. G. (1989). Stereotypes and prejudice: Their automatic and controlled components. *Journal of Personality and Social Psychology*, *56*(1), 5–18. https://doi.org/10.1037/0022-3514.56.1.5

Diamantopoulos, A., Florack, A., Halkias, G., & Palcu, J. (2017). Explicit versus implicit country stereotypes as predictors of product preferences: Insights from the stereotype content model. *Journal of International Business Studies*, *48*(8), 1023–1036. https://doi.org/10.1057/s41267-017-0085-9

Dovidio, J. F., Kawakami, K., & Beach, K. R. (2001). Implicit and explicit attitudes: Examination of the relationship between measures of intergroup bias. In R. Brown., & S. Gertner. (Eds.), *Blackwell handbook of social psychology: Intergroup processes* (4th ed., pp. 175–197). Blackwell Publishing.

Dunham, Y., Srinivasan, M., Dotsch, R., & Barner, D. (2014). Religion insulates ingroup evaluations: The development of intergroup attitudes in India. *Developmental Science*, *17*(2), 311–319. https://doi.org/10.1111/desc.12105

Dunn, E. W., & Spellman, B. A. (2003). Forgetting by remembering: Stereotype inhibition through rehearsal of alternative aspects of identity. *Journal of Experimental Social Psychology*, *39*(5), 420–433. https://doi.org/10.1016/S0022-1031(03)00032-5

Fazio, R. (2007). Attitudes as object-evaluation associations of varying strength. *Social Cognition*, *25*(5), 603–637. https://doi.org/10.1521/soco.2007.25.5.603

Fazio, R. H., & Olson, M. A. (2003). Implicit measures in social cognition research: Their meaning and use. *Annual Review of Psychology*, *54*(1), 297–327. https://doi.org/10.1146/annurev.psych.54.101601.145225

Fazio, R. H., Sanbonmatsu, D. M., Powell, M. C., & Kardes, F. R. (1986). On the automatic activation of attitudes. *Journal of Personality and Social Psychology*, *50*(2), 229–238. https://doi.org/10.1037/0022-3514.50.2.229

Fiske, S. T., Cuddy, A. J. C., Glick, P., & Xu, J. (2002). A model of stereotype content as often mixed: Separate dimensions of competence and warmth respectively follow from status and competition. *Journal of Personality and Social Psychology*, *82*(6), 878–902. https://doi.org/10.1037/0022-3514.82.6.878

Franceschini, G., Galli, S., Chiesi, F., & Primi, C. (2014). Implicit gender–math stereotype and women's susceptibility to stereotype threat and stereotype lift. *Learning and Individual Differences*, *32*, 273–277. https://doi.org/10.1016/j.lindif.2014.03.020

Gaertner, S. L., & McLaughlin, J. P. (1983). Racial stereotypes: Associations and ascriptions of positive and negative characteristics. *Social Psychology Quarterly*, *46*(1), 23–30. [Database] https://doi.org/10.2307/3033657

Gawronski, B., & De Houwer, J. (2014). Implicit measures in social and personality psychology. *Handbook of Research Methods in Social and Personality Psychology*, *2*, 283–310.

Glock, S., Kovacs, C., & Pit-ten Cate, I. (2019). Teachers' attitudes towards ethnic minority students: Effects of schools' cultural diversity. *The British Journal of Educational Psychology*, *89*(4), 616–634. https://doi.org/10.1111/bjep.12248

Gong, J., Detchkhajornjaroensri, P., & Knight, D. W. (2019). Responsible tourism in Bangkok, Thailand: Resident perceptions of Chinese tourist behaviour. *International Journal of Tourism Research*, *21*(2), 221–233. https://doi.org/10.1002/jtr.2256

Greenwald, A. G., & Banaji, M. R. (1995). Implicit social cognition: Attitudes, self-esteem, and stereotypes. *Psychological Review*, *102*(1), 4–27. https://doi.org/10.1037/0033-295x.102.1.4

Greenwald, A. G., & Banaji, M. R. (2017). The implicit revolution: Reconceiving the relation between conscious and unconscious. *The American Psychologist*, *72*(9), 861–871. https://doi.org/10.1037/amp0000238

Greenwald, A. G., McGhee, D. E., & Schwartz, J. L. (1998). Measuring individual differences in implicit cognition: The implicit association test. *Journal of Personality and Social Psychology, 74*(6), 1464–1480. https://doi.org/10.1037/0022-3514.74.6.1464

Greenwald, A. G., Nosek, B. A., & Banaji, M. R. (2003). Understanding and using the implicit association test: I. An improved scoring algorithm. *Journal of Personality and Social Psychology, 85*(2), 197–216. https://doi.org/10.1037/0022-3514.85.2.197

Greenwald, A. G., Poehlman, T. A., Uhlmann, E. L., & Banaji, M. R. (2009). Understanding and using the Implicit Association Test: III. Meta-analysis of predictive validity. *Journal of Personality and Social Psychology, 97*(1), 17–41. https://doi.org/10.1037/a0015575

Haider, A. H., Sexton, J., Sriram, N., Cooper, L. A., Efron, D. T., Swoboda, S., Villegas, C. V., Haut, E. R., Bonds, M., Pronovost, P. J., Lipsett, P. A., Freischlag, J. A., & Cornwell, E. E. (2011). Association of unconscious race and social class bias with vignette-based clinical assessments by medical students. *JAMA, 306*(9), 942–951. https://doi.org/10.1001/jama.2011.1248

Herz, M. F., & Diamantopoulos, A. (2013). Activation of country stereotypes: Automaticity, consonance, and impact. *Journal of the Academy of Marketing Science, 41*(4), 400–417. https://doi.org/10.1007/s11747-012-0318-1

Hinton, P. (2017). Implicit stereotypes and the predictive brain: Cognition and culture in "biased." *Palgrave Communications, 3*(1), 1–9. https://doi.org/10.1057/palcomms.2017.86

Hsu, C. H., & Chen, N. (2019). Resident attribution and tourist stereotypes. *Journal of Hospitality & Tourism Research, 43*(4), 489–516.

Hu, Y., Ul Hasan Abbasi, N., Wang, S., Zhou, Y., Yang, T., & Zhang, Y. (2017). Implicit and explicit attitudes of Chinese youth toward the second-generation rich. *Social Behavior and Personality: An International Journal, 45*(3), 427–440. https://doi.org/10.2224/sbp.5775

Jacobs, A. (2012, July 26). In Singapore, vitriol against Chinese newcomers. *The New York Times*. https://www.nytimes.com/2012/07/27/world/asia/in-singapore-vitriol-against-newcomers-from-mainland-china.html

Jang, D. (2016). Explicit and implicit attitudes and attitude changes toward tourism destinations: Las Vegas, USA, versus Jeongseon, South Korea. *International Journal of Tourism Research, 18*(5), 417–422. https://doi.org/10.1002/jtr.2057

Kervyn, N., Yzerbyt, V., & Judd, C. M. (2010). Compensation between warmth and competence: Antecedents and consequences of a negative relation between the two fundamental dimensions of social perception. *European Review of Social Psychology, 21*(1), 155–187. https://doi.org/10.1080/13546805.2010.517997

Kihlstrom, J. F. (1990). The psychological unconscious. In L. A. Pervin & O. P. John (Eds.), *Handbook of personality. Theory and research* (pp. 424–442). The Guilford Press.

Kim, D. Y., & Chen, Z. (2010). Are people aware of their attitudes toward destination? Understanding the implicit association test in tourism research. *Tourism Analysis, 15*(3), 299–313. https://doi.org/10.3727/108354210X12801550666060

Kim, D. Y., Chen, Z., & Hwang, Y. H. (2011). Are we really measuring what we think we're measuring? Assessing attitudes towards destinations with the implicit association test. *International Journal of Tourism Research, 13*(5), 468–481. https://doi.org/10.1002/jtr.821

King, M. F., & Bruner, G. C. (2000). Social desirability bias: A neglected aspect of validity testing. *Psychology & Marketing, 17*(2), 79–103.

Klein, C. (2020). *Confidence intervals on implicit association test scores are really rather large*. https://doi.org/10.31234/osf.io/5djkh

Koranyi, N., Grigutsch, L. A., Algermissen, J., & Rothermund, K. (2017). Dissociating implicit wanting from implicit liking: Development and validation of the Wanting Implicit Association Test (W-IAT). *Journal of Behavior Therapy and Experimental Psychiatry, 54*, 165–169. https://doi.org/10.1016/j.jbtep.2016.08.008

Kubickova, M., & Campbell, J. M. (2020). The role of government in agro-tourism development: A top-down bottom-up approach. *Current Issues in Tourism, 23*(5), 587–604. https://doi.org/10.1080/13683500.2018.1551338

Kurdi, B., Seitchik, A. E., Axt, J. R., Carroll, T. J., Karapetyan, A., Kaushik, N., Tomezsko, D., Greenwald, A. G., & Banaji, M. R. (2019). Relationship between the Implicit Association Test and intergroup behavior: A meta-analysis. *The American Psychologist, 74*(5), 569–586. https://doi.org/10.1037/amp0000364

Langer, E. J. (1989). Minding matters: The consequences of mindlessness–mindfulness. In L. Berkowitz (Ed.), *Advances in experimental social psychology* (Vol. 22, pp. 137–173). Academic Press.

Lee, K. H., & Kim, D. Y. (2013). A comparison of implicit and explicit attitude measures: An application of the implicit association test (IAT) to fast food restaurant brands. *Tourism Analysis, 18*(2), 119–131. https://doi.org/10.3727/108354213X13645733247576

Lee, K. H., & Kim, D. Y. (2017). Explicit and implicit image cognitions toward destination: Application of the Single-Target Implicit Association Test (ST-IAT). *Journal of Destination Marketing & Management, 6*(4), 396–406.

Lepore, L., & Brown, R. (1997). Category and stereotype activation: Is prejudice inevitable? *Journal of Personality and Social Psychology, 72*(2), 275–287. https://doi.org/10.1037/0022-3514.72.2.275

Maison, D., Greenwald, A. G., & Bruin, R. H. (2004). Predictive validity of the Implicit Association Test in studies of brands, consumer attitudes, and behavior. *Journal of Consumer Psychology*, *14*(4), 405–415. https://doi.org/10.1207/s15327663jcp1404_9

Maruyama, N., & Woosnam, K. M. (2015). Residents' ethnic attitudes and support for ethnic neighborhood tourism: The case of a Brazilian town in Japan. *Tourism Management*, *50*, 225–237. https://doi.org/10.1016/j.tourman.2015.01.030

McFarland, S. G., & Crouch, Z. (2002). A cognitive skill confound on the Implicit Association Test. *Social Cognition*, *20*(6), 483–510. https://doi.org/10.1521/soco.20.6.483.22977

Milman, A., Reichel, A., & Pizam, A. (1990). The impact of tourism on ethnic attitudes: The Israeli-Egyptian case. *Journal of Travel Research*, *29*(2), 45–49. https://doi.org/10.1177/004728759002900207

Ming, H. (2018). Cross-cultural difference and cultural stereotypes in tourism – Chinese tourists in Thailand. *Journal of Hotel & Business Management*, *1*, 1–5.

Monterrubio, C. (2018). Tourist stereotypes and servers' attitudes: A combined theoretical approach. *Journal of Tourism and Cultural Change*, *16*(1), 57–74. https://doi.org/10.1080/14766825.2016.1237518

Moufakkir, O. (2011). The role of cultural distance in mediating the host gaze. *Tourist Studies*, *11*(1), 73–89. https://doi.org/10.1177/1468797611412065

Neely, J. H. (1977). Semantic priming and retrieval from lexical memory: Roles of inhibitionless spreading activation and limited-capacity attention. *Journal of Experimental Psychology: General*, *106*(3), 226–254. https://doi.org/10.1037/0096-3445.106.3.226

Nosek, B. A., & Banaji, M. R. (2001). The go/no-go association task. *Social Cognition*, *19*(6), 625–666. https://doi.org/10.1521/soco.19.6.625.20886

Nosek, B. A., Greenwald, A. G., & Banaji, M. R. (2005). Understanding and using the Implicit Association Test: II. Method variables and construct validity. *Personality & Social Psychology Bulletin*, *31*(2), 166–180. https://doi.org/10.1177/0146167204271418

Nowicki, E. A., & Lopata, J. (2017). Children's implicit and explicit gender stereotypes about mathematics and reading ability. *Social Psychology of Education*, *20*(2), 329–345. https://doi.org/10.1007/s11218-015-9313-y

Nunnally, J. C. (1978). *Psychometric theory* (2nd ed.). McGraw-Hill.

Olson, K. R., & Dunham, Y. (2010). The development of implicit social cognition. In B. Gawronski & B. K. Payne (Eds.), *Handbook of implicit social cognition: Measurement, theory, and applications* (pp. 241–254). The Guilford Press.

Paolacci, G., & Chandler, J. (2014). Inside the Turk: Understanding Mechanical Turk as a participant pool. *Current Directions in Psychological Science*, *23*(3), 184–188. https://doi.org/10.1177/0963721414531598

Payne, B. K., & Cameron, C. D. (2013). Implicit social cognition and mental representation. In D. E. Carlston (Ed.), *The Oxford handbook of social cognition* (pp. 220–238). Oxford University Press.

Pickering, M. (2001). *Stereotyping: The politics of representation*. Palgrave Macmillan.

Pi-Sunyer, O. (1977). Through native eyes: Tourists and tourism in a Catalan maritime community. In V. L. Smith (Ed.), *Hosts and guests: The anthropology of Tourism* (pp. 149–155). University of Pennsylvania Press.

Piuchan, M., Chan, C. W., & Kaale, J. (2018). Economic and socio-cultural impacts of Mainland Chinese tourists on Hong Kong residents. *Kasetsart Journal of Social Sciences*, *39*(1), 9–14. https://doi.org/10.1016/j.kjss.2017.11.004

Pizam, A., Fleischer, A., & Mansfeld, Y. (2002). Tourism and social change: The case of Israeli ecotourists visiting Jordan. *Journal of Travel Research*, *41*(2), 177–184. https://doi.org/10.1177/004728702237423

Plaza, M., Boiché, J., Brunel, L., Ruchaud, F. (2017). Sport = male… But not all sports: Investigating the gender stereotypes of sport activities at the explicit and implicit levels. *Sex Roles*, *76*(3–4), 202–217. https://doi.org/10.1007/s11199-016-0650-x

Prendergast, G. P., Lam, H. S., & Ki, Y. P. (2016). Local residents' perceptions of an influx of tourists: A Hong Kong case study. *Journal of International Consumer Marketing*, *28*(4), 283–293. https://doi.org/10.1080/08961530.2016.1165160

Project Implicit. (2020). *Project implicit*. https://implicit.harvard.edu/implicit/aboutus.html

Qiu Zhang, H., Fan, D. X., Tse, T. S., & King, B. (2017). Creating a scale for assessing socially sustainable tourism. *Journal of Sustainable Tourism*, *25*(1), 61–78. https://doi.org/10.1080/09669582.2016.1173044

Rohmer, O., & Louvet, E. (2018). Implicit stereotyping against people with disability. *Group Processes & Intergroup Relations*, *21*(1), 127–140.

Rowen, I. (2016). The geopolitics of tourism: Mobilities, territory, and protest in China, Taiwan, and Hong Kong. *Annals of the American Association of Geographers*, *106*(2), 1–393. https://doi.org/10.1080/00045608.2015.1113115

Rozin, P., & Royzman, E. B. (2001). Negativity bias, negativity dominance, and contagion. *Personality and Social Psychology Review*, *5*(4), 296–320. https://doi.org/10.1207/S15327957PSPR0504_2

Rubinstein, R. S., Jussim, L., & Stevens, S. T. (2018). Reliance on individuating information and stereotypes in implicit and explicit person perception. *Journal of Experimental Social Psychology*, *75*, 54–70. https://doi.org/10.1016/j.jesp.2017.11.009

Ryan, M. (2017). *Literary theory: A practical introduction* (3rd ed.). Wiley.

Schneider, W., & Shiffrin, R. M. (1977). Controlled and automatic human information processing: I. Detection, search, and attention. *Psychological Review*, *84*(1), 1–66. https://doi.org/10.1037/0033-295X.84.1.1

Sciarini, M., Beck, J., & Seaman, J. (2012). Online learning in hospitality and tourism higher education worldwide: A descriptive report as of January 2012. *Journal of Hospitality & Tourism Education*, *24*(2–3), 41–44. https://doi.org/10.1080/10963758.2012.10696668

Sheldon, P. J., & Var, T. (1984). Resident attitudes to tourism in North Wales. *Tourism Management*, *5*(1), 40–47. https://doi.org/10.1016/0261-5177(84)90006-2

Shen, H., Luo, J., & Zhao, A. (2017). The sustainable tourism development in Hong Kong: An analysis of Hong Kong residents' attitude towards mainland Chinese tourist. *Journal of Quality Assurance in Hospitality & Tourism*, *18*(1), 45–68.

Shen, X., Lv, X., Lin, S., & Li, C. (2019). Application of the stereotype content model to destination image: Evidence from residents of mainland China. *Journal of Destination Marketing & Management*, *14*, 100375.

Sim, D. (2019, September 9). Where are Chinese tourist going if they're giving protest-hit Hong Kong a miss? *South China Morning Post*. https://www.scmp.com/week-asia/economics/article/3027939/where-are-chinese-tourists-going-if-theyre-giving-protest-hit

Siswandio, P. (2019, March 18). Netizen's reactions to rude Chinese tourist is exactly why payback isn't the answer. *The Pride*. https://pride.kindness.sg/netizens-reactions-to-rude-chinese-tourist-is-exactly-why-payback-isnt-the-answer/

So, K. K. F., King, C., & Sparks, B. (2014). Customer engagement with tourism brands: Scale development and validation. *Journal of Hospitality & Tourism Research*, *38*(3), 304–329.

Suess, C., Woosnam, K. M., & Erul, E. (2020). Stranger-danger? Understanding the moderating effects of children in the household on non-hosting residents' emotional solidarity with Airbnb visitors, feeling safe, and support for Airbnb. *Tourism Management*, *77*, 103952. https://doi.org/10.1016/j.tourman.2019.103952

Tajfel, H. (1969). Cognitive aspects of prejudice. *Journal of Biosocial Science*, *1*(S1), 173–191. https://doi.org/10.1017/S0021932000023336

Taylor, D. M., Ruggiero, K. M., & Louis, W. R. (1996). Personal/group discrimination discrepancy: Towards a two-factor explanation. *Canadian Journal of Behavioural Science/Revue Canadienne Des Sciences du Comportement*, *28*(3), 193–202. https://doi.org/10.1037/0008-400X.28.3.193

Tourism Commission. (2019). *Tourism performance*. Commerce and Economic Development Bureau. https://www.tourism.gov.hk/english/statistics/statistics_perform.html

Tse, S., & Tung, V. W. S. (2020). *Residents' discriminatory actions against tourists* [Paper presentation]. 2020 APacCHRIE Conference, Hong Kong Special Administrative Region, China. https://www.polyu.edu.hk/htm/conference/apacchrie2020/submissions.html

Tse, T. S., & Qiu, H. (2016). Issues arising from the rapid growth of Mainland Chinese visitors to Hong Kong: Implications for tourism marketing. *Journal of China Tourism Research*, *12*(3–4), 291–312. https://doi.org/10.1080/19388160.2016.1267670

Tulving, E. (1972). Episodic and semantic memory. In E. Tulving & W. Donaldson (Eds.), *Organization of memory* (pp. 382–402). Academic Press.

Tulving, E. (2002). Episodic memory: From mind to brain. *Annual Review of Psychology*, *53*(1), 1–25. https://doi.org/10.1146/annurev.psych.53.100901.135114

Tung, V. W. S. (2019). Helping a lost tourist: The effects of metastereotypes on resident prosocial behaviors. *Journal of Travel Research*, *58*(5), 837–848. https://doi.org/10.1177/0047287518778150

Tung, V. W. S., & King, B. (2016). The stereotyping of tourism management students in a business school setting. *Journal of Teaching in Travel & Tourism*, *16*(1), 60–77.

Tung, V. W. S., King, B. E. M., & Tse, S. (2020). The tourist stereotype model: Positive and negative dimensions. *Journal of Travel Research*, *59*(1), 37–51.

Uleman, J. S., & Bargh, J. A. (1989). *Unintended thought*. The Guilford Press.

van Veelen, R., Otten, S., Cadinu, M., & Hansen, N. (2016). An integrative model of social identification: Self-stereotyping and self-anchoring as two cognitive pathways. *Personality and Social Psychology Review*, *20*(1), 3–26. https://doi.org/10.1177/1088868315576642

Vescio, T. K., & Biernat, M. (1999). When stereotype-based expectancies impair performance: The effect of prejudice, race, and target quality on judgments and perceiver performance. *European Journal of Social Psychology*, *29*(7), 961–970. https://doi.org/10.1002/(SICI)1099-0992(199911)29:7<961::AID-EJSP977>3.0.CO;2-4

Woosnam, K. M., & Norman, W. C. (2010). Measuring residents' emotional solidarity with tourists: Scale development of Durkheim's theoretical constructs. Journal of Travel Research, *49*(3), 365–380.

Woosnam, K. M., Maruyama, N., Boley, B. B., & Erul, E. (2018). Stereotypes and perceived solidarity in ethnic enclave tourism. *Journal of Tourism and Cultural Change*, *16*(2), 138–154. https://doi.org/10.1080/14766825.2016.1242596

Yan, A. (2018, August 22). Chinese tourist numbers surge as overseas travel becomes easier. *South China Morning Post*. https://www.scmp.com/news/china/society/article/2160906/chinese-tourist-numbers-surge-overseas-travel-becomes-easier

Yang, J., He, J., & Gu, Y. (2012). The implicit measurement of destination image: The application of Implicit Association Tests. *Tourism Management*, *33*(1), 50–52. https://doi.org/10.1016/j.tourman.2011.01.022

Interferences generated on the well-being of local communities by the activity of online platforms for tourist accommodation

José Maria Martín-Martín, Juan F. Prados-Castillo, Juan de Dios Jiménez Aguilera and Eva Porras González

ABSTRACT

The tourism sector can boost the economic and social development of entire cities while simultaneously triggering critical challenges to its own sustainability. Specifically, the additional stress imposed on residential neighborhoods due to the increasing number of tourist accommodations mediated online can compromise the social sustainability of tourism. This study focuses on the city of Barcelona (Spain) to shed light on crucial aspects of the impact of peer-to-peer accommodation platforms upon the well-being of its residents. A key contribution is that this work uses for the first time in tourism the Human Scale Development approach. Among the many consequential findings, of particular interest to economists and policy-makers are the fact that there is no economic sustainability without social sustainability, and that guaranteeing social cohesion and the permanence of a fixed resident population in tourist neighborhoods is essential. One additional breakthrough is the participants' strong viewpoint that a major roadblock to any progress is the lack of adequate regulation. In their opinion, any satisfactory legal framework should use participatory mechanisms to incorporate the neighbors' feedback over issues that affect their lifestyle. Last, the importance of establishing cooperation mechanisms between institutions, tourists, neighbors and businesses was also forcefully emphasized.

Introduction

The social dimension of sustainability has increasingly assumed a more prominent role in the literature over time (Dempsey et al., 2011). However, this attention is still insufficient particularly in works analyzing tourism competitiveness, which for the most part, do not include social dimension indicators (Guaita et al., 2020; Salinas et al., 2020). One reason for this is that the tourism sector has prioritized economic growth over other needs. Consequently, the problems generated by this activity such as those associated with social conflicts among stakeholders, have been routinely overlooked (Martín, Guaita, et al., 2018). Overall, Europe has chosen simple strategies to develop the tourism sector, mainly by accepting pro-growth strategies which often exclude sustainability indicators as a measure of success (Russo & Quaglieri, 2014).

Lack of attention toward the social sustainability aspect of tourism, at least as compared to that paid to environmental sustainability, contrasts with the increased scrutiny demanded by numerous collectives. Over the last decade, multiple voices have been raised pointing to the urgent need to consider the receiving communities' wants. In fact, the sustainable tourism indicator system of the UNWTO includes the social perspective explicitly (UNWTO, 2004). In 1987, the World Commission on Environment and Development (WCED) already defined Sustainable Development as the "development that meets the needs of the present without compromising the ability of future generations to meet their own needs" (WCED, 1987, p. 43). This document attributed equal importance to the impacts that development has on the economy, on the society and on the natural environment. This perspective inspired the political guidelines agreement at the UN conference on Environment and Development in Rio de Janeiro in 1992, as well as those of ensuing symposiums on sustainability organized within the framework of the UN (Adams, 2009; UN, 1992, 2012). Regarding this sector's competitiveness, one of the most widespread indicators is The Travel and Tourism Competitiveness Report (TTCR) published by the World Economic Forum. This work already singled out the importance of promoting sustainable development models capable of guaranteeing respect for both the environment and the local communities that depend on tourism in its 2017 edition (WEF, 2017).

The institutional recognition of social sustainability as a relevant objective (Guaita et al., 2019), and the increased pressure on tourist destinations, which often spur tourism-phobia feelings among the receiving communities (Martín, Guaita, et al., 2018; Martín, Rodríguez, et al., 2018), have resulted in growing attention to these values and dynamics. Given that peer-to-peer online platforms for tourist accommodation introduce tourist flows into residential environments, this tension has intensified concurrently with the penetration and use of these technological tools (Gravari-Barbas & Guinand, 2017; Martín, Ostos, et al., 2019). These platforms are a part of the set of activities known as the "sharing economy" (SE), which refers to the shared use of underutilized vehicles, spaces and other assets (Botsman & Rogers, 2010; Geron, 2013; Sacks, 2011). Guttentag (2015, p. 1193) describes them as: "essentially an online platform through which ordinary people rent out their spaces as accommodation for tourists".

A considerable number of studies have looked into the far-reaching effects that this type of tourist intermediation systems have on tourist destination communities (see, for example, Gunter & Önder, 2018; Gutierrez et al., 2017; Quattrone et al., 2016). It is significant that many of these works have denounced meaningful alterations on the lives and local culture of residents in tourist areas, such as loss of social cohesion (Cócola, 2016; Gallagher, 2017; Martín, Guaita, et al., 2018). However, even though the number of studies that describe the different impacts derived from this type of platforms grows, the literature on the perception of residents in tourist areas with respect to the interferences generated in their lives is scarce. Nonetheless, as an exception, recent works by Mody et al. (2020) and Yeager et al. (2020) have looked into the perception that residents have on the impacts of this type of tourism. In their works, the authors stress the need to increase the number of theoretical contributions on this type of analysis.

Several authors have highlighted the importance of closing this research gap (Gutierrez et al., 2017; Lyons & Wearing, 2015; Richardson, 2015). In this line, both Guttentag (2015) and Gallagher (2017) have asserted that the disruptive innovation in the tourism sector caused by this category of rental platform deserves to be studied further, particularly considering its unpredictable outcomes. Among the many other plausible benefits of further analyses, Cheng (2016) asserts that a better understanding of this plight could help public authorities improve legislation and design a roadmap to take advantage of the opportunities that may arise. This is even more likely considering the general assumption of material wealth and employment creation potential that is associated with the SE (OECD, 2016). Furthermore, this potential would be reinforced should the established regulations be capable of promoting an orderly growth that respects the interests of the various stakeholders involved (Martín, Ostos, et al., 2019). The fact that these activities provoke numerous conflicts among stakeholders implies the sector would greatly benefit from the additional attention the academic research community could grant (Cohen & Munoz, 2016). Therefore, knowledge of these interactions and conflicts should result in an active line of academic research.

This study attempts to cover the mentioned research gap and evaluate the perceptions of residents in tourist areas with respect to the impacts that the intense growth on tourist rental accommodations has on their social well-being. The need to shed light on the residents' perceptions has been highlighted by Guttentag (2015), McGehee and Andereck (2004) and Suess, Woosnam, Mody, et al. (2020). These authors also underline the benefits of learning whether these perceptions are conditioned by the financial links between residents and the tourism sector. Thus, to fulfill these objectives and for the first time in the tourism sector, we use the Human Scale Development (HSD) approach (Max-Neef, 1991). This theoretical framework puts human needs at the center of development, and grants equal importance to economic, social and environmental dimensions, emphasizing the interdependence among all three. HSD focuses on the importance of balanced relationships between society, the economy and the environment. The satisfaction of human needs is at the core of this conceptual framework which proposes a participatory methodology to help communities and public authorities plan their development (Guillén-Royo et al., 2017).

This work offers a double contribution to the academic literature. First, it sheds light on the problems created for residents' well-being by the growth of tourist accommodation mediated through online platforms. And second, it chooses a novel procedure that takes human needs as a starting point, a methodology that has rarely been used for this purpose before. Specifically, the first contribution indicated was channeled through three research questions. RQ1: How do residents assess the changes in their well-being associated with tourist rental platforms? RQ2: What are the underlying factors that condition the attitude of the residents? RQ3: What actions could limit the negative effects on the lives of the residents?

This study centers around needs-based workshops in which residents living in the tourist center of Barcelona participated. Barcelona was chosen because it is one of the prime tourist hotspots in Spain and Europe, being a city that suffers from high tourist pressure in its historical center. This case study illustrates how the HSD approach is used to reveal the types of interactions that harm social sustainability. Also, this methodology helps to uncover possible interventions that can contribute to meet the needs of the neighbors, preserving social sustainability and ensuring the viability of the tourist activity linked to these platforms. Therefore, in disclosing our findings, we contribute a unique perspective using a novel methodology applied to the tourism industry, which may be replicated in other destinations.

Tourist development and social sustainability

Growing pressure upon receiving communities results from the ease with which interactions are generated when tourist accommodation platforms are used (Russo & Quaglieri, 2014). One important side effect is the additional strain that can push residents to abandon the main tourist areas seeking more livable spaces (Russo, 2002). In extreme cases, these gentrification processes can alter the profile of the residents and business activities located in these hard-hit neighborhoods (Gotham, 2005). Cócola (2015, p. 4) defines this phenomenon as "a process of socio-spatial change in which neighborhoods are transformed according to the needs of affluent consumers, residents and visitors alike". The main effects associated with this process are the displacement of long-term tenants, rent increases, a shortage of rental property and the loss of local identity (Edelman & Geradin, 2016; Gurran & Phibbs, 2017; Guttentag, 2015; Postma & Schmuecker, 2017; Richards et al., 2019; Wegmann & Jiao, 2017). Therefore, the excessive concentration of short-term rental apartments in some areas and the increase in rental prices that is associated with it, can serve as a trigger or enhancer of the aforementioned gentrification (Richards, 2016). This would be the expression of the phenomenon known as "airbnbification" (Richards, 2016). The recognition of these phenomena has lead observers to infer that the activities related to tourism have great capacity to transform city centers, both in tangible and intangible ways (Gutierrez et al., 2017; Martín, Salinas, et al., 2019).

Tourism generates complex and varied interactions with the environment in which it develops. In the academic literature, numerous studies have identified both positive and negative impacts associated with this activity (Puczkó & Rátz, 2000). With respect to the negative aspects, Puczkó and Rátz (2000) point out that an inadequate tourism development usually implies an increase in pressure on destinations, as well as negative changes in the socio-cultural and environmental characteristics of these (Martín, 2019). With respect to the positive ones, the intrinsic virtues in tourism are highlighted. As an example, this activity is considered environmentally benign and a viable economic alternative to other more damaging options (Doswell, 1997). Furthermore, tourism often draws attention to aspects that stimulate environmental conservation initiatives (Doswell, 1997), hence, potentially contributing to Sustainable Development. However, there are also negative impacts inherent to the social and natural environment that should be identified and contained in as much as possible (Puczkó & Rátz, 2000). The final balance and the magnitude of the impacts will depend on variables such as the volume of tourists, the activities carried out by them, the fragility of the environment, or the strength of the local culture (Roberts & Hall, 2001). Equally important are the way the tourism activity system is organized, the regulation of the sector, and the structured mechanisms available to adjust this activity to the needs of each of the stakeholders involved (Fang et al., 2016; Martín, Ostos, et al., 2019).

Indeed, the development of activities related to tourism can lead to positive and negative impacts on local communities (Youell, 1998). These are the so-called "tourism impacts" (Mathieson & Wall, 1982). From a broad perspective, the impacts can derive positive aspects for local communities, such as the generation of new employment opportunities for residents (Andereck & Nyaupane, 2011); the improvement and strengthening of the business network, the contribution to valuing and preservation of the local heritage (Andereck et al., 2005); or improvements in local identity and pride (Andereck et al., 2005). On the other hand, the impacts can also derive negative heterogeneous aspects. These incorporate a wide range of phenomena grouped into environmental, socio-cultural and economic categories (Martín et al., 2014). Examples include: the effects on infrastructure congestion, price growth, increased alcohol consumption, environmental damage, greater generation of waste and the alteration of the lifestyle of local communities (Martín et al., 2017).

The new ways in which tourism is organized have contributed unique opportunities, but have also brought interactions with local communities leading to novel impacts or variants of those observed traditionally (Ioannides et al., 2019). Airbnb, Couchsurfing and Homeway are the best-known examples of this type of tourist accommodation platforms. Couchsurfing started in 2004 as a non-profit organization; however, in 2011, it changed its legal form to become a for-profit business. In 2008, Airbnb was founded as an online platform that charged a fee to enable users the opportunity to share spare spaces and all sorts of tourist accommodation solutions. According to Merrill Lynch, by 2020, Airbnb could account for up to 1.2% of the hotel offering and 3.6%–4.3% of the inventory with an estimated 40%–50% growth in annual listings (Heo, 2016). The expansion of online tourist accommodation platforms has generated debates about the legitimacy of companies such as Airbnb in several of the most-affected cities (Bort, 2014; Brustein, 2014).

The academic literature has described various benefits associated with this type of accommodation intermediation models. For example, visitors can enjoy a more authentic experience (Forno & Garibaldi, 2015; OECD, 2016; Russo & Quaglieri, 2016; Sigala, 2017; Tussyadiah & Pesonen, 2018), and it becomes more plausible to improve the interaction with locals (Belarmino et al., 2019). Also, these platforms expand the variety of accommodation options available at affordable prices (Ioannides et al., 2019; Juul, 2015; Shaheen et al., 2012), and this reduction in costs may result in an increase in tourism (Zervas et al., 2014). Three other positive aspects are: that this alternative drives tourist spending to neighborhoods which have not benefited from these earnings before (Porges, 2013); that it is easier to start up a business in the framework of the collaborative economy (Nadler, 2014); and that this new type of accommodation offer increases the lodging capacity of destinations in peak times, thus completing traditional services (Juul, 2015).

However, even though the positive impacts are well known, there is an increasing concern (Dredge & Gyimóthy, 2015; Queensland Tourism Industry Council, 2014) possibly due to the lack of planning associated with the disruptive and poorly regulated economic model linked to this sort of activity (Martín, Ostos, et al., 2019; Nieuwland & van Melik, 2020). The analyses of the impacts derived from the SE are incomplete, particularly those related to these types of lodging services (Guttentag, 2015). As pointed out by Cheng (2016, p. 67): "there appears scope for more research into the ecological, economic, and social impacts of SE". The inquiry on Sustainable Development linked with these services has earned relatively little attention in the past. Nonetheless, in recent years the academic community has shown a growing interest in studying innovation processes in the collaborative economy and the sustainability realms (Martín & Upham, 2016). As a result, new works have been published contrasting the benefits and the risks inherent to these platforms for intermediated accommodation. Some of the findings point to the fact that even though the collaborative economy can increase the income of residents in tourist areas, it can also lead to the degradation of working conditions when the locals' earning capacity depends exclusively on these types of activities (Lyons & Wearing, 2015; Schor & Fitzmaurice, 2015). Evidence points to its impact on the salary level of hotel employees resulting in lower incomes (Suciu, 2016) and in the reduction of the occupation level of hotel accommodations generating layoffs that are not balanced out by the hiring in tourist dwellings (Fang et al., 2016). Recent studies have described other problems such as the increase in residential housing prices, evictions of long-term tenants and lack of residential housing in tourist areas (Edelman & Geradin, 2016; Jefferson-Jones, 2014; Lines, 2015). The latter points to the evidence that investors groups purchase residential housing to convert it into tourist accommodation (Gurran & Phibbs, 2017). There have also been reports of discomfort and loss of social cohesion in traditional neighborhoods (Cócola, 2016; Gallagher, 2017) as well as added traffic, more noise in residential buildings, the appropriation of public space and the congestion of public areas (Gallagher, 2017; Gurran & Phibbs, 2017; Martín et al., 2017). In addition, changes have been reported in the perception of safety in residential neighborhoods (Suess, Woosnam, et al., 2020). Additional consequences with impacts on society as a whole are tax evasion and unfair competition (Lyons & Wearing, 2015; Oskam & Boswijk, 2016).

The competition exerted on the traditional hotel sector has also been analyzed (Choi et al., 2015; Zervas et al., 2014). The creation of large companies capable of significantly increasing the flow of tourists implies an increasing power of influence in defining the legal framework (Martín, Ostos, et al., 2019). New risks have also emerged as, under this new business model, it is now more complex to guarantee the personal safety of travelers and ensure problem-free economic transactions (Sigala, 2017). The reason for this is due to the fact that traditional roles of consumers and suppliers have been redefined, as well as the context in which the overnight stay itself is carried out (Cheng, 2016).

The need to expand research on the mechanisms of interaction between the users of the accommodations described and the communities in which they are located, as well as the importance of evaluating the impacts derived from these new business models, have been highlighted by numerous authors (Martín, Guaita, et al., 2018). The fact that these platforms have introduced tourist activity in residential buildings and areas demands a detailed study of the social impacts generated. In this context, Martín et al. (2015) point out the importance of establishing communication channels among members of the community to build resilient networks through empowerment. These interactions do not need to be solely negative. As John (2013) indicates, the introduction of activities linked to the collaborative economy in local communities could contribute to promoting values such as equality, mutuality, honesty, openness, empathy and an ethic of care. It has also been postulated that these activities and the feeling of belonging to the community generates could help build social capital as people interact in the process of sharing through communication and could allow a more equitable distribution of goods and services (Martín et al., 2015).

In summary, it could be concluded that the sustainable development in the tourism sector should establish goals of improving the quality of life of residents, optimizing the economic benefits perceived by the local communities and protecting the environment while offering a quality experience to the visitor (Bramwell & Lane, 1993; Martín, Salinas, et al., 2019; McIntyre, 1993; Park et al., 2008; Park & Yoon, 2009). Tourism development must be economically viable but also socially and environmentally sensitive (Puczkó & Rátz, 2000). The participation and support of residents in the tourist area are essential for the sustainability of the tourism industry in any destination (Gursoy et al., 2010). This support depends on the perception that citizens have with respect to the negative and positive impacts linked to tourism (Martin, 2019). Therefore, it is essential to learn about and understand the point of view of the local residents in relation to the negative impacts generated, so that through appropriate strategies, the community can be most supportive of tourism (Prayag et al., 2013).

Methodology and data

Conceptual framework

The HSD approach was proposed in the 1980s by economist Manfred Max-Neef and other experts to support endogenous development processes. This approach is based on three pillars: self-reliance, balanced relationships and human needs satisfaction (Guillén-Royo et al., 2017). The first point refers to the concept of centrality of communities and the need to activate their endogenous development. The second involves the necessity of maintaining balanced horizontal relationships between levels or dimensions of human activity. For example, among public powers, institutions and the economic sectors, or between the technology, the economy and nature. The third pillar highlights the urgency of respecting and meeting human needs in any development process. This conceptual framework is associated with a methodological system based on participatory workshops (Max-Neef, 1991).

Human needs are not only understood as requirements for a good life, but also represent opportunities for personal and social mobilization that can support processes of social change. They are considered to have a socio-universal character (Alkire, 2002), meaning that they are shared by different cultures over time, even if they are not felt with the same degree of intensity at any given moment (Cruz et al., 2009). Fundamental human needs are defined as the axiological needs for subsistence, protection, affection, understanding, participation, leisure, creation, identity and freedom, as well as with the existential needs of being, having, doing and interacting. As per Max-Neef, there is no hierarchy of needs although some can logically precede others, as could be the case of the need for subsistence. Existential needs represent the ways in which needs are expressed. Therefore "being" refers to the attributes of individuals or groups, "having" concerns institutions, values, tools and forms of organization, "doing" identifies collective and personal actions, and "interacting" links the characteristics of spaces and environments. The HSD needs approach is usually represented by a matrix in which the first column characterizes the nine axiological needs and the first row characterizes the four existential needs.

The 36 cells resulting from the intersection of axiological and existential needs identify the satisfiers: the ways of being, having, doing and interacting associated with the fulfillment of needs. Satisfiers are the values, attitudes, norms, laws, institutional agreements, organizations, actions, and ways of using space, resources, and nature that define the needs for satisfaction in a specific context and that vary throughout time and cultures. For example, in relation to the fundamental need for subsistence, a sustainable community can be characterized by satisfiers, such as being cooperative, supportive and caring (being); basic income schemes and organic farming activities (having); volunteering, respecting other community members and contributing to local initiatives (doing); and the availability of communal land and open flexible spaces for gatherings (interacting) (Guillen-Royo, 2016).

Modern capitalist societies, by contrast, may require sets of interlinked satisfiers different from those listed above. As Cruz et al. (2009) point out

> the rise of modern free-market society, (as a new interacting milieu), requires for the members of society a full range of new satisfiers at the having level (money, property, credit, etc.), of being (consumer, owner, free to buy and sell, etc.) and doing level (shopping, acting "rationally" in chrematistic terms, etc.) in order to satisfy their fundamental needs. (Cruz et al., 2009, p. 2023)

A wide range of satisfiers can be found in societies, and these may have different capacities to meet needs. To clarify this point, Max-Neef proposed a classification with five groups of satisfiers. The first group consists of satisfiers that focus on meeting only one out of the nine fundamental needs (singular); the second considers satisfiers that simultaneously support the updating of more than one need (synergic); the third involves satisfiers which over-satisfy a particular need while they reduce the ability to meet other needs (inhibiting); the fourth encompasses satisfiers that confuse people into believing that a need is satisfied while in the long run, the effect is the opposite (pseudo-satisfier); and the fifth group represents satisfiers that prevent fulfilling a long-term need at the same time that they prevent reaching other needs (destroyers or violators) (Max-Neef, 1991). When societies are characterized by the satisfiers described in the last three groups, the fulfillment of personal and social needs is nullified, and the conservation and protection of the environment is threatened.

The relationship between destroyers, inhibiting satisfiers and pseudo-satisfiers and environmental degradation was not addressed in Max-Neef's initial work. Recent studies have suggested that satisfiers such as pollution of water and soil sources, the effects of global warming in terms of droughts and floods, the loss of biodiversity and the progressive erosion of green areas are related to other satisfiers. Some examples are authoritarianism, consumerism and overconsumption materialistic values, hectic lifestyles, marginalization, lack of institutional transparency and limited political participation (see Guillen-Royo, 2016; Smith & Max-Neef, 2011 for further references). Therefore, from the needs-based perspective, the satisfiers that characterize the economic, social or environmental sustainability, or lack of sustainability, cannot be understood in isolation, but should be explored in terms of their interconnections. In this context, decisions should lead to improving aspects such as energy efficiency, the way in which decisions are made, the rhythm of daily life, and the values that inspire personal development and social coexistence (Guillen-Royo, 2016). Even though this conceptual scheme is solid both conceptually and methodologically, Max-Neef's concern was that this theory should not become static but that it would stay flexible to generate frameworks fitted to the professional requirements of each task (Guillén-Royo et al., 2017).

Max-Neef upheld the belief that collaborating in workshops designed to support participatory processes in communities was potentially enriching for the members of such communities. This idea was sustained by surmounting worldwide evidence (Smith & Max-Neef, 2011). The authors suggested the use of empty matrices in the participatory workshops designed to boost these participatory processes in the hopes of improving the situation of these communities and their sustainability. An example of such matrix is represented in Table 1.

Study context

The city of Barcelona is located in northeastern Spain. With a population of 1.6 million (National Statistics Institute, 2020), and a metropolitan area of 5.5 million inhabitants, Barcelona is the

Table 1. Matrix of needs and satisfiers.

	Being	Having	Doing	Interacting
Subsistence				
Protection				
Affection				
Understanding				
Participation				
Idleness				
Creation				
Identity				
Freedom				

Source: Max-Neef (1991, pp. 32–33).

second-most populated city in the country, representing a highly crowded area with a density of 158.3 inhabitants per hectare (Gutierrez et al., 2017). Barcelona is the capital of one of the richest regions in Spain, where tourism is at the core of the economy representing 15% of its GDP and 9% of its employment (Barcelona City Council, 2017). Tourism in Barcelona grew considerably after the celebration of the 1992 Olympic Games. In 1990, the city was visited annually by 1.73 million tourists, while in 2019 the number of people staying at hotels exceeded 9.4 million (National Institute Statistics, 2020). Nonetheless, if we add those lodged in other types of establishments, the arrivals sum to 13.9 million per year in the metropolitan area, and 20.2 million in the area of the joint tourist destination. These figures only include visitors who stay overnight (Barcelona City Council, 2020). Considering international arrivals alone, Barcelona is the 4th most visited city in Europe (Lonely Planet, 2020) and the 17th in the world (Hosteltur, 2020). The success of this city as a tourist attraction is based on the wide range of resources it can offer including cultural tourism, conferences and meetings, sun and beach, shopping, sports, and so on.

In recent years, the activity linked to tourist accommodation platforms has significantly increased the city's lodging capacity. Two elements have come together to alter the organization of tourism in this city: the expansion of tourist lodging intermediated by individuals, and the growth of low-cost flights (Martín, Guaita, et al., 2018). These have generated a new visitor profile, which interacts with the city in a different way (Abril-Sellarés et al., 2015).

Together with Berlin, Barcelona has been the city with the highest growth in the number of tourist accommodations (Europa Press, 2017). In 2019, the area that delimits the tourist destination of Barcelona had 3,31,747 lodgments: 58,583 of the 1,49,467 that concentrated in the city were online intermediated tourist accommodations (Barcelona City Council, 2020). In this city, only 50.5% of the tourists who stayed overnight used hotels or apart-hotels according to the Barcelona's Tourism Activity Report (Barcelona City Council, 2016). Therefore, the new models of tourism organization coincide with an increase in the pressure exerted upon the city by tourism (Gutierrez et al., 2017). In terms of overnight stays, Barcelona registers a ratio of 9807 nights per 1000 residents, almost twice the European Union (EU) average (5209 nights per 1000 inhabitants) (Barcelona City Council, 2016). As a result of the stress that these trends cause upon the resident population, an area highly dependent on tourism has developed a strong feeling of rejection toward this sector. Barcelona, Berlin and Venice are the three European cities suffering the worse overcrowding problems. This has provoked feelings of aversion from the local population, not only due to the pressure of tourism but also because of its impact in residential areas (Europa Press, 2017). With respect to Barcelona, many factors have contributed to this rejection or "tourism-phobia"; one example is the increase in the price of housing. Between 2013 and 2018, the average price of rentals in Barcelona grew by 36.4% (Martín, Guaita, et al., 2018). Also, Cócola (2016) points out to the loss of social cohesion in some of the city's neighborhoods and even an alteration of the local culture, a trend shared by other cities throughout the world (Gallagher, 2017). In the 2017 Barcelona's biannual barometer, its residents singled out tourism as the city's biggest problem. Furthermore, according to the 2017 Barcelona's Tourism Activity Report (Barcelona City Council, 2017), the number of residents claiming that tourist activity in

Table 2. Main characteristics of the participants in the study.

Characteristics	Survey participants (%)	Workshops participants (%)
Men	42.0	45.0
Women	58.0	55.0
18–30	27.1	30.0
31–45	23.9	22.0
46–65	31.2	28.7
Over 65	17.8	19.3
Primary education	5.2	2.90
Secondary education	45.8	44.2
Higher education	49.0	52.9
Unemployed	5.4	7.2
Monthly family income <1000€	4.9	6.9
1000–2000€	48.3	45.1
2000–5000€	36.2	37.8
>5000 €	10.6	10.2
Born in Barcelona	54.3	61.1
Resident in Barcelona >10 years	89.4	91.6
His/her income depends in part on the tourism sector	15.4	17.7

Source: Own elaboration. April 2020.

the city has peaked went from 25% in 2012 to 48.9% in 2016. In parallel, acts of vandalism linked to the rejection of tourism has begun to occur. Among others, these included: graffiti, assaults on tourist buses, and demonstrations which have occurred mostly in Barcelona but also in the Balearic Islands (Martín, Rodríguez, et al., 2018).

Methodology and data

This study proposes the use of a novel framework to achieve the described purpose. As detailed earlier, this framework expands the knowledge of the interactions between tourists and locals considering the new forms of tourism. The data collection in Barcelona was carried out in two phases during September 2019. The first phase consisted of a telephone survey questionnaire to city center residents. This inquiry included questions on demographic and socioeconomic characteristics. In this questionnaire, the citizens were also asked about their willingness to participate in one of the HSD workshops. A total of 269 participants responded to the questionnaire and 39 of those answered positively to participating in the workshops. Following earlier studies using the same methodology (see Guillen-Royo, 2016), the objective of recruiting between 30 and 40 participants was set, so that each of the two initial workshops consisted of 15–20 individuals. That is because a larger number of participants would create operational problems. The profile of those attending the HSD workshops did not differ from the profiles obtained in the prior survey (Table 2). Therefore, no additional adjustments in the sample obtained were needed. The volunteers became the members of the working groups. Had it been required, a representative sample of the population profile would have been randomly generated. However, this was not necessary. The neighborhoods of Barcelona in which the telephone survey was applied, and thus the areas of residence of the participants in the workshops are: El Raval, Barrio Gótico, La Barceloneta, Sant Pere, Santa Caterina i la Ribera, Sagrada Familia, Poble Nou, Vila de Gràcia, Dreta de l'Eixample, Poble Sec, Sagrada Familia, l'Antiga, Sant Antoni, Nova Esquerra and Fort Pienc. The selection of these neighborhoods was made taking into account the presence of homes offered on Airbnb in the city of Barcelona, so that the neighborhoods with the highest activity were considered. Volunteers' participation was encouraged and thanked with the reimbursement for their metro and/or bus traveling expenses. In addition, they were offered catering services in the break prior to the third meeting, and they were granted a certificate of participation. Most of these volunteers declared that their main motivation to cooperate was the opportunity to comment on matters that affected their lives, that of their family members and, in general, the society in which they lived.

The second phase of data collection revolved around three Human Needs-Based participatory workshops. These workshops were planned aiming to: (1) generate a negative matrix – a matrix including those satisfiers labeled as inhibiting, destroyers and pseudo-satisfiers, which hamper the fulfillment of needs; (2) define a utopian matrix – a matrix including those synergic and singular satisfiers that can promote the optimal fulfillment of needs; and (3) identify the bridge that would allow society to progress toward the utopian scenario (Guillen-Royo, 2016). Participants had to choose between joining the first or second workshop, but everyone had to participate in the third. The reason was to prevent people in the second workshop using the information from the first to construct a utopian matrix that was simply the opposite of the negative one.

The first step consisted of defining the negative matrix. Here, a poster-size copy of the matrix was used in such a way that little by little stickers with specific proposals were added to each of the 36 cells. This process was supported by "facilitators" who guided participants with respect to the meaning of each cell in accordance with the theory, but who could not make their own proposals or formulate ideas. Upon completion of this step, the facilitators coordinated the participants so they reached a consensus on which were the one or two most representative items for each cell. This process was repeated to build the utopian matrix. Following Guillen-Royo (2016), the research team analyzed the utopian matrix prior to the last workshop to identify common categories that could summarize the synergic and singular satisfiers suggested by participants. These themes or categories of satisfiers were then proposed as those that define a society with optimal needs satisfaction and were used as the grounds for the discussion in the third workshop. This analysis was performed during the break between sessions, when no participants were present. The objective of this preliminary analysis was to provide general guidelines to conducting the last session. Copies of the negative and utopian matrices were distributed among the participants of the third workshop, to allow a choice of information to be made when they analyzed the classification and offered their opinion on the suggested categories. To promote an in-depth discussion on synergic bridging satisfiers that would allow society to progress toward a utopian scenario, the 39 participants in the latter workshop were divided into two groups, both of which addressed two of the proposed categories. In this case, the focus was on the formulas proposed to reflect mechanisms, supports and practical tools – endogenous and exogenous – available and realistic, defined as means to reach the ideal situation. Both groups used a system similar to that described above. With a poster-size matrix representation, the initial ideas were added by placing stickers. Then, a discussion followed and work was done to reach consensus. In order to generate strategies that could make the synergic satisfiers emerge, this debate was articulated around the forms of Being, Having, Doing and Interacting. In the following section, we present the analysis of the satisfiers analyzed in the workshops.

In this process, the work of the facilitators is essential. At first, they presented the objectives of the fieldwork in a simple way and the nine fundamental human needs (i.e. protection, subsistence, affection, participation, understanding, idleness, creation, identity and freedom) as well as the four existential categories (i.e. having, being, doing and interacting), through examples applied to an area other than tourism (in this case, the examples referred to environmental pollution). Following the methodological basis of this proposal,

> Being would be identified by adjectives (e.g. chauvinist, authoritarian, compassionate, inclusive, open, etc.), Having by nouns concerning values, laws, traditions, tools or institutional agreements (e.g. basic income, greed, formal education, repressive police forces, non-independent media), Doing by verbs (e.g. cooperating, excluding, sharing, discriminating, etc.) and Interacting by the characteristics of spaces or environments (e.g. free public parks, surveillance cameras, information in indigenous language, sports facilities, spaces for creativity, etc.). (Guillen-Royo, 2016, p. 65)

In a second phase, the facilitators presented the objective of each matrix, and last they presented the information that should be expressed in each cell. The latter was done right before

Table 3. Negative matrix.

	Being	Having	Doing	Interacting
Subsistence	Individualist, closed	System that benefits large investors	Individualistic behaviors	Loss of population and local commerce
Protection	Fearful of interactions	Evictions of residents	Lack of proximity police	Police indifference to street and building noise
Affection	Individualist	Lack of neighborhood associations	The network of affection between neighbors is lost	Interactions become temporary and short-term
Understanding	Vulnerable	Neighbors do not intervene in regulation	Unproductive protests and acts of vandalism	Disconnection of regulation with the needs of neighbors
Participation	Selfishness	There are no channels to create community	Scarce organization capacities of residents to maintain the essence	Superficial relationships with tourists
Idleness	Dissatisfied	Lack of resident-oriented spaces and activities	Reduction of social leisure	The spaces are geared to tourists rather than to citizens
Creation	Standardization	Loss of cultural identity	There is no implication in the definition of activities	The authentic is replaced by the commercial
Identity	Independence	Loss of local identity	Places are stereotyped	The commercial replaces the social
Freedom	Depression	Loss of lifestyle	Alteration of social environments	Discomfort in community life

Source: Own elaboration. April 2020.

starting to work with a new cell in order to keep concepts fresh. The facilitators are researchers who know this methodology and its conceptual framework in depth. The definition of the consensus was carried out after a brief debate, once several ideas had emerged to complete each cell. Therefore, it was required that the participants reach a consensus before completing the work associated with each cell, trying to minimize interference from researchers.

Results

Identification of satisfiers that limit and enhance sustainable development

This section presents the analysis of the satisfiers gathered from the negative and utopian synthesis matrices, taking into account the three contexts suggested by Max-Neef (1991, p. 18). The contexts are: (a) oneself (Eigenwelt – the individual level); (b) the social group (Mitwelt – the community level); and (c) the environment (Umwelt – the societal or "governance" level). These three levels will be considered to illustrate the potential focus of public policies (Jolibert et al., 2014). However, this work does not center on the analysis between levels or contexts, but on the interconnections between satisfiers.

The negative matrix summarized in Table 3 shows the satisfiers that hamper the fulfillment of the residents' needs, according to their own reports. This matrix describes a society characterized by the loss of quality of life and by the degradation of the social structures due to the appropriation of neighborhood spaces by tourists and the activities aimed at these. At the individual level, individualism, lack of cooperation or even certain fears condition the citizen's vision of the process of degradation of their environment. These individual feelings connect with a social perspective in which three forces of transformation stand out: economic interests distant from social dynamics, the disconnect of citizens and the regulation and planning processes that affect them, and the loss of neighborhood cohesion. The environmental aspect was introduced as an element

Table 4. Utopic matrix.

	Being	Having	Doing	Interacting
Subsistence	Proactive	Preserve the family residency	Contain the increase in housing prices	Maintain resident population
Protection	Solidary	Avoid speculation	Commercial activities in residential areas	Generate social cohesion
Affection	Empathic	Neighbors and family networks	Improvement of public spaces	Interaction activities with tourists
Understanding	Involved	Neighborhood information	Consider residents' opinion	Interaction with tourists
Participation	Active	Spaces for communal activities	Participation in the making of the regulation	Effective channels of participation
Idleness	Committed	Maintain resident-focused services	Local leisure opportunities	Leisure proposals from the community
Creation	Happy	Have cultural spaces	Creation as a link in the community	Visitor attraction based on culture
Identity	Proud	Maintain the essence of the neighborhood	Keep the neighborhood local commerce	Introduce tourists to local commerce
Freedom	Satisfied	Possibility of choosing to live in the neighborhood	Guarantee the generational replacement	Maintain daily routines

Source: Own elaboration. April 2020.

of debate in the dynamics of the workshops. However, the participants did not clearly associate the growth processes of tourism with the environmental pressure. This connection only occurred when the lack of quality public spaces at the residents' service that had not been appropriated by tourism and in which there was a damaged social interaction, was pointed out to them.

With respect to the singular and synergic satisfiers discussed in the second workshop, the analysis carried out by the researchers uncovered three categories of satisfiers. These are presented in the summary utopian matrix in Table 4. The first category of satisfiers relates to the need of keeping a cohesive community and to the creation of meeting spaces. The second refers to the need of maintaining the identity of the neighborhood. And the third, looks into the interconnection processes, either between citizen and legislators or between citizen and tourists, in a framework of respect to the needs of each party and to the benefits that each stakeholder can derive. These categories encompass the vision of a society that balances the respect for the life of the locals and the openness to an economic activity generally valued as positive.

The first category of satisfiers turned out to be by far the ones that raised the most concerns among residents. From an individual point of view, the need to generate collaborative, caring and committed attitudes that translate into greater fulfillment and personal satisfaction is highlighted. Part of this process is associated with guaranteeing the permanence of neighbors, ensuring the generational replacement and maintaining a community based on daily routine relationships, interactions among individuals and between these and neighborhood commerce. An important component in this category is the need to ensure the availability of meeting areas at the service of the citizen. These spaces would guarantee satisfying that the individual needs for interaction and accompaniment are fulfilled through the social activity that secures them. As a key fact, residents point out that the planning of these spaces should not focus on tourists, and that their use is not privatized with this excuse, but rather that it responds to neighborhood needs, even if the spaces are then shared with the visitors.

The second category concerns the identity of the neighborhood, and particularly stresses the commercial aspect of the activities proposed to be displayed in its environment. Here, the participants defend the need to protect local commerce, to maintain the identity elements of the neighborhood, to avoid the replacement of traditional activities, and even to prevent residents actions leading to a "theme park" model that results in a facade without real human content.

Table 5. Summary of harmful, synergic and synergic bridging satisfiers in Barcelona.

Workshop 1. Negative matrix (inhibits, pseudo-satisfiers and destroyers)	Workshop 3. Synergic bridging satisfiers	Workshop 2. Utopian matrix (synergic satisfiers and singular satisfiers)
Population loss as a result of speculation and the conversion of homes into tourist accommodation	Improvement of the regulation agreed upon with neighborhood associations	Maintain cohesive local communities, guaranteeing generational replacement
Loss of tranquility in residential environments and alteration of quality of life	Reduce the pressure of tourist housing in residential areas	Foster the identity of the neighborhood and proximity activities
Loss of social cohesion and interactions that generate the social network	Create codes of conduct for tourists. And promote more respectful tourism models	Interaction with tourists based on mutual respect mechanisms
Lack of spaces, commerce and policies at the service of the citizens	Promote social interaction	Participation infrastructures, socialization that include tourists
Disconnect between neighborhood needs and regulation	Creation of public spaces at the service of residents' needs	Involvement of the residents in decision-making processes

Source: Own elaboration. April 2020.

The third category deals with the need to guarantee interactions that mobilize citizens. In this dimension, the first concern centers on the importance of considering the voice of citizens structured through neighborhood associations. This voice is expected to be particularly relevant in planning and legislation processes; specifically with respect to the regulation of tourist homes, in the creation of common spaces, in the definition of mobility, and in the promotion of local commerce. In the second place, the importance of tourism in economic terms and in cultural enrichment is recognized, and for this reason, it is pointed out that interaction between residents and tourists based on mutual respect, should be encouraged.

Synergic bridging satisfiers: toward socially sustainable development

The last workshop dealt with specific synergic bridging satisfiers that could make the connection between the negative matrix and the utopian matrix. Following Max-Neef (1991) the participants in this workshop discussed these bridging satisfiers in terms of either their endogenous or their exogenous character. This process required the participants' assessment of the community's capacity to propose satisfiers without external help (endogenous). If the local groups turned out to be incapable of performing such task, then it was proposed that experts, policy-makers or organizations that can contribute to the design of satisfiers (exogenous) should be identified and engaged. In the analysis that follows (Table 5), we differentiate the individual, community and social context in which satisfiers are expressed.

Starting with the individual context, participants believed that to achieve the proposed goals each resident should first adjust its own attitudes and behaviors; the reason is that the sum of the individuals' behaviors is the main axis and the key engine of social transformation. Participants pointed out that promoting social participation and interaction is essential to unite the community. To this end, they suggested that neighborhood associations receive funds directly from the city council. In this way, these organizations can run their own activities, meetings, projects to better the neighborhood, and so on. This would contribute to improving the socialization of citizens. In this sense, external help is required to define the correct control and participation mechanisms.

The community context is directly related to the issues discussed above. In this case, the bridges defined are based on the creation of spaces available to citizens and not privatized with commercial excuses (bar terraces in public squares, paid museums, spaces designed for passing individuals, etc.). It is proposed that these elements become the scenarios of associative policies. The participants vehemently exhort the need to establish a permanent population that will promote a natural generational change in the community. To this end, they suggest setting maximum quotas for tourist homes in each neighborhood as well as giving voice to residents' associations in tourism legislation. Furthermore, albeit unsure of how to go about this goal, another recommendation is to structure policies that promote local businesses and commerce of proximity that is not geared to tourists' needs. With respect to the regulation of tourist accommodation, several novel ideas are proposed: limiting tourist apartments to the first floor of buildings to minimize discomfort; a system of points that communities can use to report noise and discomfort and that will help promote the automatic withdrawal of the tourist housing license; and the possibility that tourist lodgings are only permitted in buildings intended for that purpose exclusively.

Last, with respect to the society dimension, the proposals are very focused on the tourism model. Consistently, three lines of action are proposed to help move from a negative model to the utopian matrix. The first of these proposals relies on the clear commitment of the public sector to promote more respectful models of tourism. As an example, citizens directly point to alcohol and party tourism. They also indicate that load capacities should be specified at the neighborhood level, and that maximum inflows distributed throughout the year should be set. They advocate that the neighborhoods should be treated as monuments, in which the number of visits – in this case, overnight stays – are controlled to avoid deterioration. Lastly, they propose that codes of good conduct for tourists are created and that resident-tourist interaction events are promoted to generate greater empathy between both groups.

Discussion and conclusions

Contrary to what could have been expected of a context highly dependent on tourism, the promotion of this activity or the need to preserve it was barely been taken into account. Even though the income of a large proportion of the participants in the workshops depended directly on tourism, the problems they prioritized were related to the direct damage that these activities provoked in their communities, preventing basic needs from being met. This is in line with Martín, Guaita, et al. (2018) conclusions. Following the simile of environmental sustainability, the participants stressed that if tourism is not socially viable, the climate of conflict generated will damage this activity at the end. From the discussions generated in the workshops, we can extract several distinctly significant conclusions. First, it is particularly relevant that in improving their social situation residents credit a key role to the personal behavior, attitudes and individual involvement of locals. In this context, they attribute a predominant role to the maintenance of the cohesion of the neighborhoods, in their different aspects. The aforementioned aspects affect the dimensions of sustainable development – economic, social and environmental – although the latter has less importance in this urban environment. Therefore, with respect to RQ1: "How do residents assess the changes in their well-being associated with the use of tourist rental platforms?" some conclusions can be highlighted. First and foremost, the participants clearly point out to the damage done to social structures, which leads to changes in their well-being. These changes imply a loss of cohesion in society and the promotion of individualism. The importance that residents attribute to social cohesion has been described in previous works (e.g. Yeager et al., 2020). The findings showed a more intense concern with respect to the social damage than to the economic effects of the platforms, in part because the latter are considered triggers for the social damage that ultimately occurs.

RQ2 asks: What are the underlying factors that condition the attitude of residents? The results show how the rejection of tourism is rather a derivative of a lack of planning, and a lack of containment of the negative impacts it generates, therefore residents ask for more involvement from policy-makers, albeit taking into account the voice of local communities. Another conditioning factor of their attitude mentioned is the breakdown of the balance between legitimate economic interests and respect for the life of the locals. Citizens have acknowledged and valued the positive impacts associated with this kind of accommodation service. Despite this, they also call for a fairer balance between their personal interests as neighbors and those of the owners. This is in line with the findings of Nunkoo and So (2016), Mody et al. (2019) and Suess, Woosnam, et al. (2020). A change in attitude could be achieved by promoting citizen participation in the regulatory processes if this implies the creation of regulation models based on mutual respect, and on the preservation of the local identity, traditional activities and the cohesion in the neighborhoods. It has been highlighted that when residents perceive harm to themselves or their communities' well-being, there will be a rejection of both these platforms and the tourist activity as a whole. This is in line with Garau-Vadell et al. (2019), Suess et al. (2018) and Uysal et al. (2016) findings.

Finally, RQ3 asks: What actions could limit the negative impacts on the lives of residents? The workshop geared to the definition of corrective measures was really productive, and aside from the information summarized in this work, many nuances, ideas, and useful perspectives for public planning were uncovered. Hence, this methodology should be replicated in tourist environments suffering from high pressure. Following Andersen and Siim (2004, p. 3), citizens' empowerment can be defined as "the process of awareness and capacity-building, which increases the participation and decision-making power of citizens and may potentially lead to transformative action". It seems that this awareness is clear among the citizens of Barcelona; that consciousness justifies the demand that the voices of the associations become more relevant, even though the need to improve neighborhood cohesion is also pointed out as a previous step. A number of studies support the idea that the empowerment of residents and their active participation in decision-making will condition their support for these activities and their perception of the impact of those on their lives (Mody et al., 2020; Yeager et al., 2020). Institutionalizing neighborhood participation in decision processes is a synergic satisfier supporting people's empowerment. In the process of containing the problems derive from excessive tourism pressure, cohesion and neighborhood participation are found to be as important as the regulation of the tourism sector itself. It is interesting to analyze the way in which satisfiers across contexts (personal, societal and environmental) and sustainability dimensions (economic, social and environmental) are connected. Understanding these connections, and how they should be cared for and delivered, is a basic aspect of solving problems that affect local communities. It is interesting to highlight some specific proposals or needs mentioned. For example, great importance has been attached to maintaining neighborhood routines, generational replacement and the interactions among people in public spaces. In summary, the increase in tourist pressure on urban centers and the increasing influx of tourists into residential environments that tourist accommodation brokerage platforms enable, has generated problems of considerable importance for citizens. This obviously complex situation requires the consensus of the agents involved in the development of tourism. Methodologies such as the one presented here have helped to systematize citizen sentiment, and to define lines of improvement and action to achieve a sustainable situation for local communities that guarantees the viability of an economic activity as important as this one. Therefore, a clear public policy recommendation is offered: in order to systematically collect citizen sentiment communication channels with local communities should be improved. Improving neighborhood cooperation networks would be the first step on a path that would culminate in an update of the regulation so that it integrates the needs of citizens. In this sense, residents have expressed the need for some type of institutional external help that allows the creation and maintenance of structures of cooperation and participation in the regulatory processes. Citizens have paid special attention to the final expression of some problems. For example, they have expressed the importance of guaranteeing generational change and the routines within the neighborhood. This implies reflecting on policies for access to housing, controlling the density of tourist apartments in each neighborhood and promoting the preservation of traditional activities. These concerns are in line with Martín, Guaita, et al.'s (2018) findings. They have also expressed the importance of guaranteeing coexistence in residential spaces. In this sense, imaginative solutions are proposed such as point systems linked to the continuity of a license, limiting tourist dwellings to the first floor of buildings or to independent buildings. Public authorities should take up some of these ideas and proposals in future regulatory processes. This study should be replicated in different contexts, since it is understood that the particular nature of each social context can influence the results obtained. The repetition of the study will help to understand the different citizen attitudes toward this context and the influence of the environment on this attitude.

The interpretation of the results must take into account one key limitation of this study which should be highlighted: the context. That is because one ought to keep present the environment of high tourist pressure as well as the specific model of social relations. This study should be replicated in different contexts, since it is understood that the particular nature of each social context can influence the results obtained. Therefore, the main limitation associated with the current study offers an opportunity for future research. It is recommended then, that other fieldworks of equivalent characteristics are performed in cities with less tourist pressure, in rural environments, coastal destinations and in tourist destinations with a different model of social relations than the one ruling in Barcelona. The repetition of the study will help to understand the different citizen attitudes toward this context and the influence of the environment on this attitude. In addition, this study assumes three constraints associated with the methodology used (Guillen-Royo, 2016). First, it is the "inner dimension", which refers to the excessive influence that certain marginalized portions of the population could have when trying to reach groups' conclusions. The second limitation refers to the anticipated frustration that may condition working groups when they expect their proposals will not be considered or will be diluted. Finally, there are problems in articulating public policy recommendations from the point of view of the HSD perspective. No specific "recipes" have been developed to help articulate different levels of governance to organize a transition toward sustainable development around the indicated principles.

In addition to replicating this work scheme in other contexts, a specific analysis of some of the conclusions and insights issued by the work groups could shed further light over different aspects of this work. This extension should use quantitative techniques, so that the number of subjects could be expanded to reinforce or refute the conclusions of this work. We do not recommend a direct translation of the methodology used in this work to a quantitative analysis, as an adapted framework has not yet been developed. This is in itself a challenge for the future. Applying this methodology through a survey-based process could bring some advantages, even though the problems to be resolved in this pursuit are numerous.

Disclosure statement

No potential conflict of interest was reported by the authors.

ORCID

José Maria Martín-Martín http://orcid.org/0000-0002-5639-2213

References

Abril-Sellarés, M., Azpelicueta, M. C., & Sánchez-Fernandesz, M. D. (2015). Turismo sostenible: Lugareños frente a turistas. El caso de la ciudad de Barcelona. *HOLOS*, *3*, 331–337. https://doi.org/10.15628/holos.2015.2809

Adams, W. M. (2009). *Green development: Environment and sustainability in a developing world*. Routledge.

Alkire, S. (2002). *Valuing freedoms: Sen's capability approach and poverty reduction*. Oxford University Press.

Andereck, K. L., & Nyaupane, G. P. (2011). Exploring the nature of tourism and quality of life perceptions among residents. *Journal of Travel Research*, *50*(3), 248–260. https://doi.org/10.1177/0047287510362918

Andereck, K. L., Valentine, K. M., Knopf, R. C., & Vogt, C. A. (2005). Residents' perceptions of community tourism impacts. *Annals of Tourism Research*, *32*(4), 1056–1076. https://doi.org/10.1016/j.annals.2005.03.001

Andersen, J., & Siim, B. (2004). *Introduction: The politics of inclusion and empowerment: Gender, class and citizenship*. In J. Andersen & B. Siim (Eds.), *The politics of inclusion and empowerment: Gender, class and citizenship* (pp. 1–18). Palgrave Macmillan.

Barcelona City Council. (2016). *Barcelona's Tourism Activity Report 2015*. Ayuntamiento de Barcelona.

Barcelona City Council. (2017). *Barcelona's Tourism Activity Report 2016*. Ayuntamiento de Barcelona.

Barcelona City Council. (2020). *Barcelona's Tourism Activity Report 2019*. Ayuntamiento de Barcelona.

Belarmino, A., Whalen, E., Koh, Y., & Bowen, J. T. (2019). Comparing guests' key attributes of peer-to-peer accommodation and hotels: Mixed-methods approach. *Current Issues in Tourism*, *22*(1), 1–7. https://doi.org/10.1080/13683500.2017.1293623

Bort, J. (2014). *San Francisco makes Airbnb legal at last*. Business Insider. Retrieved February 18, 2015, from https://www.businessinsider.com/san-francisco-makes-airbnb-legal-at-last-2014-10?IR=T

Botsman, R., & Rogers, R. (2010). *What's mine is yours: The rise of collaborative consumption*. Harper Business.

Bramwell, B., & Lane, B. (1993). Sustainable tourism: An evolving global approach. *Journal of Sustainable Tourism*, *1*(1), 1–5. https://doi.org/10.1080/09669589309450696

Brustein, J. (2014). *An Airbnb rival challenges San Francisco's Airbnb law*. Bloomberg Businessweek. Retrieved February 16, 2017, from http://www.bloomberg.com/bw/articles/2014-11-03/san-francisco-faces-a-lawsuit-over-its-airbnb-law

Cheng, M. (2016). Sharing economy: A review and agenda for future research. *International Journal of Hospitality Management*, *57*, 60–70. https://doi.org/10.1016/j.ijhm.2016.06.003

Choi, J. H., Ju, S., Kim, K.-S., Kim, M., Kim, H. J., & Yu, M. (2015). The relationship between Airbnb and the hotel revenue: In the case of Korea. *Indian Journal of Science and Technology*, *8*(S8), 1–8. https://doi.org/10.17485/ijst/2015/v8iS8/64705

Cócola, A. (2015). Tourism and commercial gentrification [Paper presentation]. RC21 International Conference on "The ideal city: Between myth and reality. Representations, policies, contradictions and challenges for tomorrow's urban life", Urbino, Italy, August 27–29. Retrieved May 8, 2017, from http://www.rc21.org/en/wp-content/uploads/2014/12/E4-C%C3%B3cola-Gant.pdf

Cócola, A. (2016). Holiday rentals: The new gentrification battlefront. *Sociological Research Online*, *21*(3), 1–9.

Cohen, B., & Munoz, P. (2016). Sharing cities and sustainable consumption and production: Towards an integrated framework. *Journal of Cleaner Production*, *134*, 87–97. https://doi.org/10.1016/j.jclepro.2015.07.133

Cruz, I., Stahel, A., & Max-Neef, M. (2009). Towards a systemic development approach: Building on the Human-Scale Development paradigm. *Ecological Economics*, *68*(7), 2021–2030. https://doi.org/10.1016/j.ecolecon.2009.02.004

Dempsey, N., Bramley, G., Power, S., & Brown, C. (2011). The social dimension of sustainable development: Defining urban social sustainability. *Sustainable Development*, *19*(5), 289–300. https://doi.org/10.1002/sd.417

Doswell, R. (1997). *How effective management makes a difference*. Butterworth-Heinemann.

Dredge, D., & Gyimóthy, S. (2015). The collaborative economy and tourism: Critical perspectives, questionable claims and silenced voices. *Tourism Recreation Research*, *40*(3), 286–302. https://doi.org/10.1080/02508281.2015.1086076

Edelman, B. G., & Geradin, D. (2016). Efficiencies and regulatory shortcuts: How should we regulate companies like Airbnb and Uber? *Stanford Technology Law Review*, *19*(2), 293–328.

Europa Press. (2017). Barcelona y Berlín son las Ciudades Europeas con Mayor Crecimiento de Viviendas de Uso Turístico. 2017. Retrieved September 19, 2017, from http://www.europapress.es/turismo/nacional/noticia-barcelona-berlin-son-ciudades-europeas-mayor-crecimiento-viviendas-uso-turistico-20170918142451.html

Fang, B., Ye, Q., & Law, R. (2016). Effect of sharing economy on tourism industry employment. *Annals of Tourism Research*, *57*, 264–278. https://doi.org/10.1016/j.annals.2015.11.018

Forno, F., & Garibaldi, R. (2015). Sharing economy in travel and tourism: The case of home-swapping in Italy. *Journal of Quality Assurance in Hospitality & Tourism*, *16*(2), 202–220.

Gallagher, L. (2017). *The Airbnb story: How three ordinary guys disrupted an industry, made billions … and created plenty of controversy*. Houghton Mifflin Harcourt.

Garau-Vadell, J. B., Gutiérrez-Taño, D., & Díaz-Armas, R. (2019). Residents' support for P2P accommodation in mass tourism destinations. *Journal of Travel Research*, *58*(4), 549–565. https://doi.org/10.1177/0047287518767067

Geron, T. (2013). *Airbnb and the unstoppable rise of the share economy*. Forbes. http://www.forbes.com/sites/tomiogeron/2013/01/23/airbnb-and-the-unstoppable-rise-of-the-share-economy/

Gotham, K. (2005). Tourism gentrification: The case of New Orleans' Vieux Carre (French quarter). *Urban Studies*, *42*(7), 1099–1121. https://doi.org/10.1080/00420980500120881

Gravari-Barbas, M., & Guinand, S. (2017). *Tourism and gentrification in contemporary metropolises international perspectives*. Routledge.

Guaita, J. M., Martín, J. M., & Salinas, J. A. (2020). Innovation in the measurement of tourism competitiveness. In M. A. Galindo-Martín, M. T. Mendez-Picazo, & M. S. Castaño-Martínez (Eds.), *Analyzing the relationship between innovation, value creation, and entrepreneurship* (pp. 268–288). IGI Global.

Guaita, J. M., Martín, J. M., Salinas, J. A., & Mogorrón-Guerrero, H. (2019). An analysis of the stability of rural tourism as a desired condition for sustainable tourism. *Journal of Business Research*, *100*, 165–174. https://doi.org/10.1016/j.jbusres.2019.03.033

Guillen-Royo, M. (2016). *Sustainability and wellbeing: Human-scale development in practice*. Routledge.

Guillén-Royo, M., Guardiola, J., & García-Quero, F. (2017). Sustainable development in times of economic crisis: A needs-based illustration from Granada (Spain). *Journal of Cleaner Production*, *150*, 267–276. https://doi.org/10.1016/j.jclepro.2017.03.008

Gunter, U., & Önder, I. (2018). Determinants of Airbnb demand in Vienna and their implications for the traditional accommodation industry. *Tourism Economics*, *24*(3), 270–293. https://doi.org/10.1177/1354816617731196

Gurran, N., & Phibbs, P. (2017). When tourists move in: How should urban planners respond to Airbnb? *Journal of the American Planning Association*, *83*(1), 80–92. https://doi.org/10.1080/01944363.2016.1249011

Gursoy, D., Chi, C. G., & Dyer, P. (2010). Locals' attitudes toward mass and alternative tourism: The case of Sunshine Coast. *Journal of Travel Research*, *49*(3), 381–394. https://doi.org/10.1177/0047287509346853

Gutierrez, J., García-Palomares, J. C., Romanillos, G., & Salas-Olmedo, M. H. (2017). The eruption of Airbnb in tourist cities: Comparing spatial patterns of hotels and peer-to-peer accommodation in Barcelona. *Tourism Management*, *62*, 278–291. https://doi.org/10.1016/j.tourman.2017.05.003

Guttentag, D. (2015). Airbnb: Disruptive innovation and the rise of an informal tourism accommodation sector. *Current Issues in Tourism*, *18*(12), 1192–1217. https://doi.org/10.1080/13683500.2013.827159

Heo, Y. (2016). Sharing economy and prospects in tourism research. *Annals of Tourism Research*, *58*, 166–170. https://doi.org/10.1016/j.annals.2016.02.002

Hosteltur. (2020). Las 20 ciudades del mundo con mayor tirón para los turistas. https://www.hosteltur.com/131020_las-20-ciudades-del-mundo-con-mayor-tiron-para-los-turistas.html

Ioannides, D., Röslmaier, M., & Van der Zee, E. (2019). Airbnb as an instigator of 'tourism bubble' expansion in Utrecht's Lombok neighbourhood. *Tourism Geographies*, *21*(5), 822–840. https://doi.org/10.1080/14616688.2018.1454505

Jefferson-Jones, J. (2014). Airbnb and the housing segment of the modern sharing economy: Are short-term rental restrictions an unconstitutional taking. *Hastings Constitutional Law Quarterly*, *42*(3), 557–575.

John, N. A. (2013). The social logics of sharing. *The Communication Review*, *16*(3), 113–131. https://doi.org/10.1080/10714421.2013.807119

Jolibert, C., Paavola, J., & Rauschmayer, F. (2014). Addressing needs in the search for sustainable development: A proposal for needs-based scenario building. *Environmental Values*, *23*(1), 29–50. https://doi.org/10.3197/096327114X13851122269007

Juul, M. (2015). *The sharing economy and tourism*. European Parliament. http://www.europarl.europa.eu/RegData/etudes/BRIE/2015/568345/EPRS_BRI(2015)568345_EN.pdf

Lines, G. E. (2015). Hej, not hej då: Regulating Airbnb in the new age of Arizona vacation rentals. *Arizona Law Review*, *57*(4), 1163–1182.

Lonely Planet. (2020). https://www.lonelyplanet.es/blog/las-ciudades-de-europa-mas-visitadas

Lyons, K., & Wearing, S. (2015). *The sharing economy issues, impacts, and regulatory responses in the context of the NSW visitor economy*. NSW Business Chamber.

Martín, J. M. (2019). Impacts of the tourist activity and citizens' evaluation about the necessity for resting periods. In M. A. Camilleri (Ed.), Strategic perspectives in destination marketing (pp. 81–112). IGI Global, Pennsylvania, United States. https://doi.org/10.4018/978-1-5225-5835-4.ch004

Martín, C. J., & Upham, P. (2016). Grassroots social innovation and the mobilisation of values in collaborative consumption: A conceptual model. *Journal of Cleaner Production*, *134*, 204–213. https://doi.org/10.1016/j.jclepro.2015.04.062

Martín, C. J., Upham, P., & Budd, L. (2015). Commercial orientation in grassroots social innovation: Insights from the sharing economy. *Ecological Economics*, *118*, 240–251. https://doi.org/10.1016/j.ecolecon.2015.08.001

Martín, J. M. (2019). *Impacts of the tourist activity and citizens' evaluation about the necessity for resting periods*. In M. A. Camilleri (Ed.), *Strategic perspectives in destination marketing* (pp. 81–112). IGI Global.

Martín, J. M., Guaita, J. M., & Salinas, J. A. (2018). An analysis of the factors behind the citizen's attitude of rejection towards tourism in a context of overtourism and economic dependence on this activity. *Sustainability*, *10*(8), 2851. https://doi.org/10.3390/su10082851

Martín, J. M., Jiménez, J. D., & Molina, V. (2014). Impacts of seasonality on environmental sustainability in the tourism sector based on destination type: An application to Spain's Andalusia region. *Tourism Economics*, *20*(1), 123–142. https://doi.org/10.5367/te.2013.0256

Martín, J. M., Rodríguez, J. A., Zermeño, K. A., & Salinas, J. A. (2018). Effects of vacation rental websites on the concentration of tourists—Potential environmental impacts. An application to the Balearic Islands in Spain. *International Journal of Environmental Research and Public Health*, *15*(2), 347.

Martín, J. M., Ostos, M. S., & Salinas, J. A. (2019). Why regulation is needed in emerging markets in the tourism sector. *The American Journal of Economics and Sociology*, *78*(1), 225–254.

Martín, J. M., Salinas, J. A., Rodríguez, J. A., & Jiménez, J. D. (2017). Assessment of the tourism's potential as a sustainable development instrument in terms of annual stability: Application to Spanish rural destinations in process of consolidation. *Sustainability*, *9*(10), 1692.

Martín, J. M., Salinas, J. A., Rodríguez, J. A., & Ostos, M. S. (2019). Analysis of tourism seasonality as a factor limiting the sustainable development of rural areas. *Journal of Hospitality & Tourism Research*, *44*(1), 45–75.

Mathieson, A., & Wall, G. (1982). *Tourism: Economic, physical and social impacts*. Addison Wesley Longman.

Max-Neef, M. (1991). *Human-scale development: Conception, application and further reflection*. Apex Press.

McGehee, N. G., & Andereck, K. L. (2004). Factors predicting rural residents' support of tourism. *Journal of Travel Research*, *43*(2), 131–140. https://doi.org/10.1177/0047287504268234

McIntyre, G. (1993). *Sustainable tourism development: Guide for local planners*. World Tourism Organization.

Mody, M., Suess, C., & Dogru, T. (2019). Not in my backyard? Is the anti-Airbnb discourse truly warranted? *Annals of Tourism Research*, *74*, 198–203. https://doi.org/10.1016/j.annals.2018.05.004

Mody, M., Woosnam, K. M., Suess, C., & Dogru, T. (2020). Hapless victims or empowered citizens? Understanding residents' attitudes towards Airbnb using Weber's Theory of Rationality and Foucauldian concepts. *Journal of Sustainable Tourism*. https://doi.org/10.1080/09669582.2020.1834567

Nadler, S. S. N. (2014). *The sharing economy: What is it and where is it going?* Massachusetts Institute of Technology.

National Statistics Institute. (2020). Encuesta de Ocupación Hotelera. https://www.ine.es/dyngs/INEbase/es/operacion.htm?c=Estadistica_C&cid=1254736177015&menu=ultiDatos&idp=1254735576863

Nieuwland, S., & van Melik, R. (2020). Regulating Airbnb: How cities deal with perceived negative externalities of short-term rentals. *Current Issues in Tourism*, *23*(7), 811–825. https://doi.org/10.1080/13683500.2018.1504899

Nunkoo, R., & So, K. K. F. (2016). Non-hosting residents' support for tourism: Testing alternative structural models. *Journal of Travel Research*, *55*(7), 847–861. https://doi.org/10.1177/0047287515592972

OECD. (2016). *Policies for the tourism sharing economy – OECD tourism trends and policies*. OECD Publishing.

Oskam, J., & Boswijk, A. (2016). Airbnb: The future of networked hospitality businesses. *Journal of Tourism Futures*, *2*(1), 22–42. https://doi.org/10.1108/JTF-11-2015-0048

Park, D., & Yoon, Y. (2009). Segmentation by motivation in rural tourism: A Korean case study. *Tourism Management*, *30*(1), 99–108. https://doi.org/10.1016/j.tourman.2008.03.011

Park, D., Yoon, Y., & Lee, M. (2008). Rural community development and policy challenges in South Korea. *Journal of the Economic Geographical Society of Korea*, *11*(4), 600–617.

Porges, S. (2013). *The Airbnb effect: Bringing life to quiet neighborhoods*. Forbes. http://www.forbes.com/sites/sethporges/2013/01/23/the-airbnb-effect-bringing-life-to-quiet-neighborhoods/

Postma, A., & Schmuecker, D. (2017). Understanding and overcoming negative impacts of tourism in city destinations: Conceptual model and strategic framework. *Journal of Tourism Futures*, *3*(2), 144–156. https://doi.org/10.1108/JTF-04-2017-0022

Prayag, G., Hosany, S., Nunkoo, R., & Alders, T. (2013). London residents' support for the 2012 Olympic Games: The mediating effect of overall attitude. *Tourism Management*, *36*, 629–640. https://doi.org/10.1016/j.tourman.2012.08.003

Puczkó, L., & Rátz, T. (2000). Tourist and resident perceptions of the physical impacts of tourism at Lake Balaton, Hungary: Issues for sustainable tourism management. *Journal of Sustainable Tourism*, *8*(6), 458–478. https://doi.org/10.1080/09669580008667380

Quattrone, G., Proserpio, D., Quercia, D., Capra, L., & Musolesi, M. (2016). Who *benefits* from the *"sharing" economy* of Airbnb? [Paper presentation]. WWW'16: Proceedings of the 25th International Conference on World Wide Web, Montreal, Canada, April 11–15 (pp. 1385–1394). https://doi.org/10.1145/2872427.2874815

Queensland Tourism Industry Council. (2014). *The sharing economy: How it will impact the tourism landscape and what businesses can do*. Queensland Tourism Industry Council.

Richards, G. (2016). El turismo y la ciudad ¿Hacia nuevos modelos? *Revista Cidob d'afers internacionals, 113*, 71–87. https://doi.org/10.24241/rcai.2016.113.2.71

Richards, S., Brown, L., & Dilettuso, A. (2019). The Airbnb phenomenon: The resident's perspective. *International Journal of Tourism Cities, 6*(1), 8–26. https://doi.org/10.1108/IJTC-06-2019-0084

Richardson, L. (2015). Performing the sharing economy. *Geoforum, 67*, 121–129. https://doi.org/10.1016/j.geoforum.2015.11.004

Roberts, L., & Hall, D. (2001). *Rural tourism and recreation: Principles to practice*. CABI Publishing.

Russo, A. P. (2002). The "vicious circle" of tourism development in heritage cities. *Annals of Tourism Research, 29*(1), 165–182. https://doi.org/10.1016/S0160-7383(01)00029-9

Russo, A. P., & Quaglieri, A. (2014). La Lógica Espacial del Intercambio de Casas: Una Aproximación a las Nuevas Geografías de lo Cotidiano en el Turismo Contemporáneo. *Scripta Nova Revista Electrónica de Geografía y Ciencias Sociales, 17*, 483.

Russo, A. P., & Quaglieri, A. (2016). *The shifting spatial logic of tourism in networked hospitality*. In A. P. Russo & G. Richards (Eds.), *Reinventing the local in tourism: Producing, consuming and negotiating place* (pp. 15–34). Channel View Publications.

Sacks, D. (2011). *The sharing economy*. Fast Company. http://www.fastcompany.com/magazine/155/the-sharing-economy.html

Salinas, J. A., Serdeira, P., Martín, J. M., & Rodríguez, J. A. (2020). Determinants of tourism destination competitiveness in the countries most visited by international tourists: Proposal of a synthetic index. *Tourism Management Perspectives, 33*, 100582.

Schor, J. B., & Fitzmaurice, C. (2015). *Collaborating and connecting: The emergence of the sharing economy*. In L. Reisch & J. Thogersen (Eds.), *Handbook of research on sustainable consumption* (pp. 410–425). Edward Elgar.

Shaheen, S. A., Mallery, M. A., & Kingsley, K. J. (2012). Personal vehicle sharing services in North America. *Research in Transportation Business & Management, 3*, 71–81.

Sigala, M. (2017). Collaborative commerce in tourism: Implications for research and industry. *Current Issues in Tourism, 20*(4), 346–355. https://doi.org/10.1080/13683500.2014.982522

Smith, P. B., & Max-Neef, M. (2011). *Economics unmasked: From power and greed to compassion and the common good*. Green Books Totnes.

Suciu, A. M. (2016). *The impact of Airbnb on local labour markets in the hotel industry in Germany*. https://ssrn.com/abstract=2874861

Suess, C., Baloglu, S., & Busser, J. A. (2018). Perceived impacts of medical tourism development on community well-being. *Tourism Management, 69*, 232–245. https://doi.org/10.1016/j.tourman.2018.06.006

Suess, C., Woosnam, K. M., & Erul, E. (2020). Stranger-danger? Understanding the moderating effects of children in the household on non-hosting residents' emotional solidarity with Airbnb visitors, feeling safe, and support for Airbnb. *Tourism Management, 77*, 103952. https://doi.org/10.1016/j.tourman.2019.103952

Suess, C., Woosnam, K., Mody, M., Dogru, T., & Sirakaya Turk, E. (2020). Understanding how residents' emotional solidarity with Airbnb visitors influences perceptions of their impact on a community: The moderating role of prior experience staying at an Airbnb. *Journal of Travel Research*. https://doi.org/10.1177/0047287520921234

Tussyadiah, I. P., & Pesonen, J. (2018). Drivers and barriers of peer-to-peer accommodation stay: An exploratory study with American and Finnish travellers. *Current Issues in Tourism, 21*(6), 703–720. https://doi.org/10.1080/13683500.2016.1141180

UN. (1992). Report of the United Nations Conference on Environment and Development, Rio de Janeiro, Brazil, June 3–14. Retrieved May 4, 2015, from http://www.un.org/documents/ga/conf151/aconf15126-1annex1.htm

UN. (2012). Report of the United Nations Conference on Sustainable Development, Rio de Janeiro, Brazil, June 20–22. Retrieved May 4, 2015, from https://sustainabledevelopment.un.org/rio20

UNWTO. (2004). *Indicators of sustainable development for tourism destinations: A guidebook*. World Tourism Organization.

Uysal, M., Sirgy, M. J., Woo, E., & Lina, H. (2016). Quality of life (QOL) and well-being research in tourism. *Tourism Management, 53*, 244–261. https://doi.org/10.1016/j.tourman.2015.07.013

WCED. (1987). *Our common future* (Report of the World Commission on Environment and Development). Oxford University Press.

WEF. (2017). The travel & tourism competitiveness report 2017: Paving the way for a more sustainable and inclusive future. Geneva. Dataset extracted from http://reports.weforum.org/travel-and-tourism-competitiveness-report-2017/downloads/

Wegmann, J., & Jiao, J. (2017). Taming Airbnb: Toward guiding principles for local regulation of urban vacation rentals based on empirical results from five US cities. *Land Use Policy*, *69*, 494–501. https://doi.org/10.1016/j.landusepol.2017.09.025

Yeager, E. P., Boley, B. B., Woosnam, K. M., & Green, G. T. (2020). Modeling residents' attitudes toward short-term vacation rentals. *Journal of Travel Research*, *59*(6), 955–974. https://doi.org/10.1177/0047287519870255

Youell, R. (1998). *Tourism: An introduction*. Addison Wesley Longman.

Zervas, G., Proserpio, D., & Byers, J. (2014). The rise of the sharing economy: Estimating the impact of Airbnb on the hotel industry (Research Paper No. 2013-16). Boston University School of Management.

Towards sustainable tourism development in a mature destination: measuring multi-group invariance between residents and visitors' attitudes with high use of accommodation-sharing platforms

Andrés Chamarro, José Cobo-Benita and María Dolores Herrero Amo

ABSTRACT

The aim of this study is to evaluate the differences between residents and visitors in their attitudes and behaviours towards sustainable tourism. The study examines the validity of the 21-item version of SUS-TAS, measuring invariance between residents' and visitors' attitudes in Madrid, a mature tourist destination with high use of accommodation-sharing platforms, in order to have a new testable tool for future sustainable planning. For multi-group invariance assessment, the Spanish version of the SUS-TAS for residents and the version for visitors are validated separately following a first-order, seven-factor model. The multi-group confirmatory factor analysis showed full configural invariance, partial metric invariance and partial scalar invariance across residents and visitors. Findings suggest that the main variations between residents and visitors were related to the "community-centered economy" and "maximizing community participation" constructs. To conclude, SUS-TAS can be used to measure residents' and visitors' attitudes towards sustainable tourism in a context with high use of accommodation-sharing platforms. Implications for future research and limitations are discussed.

Introduction

Tourism has played a central role in the economy of many EU countries (UNWTO, 2019). In 2018, total international tourist arrivals globally reached 1.4 billion and international tourist arrivals in Europe were 710 million, representing a 5% growth. Southern European countries lead this growth with 289.4 million, while Spain and Italy, two of the top five destinations, had 83 million and 62 million arrivals respectively.

After the economic crisis of 2008, new business models appeared based on the concepts of the sharing economy, to deal with the economic and social problems that society had suffered. Accommodation-sharing platforms (ASP) have disrupted tourism and hospitality sectors with companies like Airbnb and Homeaway (Huston, 2015; Zervas et al., 2017). Research on the impact of these platforms has increased in the last decade (Geissinger et al., 2019; Guttentag & Smith, 2017; Martin, 2016; Oskam & Boswijk, 2016), focusing on the economic, social and environmental aspects.

In a different but related realm of research, studies about tourism planning and development have highlighted the concept of sustainability as a major concern (Poudel et al., 2016; Stoddard et al., 2012). Over the last decades there has been an increase in studies on understanding the attitudes of residents as main stakeholders when planning sustainable community tourism (Nunkoo et al., 2013; Zhang et al., 2015) in order to solve possible problems. Authors have explained the importance of considering different stakeholders (Poudel et al., 2016; Sirakaya-Turk, Ekinci et al., 2008) for sustainable tourism: community members or residents, visitors and tourism. Further research emphasises the importance of residents' attitudes when planning tourism (Woosnam et al., 2018) with a sustainable perspective and pro-tourism behaviours (Ribeiro et al., 2017). To ensure long-term sustainable tourism development, scholars include residents as fundamental contributors to community involvement (Choi & Murray, 2010; Choi & Sirakaya, 2006). Previous studies have therefore used the Sustainable Tourism Attitude Scale (SUS-TAS) as a tool to evaluate residents' attitudes towards sustainable tourism in different contexts, mainly in mature destinations and more recently in emerging tourism destinations (Ribeiro et al., 2018). To this end, the authors considered the social exchange theory (SET) (Ap, 1992) and the new environmental paradigm (NEP) (Dunlap et al., 2000) mainly in mature destinations and lately in emerging tourism destinations (Ribeiro et al., 2013, 2018). Although the scale has been validated in cross-cultural settings (Ribeiro et al., 2018; Sirakaya-Turk, Ekinci et al., 2008; Sirakaya-Turk & Gursoy, 2013; Yu et al., 2009; Zhang et al., 2015), measurement invariance has not been tested in mature destinations considering both types of stakeholders simultaneously, but has tested only residents (Hsu et al., 2020) or residents and visitors separately in different settings (Poudel et al., 2016). Although tourism research has focused mainly on residents perceptions, recent studies (Joo et al., 2019; Su & Swanson, 2017) have focused their efforts on exploring tourists' perceptions of tourism impacts and development on a destination. Moreover, researchers have provided support that tourists know the importance of tourism impact and tourism development on a destination, specifically about community benefits or support for tourism (Joo et al., 2019) or environmental awareness and protective behaviour (Cheng & Wu, 2015; Su & Swanson, 2017).

Given the recent unmeasured growth of accommodation-sharing platforms, planning for sustainable tourism has changed in mature tourist destinations. The way these platforms have developed has created current issues in destinations, affecting different stakeholders of the tourism system (Guttentag, 2015; Guttentag & Smith, 2017). Visitors have changed the way they travel and interact with residents (Frenken & Schor, 2017; Gansky, 2012) and hotels' economic results have been affected: they claim that these platforms create unfair competition and promote tax avoidance (Oskam & Boswijk, 2016; Varma et al., 2016). For the most part, however, these impacts are currently affecting the host population in tourism destinations, with issues like the resident displacement, increasing rental prices and loss of cultural heritage as a result of the professionalization of the platforms (Öberg, 2018) and the participation of companies (Belk, 2014; Cherry & Pidgeon, 2018).

Following on from previous research papers, the aim of this study is to evaluate the differences between the attitudes and behaviours of residents and visitors towards sustainable tourism. The study consists in testing for multi-group measurement invariance (Byrne, 2004) between residents' and visitors' attitudes, using the 21-item version of SUS-TAS (Ribeiro et al., 2017). The study was conducted in Madrid, Spain, a mature tourist destination with high use of accommodation-sharing platforms, the aim being to obtain a new testable tool for future planning and development actions. As the impact of accommodation-sharing platforms could differ between visitors and residents, the study emphasises the importance of considering the attitudes to sustainable tourism development of both visitors and residents in a destination.

Theoretical background

Accommodation-sharing platforms and sustainable tourism development

The initial concept of accommodation-sharing platforms was linked to pursuing a more sustainable resource allocation and reducing costs (Botsman & Rogers, 2011). Research suggests that user motivations are not only related to profit but also to altruistic values such as reusing assets or helping others, resulting in a more sustainable lifestyle (Edbring et al., 2016). The sharing economy can perform in three different ways (Richardson, 2015): through community, access and collaboration. All three are latent in Airbnb, with its review system, which fosters interaction between users, collaborating to create a whole community around the platform, which is one of the main reasons for its exponential growth (Gansky, 2012; Ikkala & Lampinen, 2015). Furthermore, the possibility of guests becoming hosts and vice versa strengthens the community (Richardson, 2015), where trust is one of the main aspects considered (Botsman & Rogers, 2011; Finley, 2013). In terms of access, the platforms offer guests the possibility of buying short-term access instead of ownership. Even though there are different motivations to use accommodation-sharing platforms, cost and practical aspects remain the principal incentive for tourists to prefer this service (Guttentag et al., 2018) rather than the traditional accommodation offering.

Accommodation-sharing platforms have impacted on society in different ways. Economically, guests consider that using these platforms is a cheaper option than hotels, due to the reduction in staff and transaction costs (Guttentag, 2015), and the hosts' main reason for advertising their accommodation is to obtain extra monthly income (Hamari et al., 2016; Martin, 2016). However, for hotels, the negative impact has been quantified in previous studies showing that their economic results are affected (Dogru et al., 2019; Zervas et al., 2017). The industry claims there is unfair competition between the platforms and hotels due to the lack of regulation (Zervas et al., 2017). In addition, previous studies highlighted the negative impact on cities as a result of increased rental prices, displacement of residents and lower spending in the destinations (Garcia-López et al., 2020; Oskam & Boswijk, 2016). Socially, several studies have shown that although the main advantage is economic, people also decide to rent their houses in order to socialise with guests (Gansky, 2012; Guttentag, 2015; Ikkala & Lampinen, 2015). Environmentally, some studies believe that this more efficient use of resources -staying in a private room instead of a hotel, reducing supply costs- does have a positive impact (Botsman & Rogers, 2011; Schor, 2017). Conversely, other studies emphasise the complexity of this measurement and are sceptical with regard to the results released by the platforms, as they only focus on the direct effects (Frenken & Schor, 2017).

One of the main issues concerning these platforms is related to the fast evolution of technology, causing legal issues with local legislation (Varma et al., 2016). Currently, the main discussion around sustainability is whether the initial motivation of sharing is still a real phenomenon (Geissinger et al., 2019). Accommodation-sharing platforms have moved in new directions, such as the separation between access and owners, professionalization of the platforms (Dogru et al., 2020; Öberg, 2018) and hosts using platforms for profit purposes only (Belk, 2014). Consequently, commercial hosts are competing with residents on the housing market, increasing prices and causing gentrification mainly in tourist areas (Cherry & Pidgeon, 2018). Previous studies collected information on several cities like Paris, Barcelona, Berlin or Amsterdam, which have introduced new regulation policies (Oskam & Boswijk, 2016) on short-term rental, measures to protect residents, reducing long-term rentals and forbidding unregistered hosts. Therefore, to ensure long-term sustainable tourism development, residents should be considered as fundamental contributors to community involvement. (Hung et al., 2011; Nunkoo et al., 2013).

During the last years, tourism researchers have focused their studies on resident's perceptions of accommodation sharing platforms (Garau-Vadell et al., 2018; Mody et al., 2019; Yeager et al., 2020). From a quantitative approach, Garau-Vadell et al. (2018) analysed residents' perceptions of short-term vacation rentals and the differences between hosts and residents who have never hosted using accommodation sharing platforms using the Social Exchange Theory (SET) to explain residents' level of support to accommodation sharing platforms. Moreover, a combination of SET and Weber's Theory of Formal and Substantive Rationality (WTFSR) was applied to consider both economic and non-economic factors (Yeager et al., 2020) that influence residents' support or opposition to accommodation sharing platforms. Similarly, Mody et al., (2020) were guided by the WTFSR and Foucauldian concepts of power to explain residents' attitudes regarding Airbnb.

The tourism ecosystem has changed due to the rise of accommodation-sharing platforms (Dogru et al., 2020) and to ensure long-term sustainable tourism development, residents of a destination with high use of these platforms should be considered as fundamental contributors to community involvement (Hung et al., 2011; Nunkoo et al., 2013).

SUS-TAS – Sustainable Tourism Attitude Scale

Since residents have been considered as core stakeholders in sustainable tourism, researchers have focused on measuring their attitudes and behaviours. The SET has been the main theory used by tourism researchers to explain residents' attitudes, estimating their level of support for or opposition to tourism development. The Sustainable Tourism Attitude Scale (SUS-TAS) was initially developed by Choi and Sirakaya (2005). They gathering data from residents of a tourist city in Texas, creating a new tool to evaluate residents' attitudes towards sustainable tourism in different contexts. The study incorporated aspects of the NEP and the SET, and through an Explanatory Factor Analysis, seven factors were identified to measure the original 44-item SUS-TAS: (1) perceived social costs, (2) perceived environmental sustainability, (3) perceived economic benefits, (4) long-term planning, (5) ensuring visitor satisfaction, (6) community-centered economy, and (7) maximizing community participation. Researchers have refined and validated the original 44-item scale (Choi & Sirakaya, 2005) to obtain a shorter valid 21-item version with good psychometric properties (Ribeiro et al., 2018). Some researchers have focused their studies on refining the scale, revalidating shorter versions to improve model fit and parsimony (Ribeiro et al., 2018; Sirakaya-Turk, 2007; Sirakaya-Turk, Ekinci et al., 2008; Sirakaya-Turk & Gursoy, 2013; Yu et al., 2009; Zhang et al., 2015) with a majority of the data collected in mature destinations. The initial SUS-TAS was validated in a cross-cultural context using Turkish and Cypriot residents with a shorter 33-item version. Yu et al. (2009) revalidated the 44-item scale into a 27-item one with good validity and reliability indices. Following the seven-step scale-development process proposed by (Churchill, 1979), most papers focused on steps 6 and 7 to develop and validate the SUS-TAS (Choi & Sirakaya, 2005): (6) Refine scale measurements using a new sample (using confirmatory factor analysis to re-evaluate the reliability and validity of the scale), and (7) develop norms.

Sirakaya-Turk and Gursoy (2013) reduced the SUS-TAS to a 21-item scale with adequate validities, model fit, and parsimony but under cut-off values in two model fit indices. Some studies focused on applying a validated version of SUS-TAS in order to understand the attitudes of host communities and other stakeholders towards tourism. For instance, researchers used the 33-item SUS-TAS validated by Sirakaya-Turk et al. to cluster residents' profiles in three host communities in Turkey (Sirakaya-Turk, Ingram et al., 2008). Similarly, a study measured environmental behaviour variation between Swedish and Russian tourists in Cyprus applying environmental items from the 44-item SUS-TAS (Kvasova, 2011). A study in Mauritius measured hoteliers' attitudes towards sustainability, combining items from SUS-TAS and Dyer et al's perceived tourism impact scale (Prayag et al., 2010).

In a further attempt to validate a second-order factor model, Zhang et al. (2015) refined the SUS-TAS, creating a maximum parsimonious 20-item SUS-TAS with strong psychometric properties with good fit, reliability and validity across 11 different Midwestern U.S. counties. The study includes a conceptual framework for predictive validity, revising the theoretical background of the SUS-TAS, defining two dimensions of residents' attitudes toward sustainable tourism to predict their intention to participate in tourism planning (Ribeiro et al., 2018; Zhang et al., 2015). The first dimension, *perceived tourism impacts (PTI)*, incorporates three factors: Perceived Economic Benefits (PEB), Perceived Environmental Sustainability (PES) and Perceived Social Cost (PSC). The second dimension, *expected tourism sustainability (ETS)*, contains the four remaining factors: Community Centred Economy (CE), Long-Term Planning (LP), Ensuring Visitor Satisfaction (EVS) and Maximizing Community Participation (MCP). The purpose of this study was to examine global interpretations of SUS-TAS, concluding that each factor should be interpreted individually rather than as a global dimension, due to lack of significant correlation between three factors. Ribeiro et al. (2018), in their study on residents in the Cape Verde islands, developed and tested different versions of the SUS-TAS (44-item, 33-item, 27-item and 21-item), finding a shorter version of the 21-item SUS-TAS with maximum parsimony. The paper addresses the need to validate the predictive validity of a second-order factor model, so far lacking in the scientific literature, using the two hierarchical factor dimensions, PTI and ETS, presented previously.

Multi-group measurement invariance of SUS-TAS: residents and visitors in Madrid

Researchers have used multi-group measurement invariance to analyse models across different groups (Chen, 2007; Cheung & Rensvold, 2002), especially in cross-cultural research. The main objective is to test the reliability of an assessment measure across the groups (Byrne, 2004). Invariance is confirmed when members from different groups score equally on the test conducted. The levels of invariance typically measured are configural invariance or equivalence of structure, metric invariance or factor loading equivalence and scalar or item intercept equivalence (Schmitt & Kuljanin, 2008).

Schmitt and Kuljanin (2008) and Vandenberg and Lance (2000) reviewed papers that measured invariance using Confirmatory Factor Analysis (CFA) models, explaining the different levels of invariance measured, if translation was needed and if they reach full or partial invariance. Since 2000, the examination of scalar and factor mean differences is more common (Schmitt & Kuljanin, 2008).

For invariance measurement, two possible scenarios are considered depending on the previous CFA assessed: (1) the baseline model is identically specified across groups, or (2) the baseline model is differentially specified across groups. In this paper, the first scenario is used after assessing the CFA for each group separately.

Previous literature has assessed the multi-group CFA (MGCFA) for testing measurement invariance in structural equation modelling (Sass, 2011; Vandenberg & Lance, 2000) and recently within the tourism literature (Chi, 2011; Hsu et al., 2020; Kyle et al., 2005; Poudel et al., 2016). These studies focus on comparing gender, cultural, linguistic or other demographic characteristics across groups (Milfont & Fischer, 2010).

The literature on MGCFA comparing different stakeholders of sustainable tourism is limited (Kyle et al., 2005; Poudel et al., 2016). The first of these papers examines the perspective of tourists and residents regarding sustainable tourism in two protected areas, Annapurna Conservation Area and Chitwan National Park, assessing several MGCFAs with tourism impact subscales as latent factors. Residents and tourists perceived positive and negative sustainable impacts differently across the areas, including environmental, economic and social impacts. In addition, Hsu et al. (2020) tested for measurement invariance of the SUS-TAS among residents of three Taiwan archipelagos. The study follows the predictive validity methodology from previous studies (Ribeiro et al., 2018; Zhang et al., 2015) but the second-order factor model showed worse measurement model fit and the first-order model was used for the MGCFA. The study tested for measurement invariance based on the goodness-of-fit result comparison; in addition, configural, metric, scalar and factor covariance invariances are confirmed among the Pengu, Kinmen and Matsu archipelagos. For this paper, the suggested approach combined the MGCFA of the SUS-TAS, considering both types of stakeholder, residents and visitors.

Methods

Study settings

Data were collected from residents and visitors in the Spanish capital, Madrid. In terms of tourism, the purpose was to measure their attitudes towards sustainable tourism using SUS-TAS.

Madrid is the capital and most populous city of Spain, with 3.26 million residents and 1.4% annual growth (INE, 2019). During 2018, Madrid recorded 10.2 million visitors, with 21.4 million overnight stays (Madrid Destino, 2019). From the total tourists received, approximately 5.4 million were international tourists (53.3%) and accounted for 13 million overnight stays (60.9% of the total). In particular, the USA (12.5%), Italy (6.9%), France (6.8%), the UK (6.8%) and Germany (4.9%) were the top five countries with most visitors to Madrid City in 2018.

Madrid's connectivity options are mainly the airport, the train and the bus network. Adolfo Suárez Madrid-Barajas Airport is considered one of the most important connectivity hubs in Europe, with 57.9 million travellers in 2018, connecting 74 countries and 218 destinations, becoming the European airport with most connections with Latin America (Madrid Destino, 2019). The number of international passengers increased by 8.8% in 2018, reaching nearly 42 million, the fifth most important airport in the EU ranking.

In terms of accommodation infrastructure, Madrid has evolved in recent years. The city has 793 hotels registered with 43,816 rooms available. From 2018, Madrid city hall started counting rooms on accommodation-sharing platforms, including residents that rent their own apartment for at least 90 days and professionals with an activity licence. The estimated number of apartments offered on accommodation-sharing platforms in Madrid is 20,217 with an average price of €127.80 per night (Madrid Destino, 2019). For this study, data were collected from residents and visitors in Madrid, a city where the number of apartments offered (20,017) in accommodation-sharing platforms is already half of the rooms available in city hotels (43,816). In addition, Madrid is one of the cities with highest Airbnb growth rate across Europe's major cities, with an increase of 67% in 2017 and nearly 2.2 million overnight stays in Airbnb accommodations. For this reason, the study focused on the last five years to ensure the high use of these platforms (Vázquez et al., 2018).

The authors used the 21-item shorter version of SUS-TAS from Ribeiro et al. (2018), collecting two different sets of samples: 304 residents and 300 visitors to the city of Madrid, the capital of Spain as a destination. The questionnaire included three sections: (1) demographic information, (2) questions related to the level of use and impact of accommodation-sharing platforms, and (3) the 21-item SUS-TAS.

Instrument

For this study, the seven-factor 21-item shorter version of SUS-TAS from Ribeiro et al. (2018) was used. In order to measure the multi-group invariance, two samples were required and two survey instruments were designed, one for residents and one for visitors. Participants were asked to state their level of agreement with 21 questions using a five-point Likert scale for scoring (1 = strongly disagree and 5 = strongly agree). The residents' survey was translated into Spanish by two Spanish professors who are bilingual in Spanish and English and then back into English by the researchers. The translated version was checked for semantic matching, discrepancy correction, and accuracy by two research assistants. For the visitors' survey, the 21-items from Ribeiro et al. (2018) were used with only minor wording modifications. Adjustments were made to ensure that the final versions were error-free and both surveys were tested with 10 Madrid residents and 10 visitors to Madrid before data collection. The survey instrument included three sections: (1) demographic information including gender, age, length of residence for residents and time and purpose of visit for visitors, education level, occupation and tourism-related jobs, (2) three items related to accommodation-sharing platforms measuring frequency of use, level of impact in their community (housing price, resident displacement, number of tourists) and if they were a host on accommodation-sharing platforms, and (3) the 21-item SUS-TAS.

The questions related to Accommodation-sharing platforms were assessed to understand the profile towards ASP, to understand if the frequency of use and the impact in their community is latent.

Data collection and sample profile

The data were collected from a stratified random sample (Singh & Mangat, 1996) of 304 residents and 300 visitors after discards. The questionnaires were distributed with no nationality restriction to individuals who were at least 18 years old. For the residents of Madrid, no special requirements were established apart from the minimum age and place of residence. For the visitors of Madrid, the only requirement was to have visited Madrid in the last 5 years to ensure that they have experienced the growth of accommodation-sharing platforms as explained in the previous study setting.

The online questionnaires, designed with Google Forms, were administered to residents and visitors of Madrid over an eight-week period during January and February 2019, using three distribution channels: (1) through e-mail, (2) embedded links in social media (neighbourhood newspapers and community groups) and (3) approaching respondents at random streets, public parks or international conventions with the collaboration of two bilingual assistants. The number of respondents for residents (n = 304) and visitors (n = 300) complied with the recommended minimum sample size of n = 200 per group for measurement invariance analysis (Meade & Lautenschlager, 2004).

In total, 320 residents and 318 visitors were obtained, discarding 16 of the residents' answers and 18 of the visitors' answers due to missing values.

The descriptive statistics of sociodemographic profiles for each sample are illustrated in Table 1. Both samples have more female participants, with 63% in the residents' sample and 58.3% in the visitors' sample. The majority of the respondents in both samples are aged between 18 and 29, 34.2% and 46.7% respectively. Nationality differs between residents and visitors, with a diverse sample among the visitors, including 48.3% national tourists and a high representation from France (14.7%), the UK (6.3%), Germany (5.3%) and Italy (5.3%). Moreover, the visitor sample proportion of international respondents (51.7%) was considered appropriate for the study as it was similar to the proportion of international visitors to Madrid (53.3%) in 2018 (Madrid Destino, 2019).

Regarding education, 81.3% of the residents and 87.7% of the visitors have studied for at least a Bachelor's or a University degree.

In relation to accommodation-sharing platforms, a five-point Likert scale is used to measure the frequency of use and the level of impact accommodation-sharing platforms has on their community (housing price, displacement of residents, number of tourists). Regarding frequency of use, 65.1% of residents and 59.4% of visitors have used the platforms (see Table 2). The level of impact of Accommodation-sharing platforms is perceived as important at least (scale from 3 to 5) for 72.1% of the residents and 66% of the visitors, showing a certain difference between samples. Data collection did not target accommodation-sharing platform hosts specifically and the number of hosts that answered the questionnaire was 7.6% of residents and 9.3% of visitors.

Data analysis

To validate the use of the 21-item SUS-TAS from Ribeiro et al. (2018) for visitors (n = 300) and the Spanish version for residents (n = 304), two independent Confirmatory Factor Analyses (CFA) were performed using the maximum likelihood estimation (Jöreskog & Sörbom, 1996) in AMOS 25 software (Byrne, 2004) for both samples. After testing the measurement model for both samples with acceptable model-fit, convergent validity, composite reliability, and discriminant validity, they were cross-validated separately to find each baseline model to measure multi-group invariance. In addition, the results of model-fit, convergent validity, composite validity and discriminant validity are reported in section *4.1.1 Overall Fit measurement model*.

In this paper, no second-factor model was tested as the main focus was the cross-validation between residents and visitors and the descriptive analysis of the impact of accommodation-sharing platforms.

Table 1. Sociodemographic profiles of resident and visitor respondents.

	Residents (n = 304)		Visitors (n = 300)	
Demographics	Frequency (n)	Percentage (%)	Frequency (n)	Percentage (%)
Gender				
Male	110	36.2	124	41.3
Female	191	62.8	175	58.3
Prefer not to say	3		1	0.3
Age				
18-29	104	34.2	140	46.7
30-39	51	16.8	32	10.7
40-49	51	16.8	44	14.7
50-59	82	27.0	54	18.0
>=60	16	5.3	30	10.0
Nationality				
Spain	276	90.8	145	48.3
France	-*	–	44	14.7
UK	–	–	19	6.3
Germany	–	–	16	5.3
Italy	–	–	16	5.3
Others	16	9.2	60	20.0
Education				
Elementary School	0	0	7	2.3
High School	47	15.5	26	8.7
Bachelor/University Degree	148	48.7	125	41.7
Master or PhD	99	32.6	138	46.0
Others	10	3.3	4	1.3
Occupation				
Self-employed person	40	13.2	37	12.3
Employed person	177	58.2	135	45.0
Student	52	17.1	88	29.3
Domestic duties	1	0.3	6	2.0
Unemployed	22	7.2	18	6.0
Retired	12	3.9	16	5.3
Job				
Not tourism-related	282	92.8	255	85.0
Tourism-related	22	7.2	45	15.0

	Residents (n = 304)		Visitors (n = 300)		
Demographics	Frequency (n)	Percentage (%)		Frequency (n)	Percentage (%)
Time of residence (years)			Times visited Madrid		
Less than 1 year	11	3.6	1-3 times	135	45.0
1 to 5	23	7.6	4-5 times	45	15.0
6 to 10 years	10	3.3	6-10 times	40	13.3
11 to 15 years	11	3.6	>10 times	80	26.7
> 15 years	249	81.9			

*Not specified.

Regarding the multi-group invariance measurement among residents and visitors, a nested model comparison is used based on analysis of covariance (ANCOVA) structures (Byrne, 2004) and MGCFA analyses are performed (Milfont & Fischer, 2010) using AMOS 25 software. In testing for invariance, equivalencies across groups are measured by constraining the model across both groups and testing for configural, metric and scalar invariance. The theoretical model is compared with observed structures, placing parameter constraints on specific parameters to check equivalence. In these cases, data for both groups need to be analysed simultaneously (Byrne, 2004). The nested models are configured by comparing each new model in terms of model-fit with the previous one, following a hierarchical ordering, thus reducing the number of constraint parameters which increase the number of degrees of freedom (Byrne, 2004; Milfont & Fischer, 2010). If the full invariance is not met, a partial invariance could be performed.

Table 2. Accommodation-sharing platform segmentation.

	Residents (n = 304)		Visitors (n = 300)	
	Frequency (n)	Percentage (%)	Frequency (n)	Percentage (%)
Frequency of use				
Never	106	34.9	122	40.7
Rarely	57	18.8	56	18.7
Sometimes	76	25	61	20.3
Often	49	16.1	39	13
Always	16	5.3	22	7.3
Level of impact				
Unimportant	46	15.1	59	19.7
Slightly important	40	12.8	43	14.3
Important	78	25.7	78	26
Moderately important	86	27.6	86	28.7
Very important	58	18.8	34	11.3
Host				
Yes	23	7.6	28	9.3
No	281	92.4	272	90.7

Results

Validity of SUS-TAS for residents and visitors: CFA

Overall fit measurement model

As a prerequisite to testing invariance, a baseline model needs to be estimated for each group separately. Firstly, the CFA provides a test of validity and reliability for the Spanish version of the SUS-TAS and the visitor version. For this purpose, the overall fit for the measurement model was assessed using several goodness-of-fit indicators for a first-order factor model with seven correlated constructs for both samples.

To measure overall fit of CFA, the chi-square statistic (χ^2) is used (Brown, 2015). Previous studies have demonstrated the sensitivity of this goodness-of-fit index to sample size (Milfont & Fischer, 2010; Vandenberg & Lance, 2000). To solve this problem, additional indices, which are less affected by the sample size (Hu & Bentler, 1999), are considered: Comparative Fit Index (CFI), Tucker-Lewis index (TLI), root mean square error of approximation (RMSEA) and the standardized root mean square residual (SRMR) (Hair et al., 2019).

A ratio of 1 to 3 is recommended for the χ^2/df and the recommended cut-off values for an excellent model fit are TLI > 0.95, CFI > 0.95, RMSEA < 0.08 and SRMR < 0.08 (Hair et al., 2019; Hu & Bentler, 1999; Vandenberg & Lance, 2000).

The results of the goodness-of-fit for the correlated first-order seven-factor model for visitors and residents indicated an acceptable fit index to the data collected. Respectively, 21-item SUS-TAS $_{residents}$ (χ^2/df = 1.865, CFI = 0.940, TLI = 0.925, SRMR = 0.058 and RMSEA = 0.053) and 21-item SUS-TAS $_{visitors}$ (χ^2/df = 1.174, CFI = 0.983, TLI = 0.979, SRMR = 0.046 and RMSEA = 0.053) showed excellent goodness-of-fit results for all cut-off criteria in the visitors version and excellent goodness-of-fit for the RMSEA and SRMR but only acceptable goodness-of-fit for CFI and TLI indices reaching identification standards (Hu & Bentler, 1999).

In comparison to the 21-item SUS-TAS first-order model from Ribeiro et al. (2018), the overall fit of both samples (see Table 3) showed worse, but still acceptable, statistical fit for our research objective. Regarding measurement model fit, a baseline model is identified for measuring invariance.

Psychometric properties of SUS-TAS for resident and visitor versions

The next step is to analyse the psychometric properties of the SUS-TAS with both samples, assessing construct validity, including content, convergent and discriminant validity and reliability, as measured in previous studies (Ribeiro et al., 2018). Previous studies (Ribeiro et al., 2018; Yu et al., 2009; Zhang et al., 2015) had already examined content validity. For this paper, to ensure

Table 3. Model fit comparison of residents and visitors separately.

First-order model	χ^2	df	χ^2/df	CFI	TLI	RMSEA	SRMR
Residents 21-item SUS-TAS (Spanish version)	313.33	168	2.191	0.91	0.89	0.062	0.058
Visitors 21-item SUS-TAS	196.08	168	1.167	0.98	0.98	0.024	0.046
Ribeiro et al. (2018) 21-item SUS-TAS	216.23	168	1.287	0.99	0.99	0.026	0.025
Cut-off criteria (Acceptable)	–	–	>3	>0.9	>0.9	<0.08	<0.08
Cut-off criteria (Excellent)	–	–	>1	>0.95	>0.95	<0.06	<0.08

no bias due to the change of language and different context, content validity was assessed by the researchers, translating the residents' version into Spanish, which a native English speaker back-translated to ensure the correct definition of the items. Next, two tourism experts, random residents and visitors evaluated both samples to check that the scale items were sufficiently understandable (Tull & Hawkins, 1993).

Convergent validity and reliability was assessed to ensure the correlation between factors and the factor loadings of the model. The cut-off value for the loadings should be equal to or higher than 0.5 and statistically significant (Fornell & Larcker, 1981) and Table 4 shows that the factor loadings are 0.58 to 0.97 for residents and 0.57 to 0.89 for visitors, both statistically significant at two-tailed level $p < 0.001$. Convergent validity is also validated (Hair et al., 2019) calculating construct reliability (CR) and average variance extracted (AVE). To verify construct validity, AVE for all factors needs to be greater than the cut-off criterion of 0.5 (Hair et al., 2019; Fornell & Larcker, 1981) which is validated for both samples as shown in Table 4, with AVE values ranging from 0.52 to 0.67 for residents and 0.50 to 0.57 for visitors (Fornell & Larcker, 1981). Moreover, the CR should be higher than the 0.7 cut-off criterion (Hair et al., 2019) which is corroborated for both cases, ranging from 0.76 to 0.86 and 0.75 to 0.8 respectively. Both criteria are met, which reflects a strong convergent validity and reliability.

Discriminant validity examines the uniqueness of a factor in the model compared with the other factors (Hair et al., 2019). This validity was tested by comparing the square root of average variances extracted (AVE) with intercorrelations between two factors (Fornell & Larcker, 1981). As shown in Table 5, the results confirm that the discriminant validity was established for both samples, as the square of the AVE for each factor is higher than each intercorrelation (Fornell & Larcker, 1981).

The study does not test for a second-order model (Ribeiro et al., 2018; Zhang et al., 2015), which focuses on a hierarchical CFA, and no predictive validity is measured.

Construct validity for both 21-item SUS-TAS is demonstrated, as content validity, convergent validity, reliability and discriminant validity are confirmed.

Measuring invariance across residents and visitors: MGCFA

Following the sequences of steps measuring invariances of a nested model comparison (Vandenberg & Lance, 2000; Schmitt & Kuljanin, 2008; Byrne, 2004), the first step is to test for a goodness-of-fit model measurement for both groups separately, which was validated previously. Previous studies (Cheung & Rensvold, 2002; Schmitt & Kuljanin, 2008) explained the inconsistency of comparing only chi-square tests of significance when measuring invariance. To resolve this, Cheung and Rensvold (2002) compared several goodness-of-fit indices and determined that the ΔCFI was the best index of change in fit, with a minimum fit change of .01 for it to be considered significant. For this study, two types of incremental goodness-of-fit statistics are used to assess measurement invariance: chi-square difference ($\Delta\chi^2$) statistics and changes in model fit indices (ΔCFI, ΔTLI, ΔRMSEA, and ΔSRMR). In addition, modification indices are checked to detect the constraints causing significant reduction of $\Delta\chi^2$ value. The nested models are considered invariant if the incremental values are ΔCFI<.010, ΔTLI<.020, ΔRMSEA<.015, and ΔSRMR<.030 (Chen, 2007; Cheung & Rensvold, 2002).

Table 4. CFA summaries of the residents' and visitors' version of SUS-TAS.

	Residents			Visitors		
	Factor loading	CR	AVE	Factor loading	CR	AVE
Perceived environmental sustainability (PES)		0.86	0.67		0.78	0.55
PES1. Our environment must be protected now and for the future	0.77			0.73		
PES 2. Tourism must protect the community environment	0.92			0.87		
PES 3. Tourism must improve the environment for future generations	0.77			0.60		
Perceived economic benefit (PEB)		0.82	0.61		0.76	0.51
PEB4. Tourism is a strong economic contributor to the community	0.76			0.70		
PEB5. Tourism diversifies our economy	0.65			0.76		
PEB6. Tourism benefits other industries in the community	0.91			0.69		
Perceived social costs (PSC)		0.84	0.64		0.80	0.57
PSC7. My quality of life has deteriorated because of tourism	0.97			0.81		
PSC8. Tourists in our community disrupt my quality of life	0.65			0.72		
PSC9. I feel uncomfortable or unwelcome in local tourism businesses	0.75			0.73		
Long-term planning (LP)		0.80	0.58		0.76	0.52
LP10. We need to take a long-term view when planning for tourism development	0.76			0.79		
LP11. Successful management of tourism requires advanced planning strategy	0.88			0.78		
LP12. When we plan for tourism, we cannot be shortsighted	0.61			0.57		
Community-centered economy (CE)		0.76	0.52		0.75	0.50
CE13. Communities' residents should receive a fair share of benefits from tourism	0.64			0.67		
CE14. Communities' residents should be given more opportunities to invest in tourism development	0.89			0.77		
CE15. Tourism industry must contribute to community improvement	0.58			0.68		
Ensuring visitor satisfaction (EVS)		0.76	0.52		0.75	0.50
EVS16. Tourism industry must ensure good quality tourism experiences for visitors	0.74			0.78		
EVS17. It is the responsibility of tourism businesses to meet visitor needs	0.72			0.68		
EVS18. Community attractiveness is a core element of ecological "appeal" for visitors	0.69			0.65		
Maximizing community participation (MCP)		0.77	0.53		0.78	0.55
MCP19. Full participation by everyone in the community is a must for successful tourism development	0.66			0.57		
MCP20. Communities' residents should have an opportunity to be involved in tourism decision-making	0.84			0.73		
MCP21. Tourism decisions must be made by all in communities regardless of a person's background	0.68			0.89		

**Significant at the 0.1 level.

Partial measurement invariance is assessed when full measurement invariance is not complied with (Byrne, 2004; Milfont & Fischer, 2010; Schmitt & Kuljanin, 2008; Vandenberg & Lance, 2000). The process consists in constraining only a subset of the parameters which are invariant across groups and the other subset of factors change across them. For this study, partial measurement invariance is considered when some but not all the parameters are invariant across groups (Vandenberg & Lance, 2000). To test further partial invariance models, configural and at least partial metric invariance need to be validated.

Table 5. Discriminant validity matrix.

Residents Factors	PES	PEB	PSC	LP	CE	EVS	MCP
PES	**0.821**[a]						
PEB	0.232**	**0.781**					
PSC	0.024	−0.181**	**0.802**				
LP	0.371**	0.268**	−0.021	**0.76**			
CE	0.252**	0.197**	−0.017	0.159	**0.719**		
EVS	0.229**	0.211**	−0.025	0.205**	0.224**	**0.72**	
MCP	0.288**	0.06	0.130	0.344**	0.468**	0.355**	**0.73**
Visitors Factors	**PES**	**PEB**	**PSC**	**LP**	**CE**	**EVS**	**MCP**
PES	**0.744**[a]						
PEB	0.141	**0.717**					
PSC	−0.207**	−0.230**	**0.753**				
LP	−0.033	−0.093	0.037	**0.718**			
CE	−0.072	0.161	−0.052	0.12	**0.71**		
EVS	−0.072	0.051	−0.087	0.114	0.055	**0.708**	
MCP	0.057	−0.025	0.124	0.128	−0.029	−0.097	**0.742**

[a]The bold diagonal elements are the square roots of each AVE; factor correlations are shown off-diagonal.
**$p<.01$ (two-tailed).

The second step after separated CFA is to combine both groups, changing to a MGCFA in order to cross-validate the hypothesized model (Model 1). This step corresponds to the measurement of *configural invariance*; with the purpose of confirming that the same number of factor loadings and the structure are invariant for residents and visitors. Model 1 showed excellent goodness-of-fit ($\chi2 = 509.411$ df $= 336$ $\chi2/df = 1.516$, CFI $= 0.958$, TLI $= 0.947$, SRMR $= 0.0584$ and RMSEA $= 0.029$) confirming configural invariance and establishing the baseline model for testing different invariances by adding parameter constraints (Milfont & Fischer, 2010). The chi-square value ($\chi2 = 509.411$) of the configural model is the sum of the chi-square of the residents ($\chi2 = 313.33$) and the visitors ($\chi2 = 196.08$) separately. Each model invariance is measured by comparing several goodness-of-fit indices (Cheung & Rensvold, 2002; Vandenberg & Lance, 2000) with the previous less restrictive model.

Metric invariance (Model 2) is measured by constraining the factor loadings tested to equality so that they are invariant across the groups. Model 2 goodness-of-fit indices are compared with Model 1 and although the cut-off criteria of the goodness-of-fit indices are met $\Delta CFI_{M1\text{-}M2} = -.008$, $\Delta TLI_{M1\text{-}M2} = -.007$, $\Delta RMSEA_{M1\text{-}M2} = .002$, and $\Delta SRMR_{M1\text{-}M2} = -.001$, the p-value from the chi-square (χ^2) test showed significance ($p < 0.01$) and a partial metric invariance was assessed. To detect which constraints were causing the model fit, modification indices were used and the procedure from Byrne (2004) was followed. CE3 and MCP2 were freely estimated across the groups and Model 4 was tested, comparing it with Model 1, and obtaining partial invariance across both groups with excellent goodness-of-fit index comparison: $\Delta CFI_{M3\text{-}M1} = -.002$, $\Delta TLI_{M3\text{-}M1} = .000$, $\Delta RMSEA_{M3\text{-}M1} = .000$, and $\Delta SRMR_{M3\text{-}M1} = -.001$ with no p-value significance (0.114).

Scalar invariance measures whether the observed variable scores are related to the latent construct scores. Model 4 tested full scalar invariance by constraining the intercepts of items to be equivalent across groups and compared it with Model 3. The results showed inadequate goodness-of-fit index comparison: $\Delta CFI_{M4\text{-}M3} = -.021$, $\Delta TLI_{M4\text{-}M3} = -.021$, $\Delta RMSEA_{M4\text{-}M3} = .007$, and $\Delta SRMR_{M4\text{-}M3} = .000$ with p-value significance ($p < 0.01$). Therefore, Model 4 varies across groups and full invariance is not established.

Partial scalar invariance is tested with the same criteria as for Model 4, freely estimating seven of the twenty-one intercepts that cause significance ($p < 0.01$) and the reduction of goodness-of-fit: PEB2, PSC1, LP1, CE1, CE2, MCP1, MCP3. Model 5 showed adequate goodness-of-fit index comparison: $\Delta CFI_{M5\text{-}M3} = -.001$, $\Delta TLI_{M5\text{-}M3} = 0.001$, $\Delta RMSEA_{M5\text{-}M3} = .000$, and

Table 6. Measurement invariance across residents and visitors.

Model description	Comparative model	χ^2(df)	χ^2/df	$\Delta\ \chi^2$ (Δdf)	*p*-Value	CFI	TLI	RMSEA	SRMR
Configural (Model 1)	–	509.411 (336)	1.516	–	–	0.958	0.947	0.029	0.0584
Metric invariance (Model 2)	M2-M1	554.783 (350)	1.585	45.372 (14)	0.000*	0.950	0.940	0.031	0.0592
Partial metric invariance (Model 3)[a]	M3-M1	527.463 (348)	1.516	18.052 (12)	0.114	0.956	0.947	0.029	0.059
Scalar invariance (Model 4)	M4-M3	637.345 (369)	1.727	109.882 (21)	0.000*	0.935	0.926	0.036	0.0593
Partial scalar invariance (Model 5)[b]	M5-M3	547.99 (362)	1.514	20.527 (14)	0.114	0.955	0.948	0.029	0.059

[a]Factor loadings of CE3 and MCP2 are freely estimated across groups.
[b]Intercepts of items PEB2, PSC1, LP1, CE1, CE2, MCP1, MCP3 are freely estimated across groups.
$*p < .01$ (significant).

$\Delta SRMR_{M5-M3} = .000$ with p-value significance ($p < 0.01$). Therefore, Model 5, the less constrained model, established partial scalar invariance.

Table 6 summarizes the different model comparisons in terms of goodness-of-fit indices and confirms full construct invariance and partial metric and scalar invariance between groups.

Discussion

The aim of this study was to describe the validation and steps to measure invariance between residents' and visitors' attitudes toward sustainable tourism using the SUS-TAS, considering a context with high use of accommodation-sharing platforms. This paper contributes to previous literature on validating the SUS-TAS in different contexts, specifically, testing for measurement invariance among different groups. In addition, it validates the 21-item SUS-TAS for residents and for visitors separately with a CFA. This paper adopted Ribeiro et al. (2018) scale to preserve consistency with respect to the tool and ensure that the validation in a new context was possible (Choi & Sirakaya, 2005), testing the application of SUS-TAS for visitors. The results show that the 21-item SUS-TAS with a seven-factor model developed by Ribeiro et al. (2018) can be used across residents and visitors simultaneously with some freely estimated factors and item intercepts. Previous studies have validated the SUS-TAS in cross-cultural settings, comparing residents from different places (Hsu et al., 2020; Sirakaya-Turk, Ekinci et al., 2008), visitors to a certain place (Kvasova, 2011) or other stakeholders like hoteliers in Mauritius (Prayag et al., 2010). Few studies have tested measurement invariance in tourist settings to examine attitudes between residents from different places (Hsu et al., 2020) or residents and visitors from different places but testing measurement invariance separately. In line with previous studies, this study is the first attempt to measure invariance across residents and visitors using the 21-item first-order SUS-TAS to compare different stakeholders simultaneously.

The results from the separate CFA continue with the validation of the SUS-TAS in different contexts and, in this study, with the visitors' sample. The goodness-of fit model indicators were better for visitors than for residents but convergent validity was near the limit for the visitor sample as the AVE values were near the cut-off value of 0.5 (Fornell & Larcker, 1981). When comparing our results to Ribeiro et al. (2018), it must be pointed out that factor loadings for both samples were lower than in the earlier paper. The difference between these results and the previous ones can be attributable to the cultural differences of respondents' profiles. In addition, another explanation could be related to the fact that the previous study (Ribeiro et al., 2018) had the purpose of testing different versions of the SUS-TAS (44-item, 33-item, 27-item and 21-item), selecting the items with higher factor loadings to achieve maximum parsimony. In this study, the main objective was to measure invariance between residents and visitors and factor loading results were validated for both samples.

The 21 intercorrrelations among the seven factors vary between the two groups with 15 significantly correlated and six nonsignificant for the resident sample, and only two significantly correlated for the visitor sample. Among the nonsignificant intercorrelations in the resident version, five were intercorrelated with perceived social costs, indicating less intercorrelation between this factor and the other factors except with perceived economic benefit. Considering the visitor sample, only two intercorrelations were significant, showing less intercorrelation between factors in this model. Contrary to the resident sample, the significant intercorrelations were both with the perceived social costs factor, showing the differences between the models.

In the MGCFA, the results showed that the structure of the model can be used across both samples with full configural invariance. The results of the full metric invariance were close to acceptable but when checking for possible factors that decreased significantly, model-fit CE3 and MCP2 were identified and freely estimated, which showed excellent model-fit with the partial metric invariance. Regarding scalar invariance, seven of the 21 intercepts were freely estimated for a partial scalar invariance, showing more issues than the previous invariance. From the seven freely estimated intercepts, two were items from CE and two were items from MCP. It is important to highlight the fact that the main issues for full metric and scalar invariance were related to the community-centered economy and maximizing community participation factors, demonstrating that the perception of residents and visitors regarding these aspects differs significantly and needs to be considered in future planning. Therefore, the invariance results showed that residents and visitors have similar perceptions of environmental sustainability and ensuring visitor satisfaction with no factors, items or intercepts freely estimated.

The study findings show that the SUS-TAS can be used to measure residents' and visitors' attitudes toward sustainable tourism in a context with high use of accommodation-sharing platforms, considering differences in perceptions of the community-centered economy and maximizing community participation between residents and visitors.

The research findings contribute to previous studies on the use of the SUS-TAS as an appropriate scale for measuring residents' attitudes and specifically for measuring residents' and visitors' attitudes simultaneously.

Regarding the methodology, the paper explains the steps to test measuring invariance across groups with the SUS-TAS from the perspective of two stakeholders, following the previous example of Hsu et al., (2020) which only compared residents' attitudes across different settings. One of the most important theoretical contributions is that the shorter version of the 21-item SUS-TAS from Ribeiro et al. (2018) is cross-validated in a mature destination, showing that it can be used in different settings, including measuring visitors' attitudes. Moreover, a shorter version of the scale is more suitable for practical use (Hsu et al., 2020). In terms of sustainable tourist planning, the study helps to understand perspectives from both types of stakeholder and the aspects that need to be considered differently in a mature destination like Madrid with high use of accommodation sharing platform. Overall, this study would help to manage future disruptions in the sector and could be applicable in other settings like other European cities with similar characteristics.

Limitations and future research

There are some limitations that need to be considered when analysing the findings of this study. Firstly, the study only considered a first-order model not including the global interpretation of the SUS-TAS with a second-order model. In addition, no predictive validity in determining planning involvement intention was considered. Secondly, data collected in the visitor sample are mainly from European countries, but it is recommended that the model be tested with other cultural settings. Thirdly, as the main purpose was to measure invariance in a mature destination with high use of accommodation-sharing platforms, not many of the questions were related to this topic. Finally, although data were collected from a stratified random sample of residents and visitors, both profiles that answered the questionnaire were generally highly educated (81.3% of the residents and 87.7% of the visitors have studied for at least a Bachelor's or a University degree) which do not represent completely the two populations considered.

For further studies, the 21-item SUS-TAS could be used to measure invariance across residents and visitors in other contexts and cultures. Additionally, further studies should consider other stakeholders, such as hoteliers (Prayag et al., 2010) or consider hosts' perspectives (Gursoy & Rutherford, 2004) targeting them for data collection and measuring invariance across other stakeholders. In relation to the sample limitation, further studies could use a different random sample that represents a destination more completely to further substantiate our findings. Bringing a new disruption, COVID-19 has changed the ways people travel and has impacted tourism drastically; further studies should consider this new context for measuring invariance or the validity of SUS-TAS in the next few years.

Disclosure statement

No potential conflict of interest was reported by the authors.

ORCID

Andrés Chamarro http://orcid.org/0000-0002-5770-459X

References

Ap, J. (1992). Residents' perceptions on tourism impacts. *Annals of Tourism Research, 19*(4), 665–690. https://doi.org/10.1016/0160-7383(92)90060-3

Belk, R. (2014). You are what you can access: Sharing and collaborative consumption online. *Journal of Business Research, 67*(8), 1595–1600. https://doi.org/10.1016/j.jbusres.2013.10.001

Botsman, R., & Rogers, R. (2011). *What's mine is yours: The rise of collaborative consumption*. Collins.

Brown, T. A. (2015). *Confirmatory factor analysis for applied research*. The Guilford Press.

Byrne, B. B. (2004). Testing for multigroup invariance using AMOS graphics: A road less traveled. *Structural Equation Modeling: A Multidisciplinary Journal, 11*(2), 272–300. https://doi.org/10.1207/s15328007sem1102_8

Chen, F. F. (2007). Sensitivity of goodness of fit indexes to lack of measurement invariance. *Structural Equation Modeling: A Multidisciplinary Journal, 14*(3), 464–504. https://doi.org/10.1080/10705510701301834

Cheng, T.-M., & Wu, H. (2015). How do environmental knowledge, environmental sensitivity, and place attachment affect environmentally responsible behavior? An integrated approach for sustainable island tourism. *Journal of Sustainable Tourism, 23*(4), 557–576. https://doi.org/10.1080/09669582.2014.965177

Cherry, C., & Pidgeon, N. (2018). Is sharing the solution? Exploring public acceptability of the sharing economy. *Journal of Cleaner Production, 195*, 939–948. https://doi.org/10.1016/j.jclepro.2018.05.278

Cheung, G. W., & Rensvold, R. B. (2002). Evaluating goodness-of-fit indexes for testing measurement invariance. *Structural Equation Modeling: A Multidisciplinary Journal, 9*(2), 233–255. https://doi.org/10.1207/S15328007SEM0902_5

Chi, C. G.-Q. (2011). Destination loyalty formation and travelers' demographic characteristics: A multiple group analysis approach. *Journal of Hospitality & Tourism Research, 35*(2), 191–212. https://doi.org/10.1177/1096348010382233

Choi, H. C., & Murray, I. (2010). Resident attitudes toward sustainable community tourism. *Journal of Sustainable Tourism, 18*(4), 575–594. https://doi.org/10.1080/09669580903524852

Choi, H. C., & Sirakaya, E. (2006). Sustainability indicators for managing community tourism. *Tourism Management, 27*(6), 1274–1289. https://doi.org/10.1016/j.tourman.2005.05.018

Choi, H.-S C., & Sirakaya, E. (2005). Measuring residents' attitude toward sustainable tourism: Development of sustainable tourism attitude scale. *Journal of Travel Research, 43*(4), 380–394. https://doi.org/10.1177/0047287505274651

Churchill, G. A. (1979). A paradigm for developing better measures of marketing constructs. *Journal of Marketing Research, 16*(1), 64–73. https://doi.org/10.2307/3150876

Dogru, T., Mody, M., & Suess, C. (2019). Adding evidence to the debate: Quantifying Airbnb's disruptive impact on ten key hotel markets. *Tourism Management, 72*, 27–38. https://doi.org/10.1016/j.tourman.2018.11.008

Dogru, T., Mody, M., Suess, C., Line, N., & Bonn, M. (2020). Airbnb 2.0: Is it a sharing economy platform or a lodging corporation? *Tourism Management, 78*, 104049. https://doi.org/10.1016/j.tourman.2019.104049

Dunlap, R. E., Liere, K. D., Mertig, A. G., & Jones, R. E. (2000). New trends in measuring environmental attitudes: Measuring endorsement of the new ecological paradigm: A revised NEP scale. *Journal of Social Issues, 56*(3), 425–442. https://doi.org/10.1111/0022-4537.00176

Edbring, E. G., Lehner, M., & Mont, O. (2016). Exploring consumer attitudes to alternative models of consumption: Motivations and barriers. *Journal of Cleaner Production, 123*, 5–15. https://doi.org/10.1016/j.jclepro.2015.10.107

Finley, K. (2013). Trust in the sharing economy: An exploratory study. Center for Cultural Policy Studies, University of Warwick. Retrieved from https://www.heartland.org/_template-assets/documents/publications/ccps_a4_ma_gmc_kf_3.pdf

Fornell, C., & Larcker, D. F. (1981). Evaluating structural equation models with unobservable variables and measurement error. *Journal of Marketing Research, 18*(1), 39–50. https://doi.org/10.2307/3151312

Frenken, K., & Schor, J. (2017). Putting the sharing economy into perspective. *Environmental Innovation and Societal Transitions, 23*, 3–10. https://doi.org/10.1016/j.eist.2017.01.003

Gansky, L. (2012). *The mesh: Why the future of business is sharing*. Portfolio.

Garau-Vadell, J. B., Gutiérrez-Taño, D., & Díaz-Armas, R. (2018). Residents' support for P2P accommodation in mass tourism destinations. *Journal of Travel Research, 58*(4), 549–565. https://doi.org/10.1177/0047287518767067

Garcia-López, M.-À., Jofre-Monseny, J., Mazza, R. M., & Segú, M. (2020). Do short-term rental platforms affect housing markets? Evidence from Airbnb in Barcelona. *Journal of Urban Economics*. https://doi.org/10.1016/j.jue.2020.103278

Geissinger, A., Laurell, C., Öberg, C., & Sandström, C. (2019). How sustainable is the sharing economy? On the sustainability connotations of sharing economy platforms. *Journal of Cleaner Production, 206*, 419–429. https://doi.org/10.1016/j.jclepro.2018.09.196

Gursoy, D., & Rutherford, D. G. (2004). Host attitudes toward tourism: An improved structural model. *Annals of Tourism Research, 31*(3), 495–516. https://doi.org/10.1016/j.annals.2003.08.008

Guttentag, D. (2015). Airbnb: Disruptive innovation and the rise of an informal tourism accommodation sector. *Current Issues in Tourism, 18*(12), 1192–1217. https://doi.org/10.1080/13683500.2013.827159

Guttentag, D. A., & Smith, S. L. (2017). Assessing Airbnb as a disruptive innovation relative to hotels: Substitution and comparative performance expectations. *International Journal of Hospitality Management, 64*, 1–10. https://doi.org/10.1016/j.ijhm.2017.02.003

Guttentag, D., Smith, S., Potwarka, L., & Havitz, M. (2018). Why tourists choose Airbnb: A motivation-based segmentation study. *Journal of Travel Research, 57*(3), 342–359. https://doi.org/10.1177/0047287517696980

Hair, J. F., Black, W. C., Babin, B. J., & Anderson, R. E. (2019). *Multivariate data analysis*. Cengage Learning, EMEA.

Hamari, J., Sjöklint, M., & Ukkonen, A. (2016). The sharing economy: Why people participate in collaborative consumption. *Journal of the Association for Information Science and Technology, 67*(9), 2047–2059. https://doi.org/10.1002/asi.23552

Hsu, C.-Y., Chen, M.-Y., Nyaupane, G. P., & Lin, S.-H. (2020). Measuring sustainable tourism attitude scale (SUS-TAS) in an Eastern island context. *Tourism Management Perspectives, 33*, 100617. https://doi.org/10.1016/j.tmp.2019.100617

Hu, L.-T., & Bentler, P. M. (1999). Cutoff criteria for fit indexes in covariance structure analysis: Conventional criteria versus new alternatives. *Structural Equation Modeling: A Multidisciplinary Journal, 6*(1), 1–55. https://doi.org/10.1080/10705519909540118

Hung, K., Sirakaya-Turk, E., & Ingram, L. J. (2011). Testing the efficacy of an integrative model for community participation. *Journal of Travel Research, 50*(3), 276–288. https://doi.org/10.1177/0047287510362781

Huston, C. (2015, August 13). *As Airbnb grows, hotel prices expected to drop.* Retrieved from Marketwatch https://www.marketwatch.com/story/as-airbnb-grows-hotel-prices-expected-to-drop-2015-08-13

Ikkala, T., & Lampinen, A. (2015). *Monetizing network hospitality* [Paper presentation]. Proceedings of the 18th ACM Conference on Computer Supported Cooperative Work & Social Computing – CSCW, *15.* https://doi.org/10.1145/2675133.2675274

INE. (2019). *Population census data.* Madrid City.

Joo, D., Cho, H., & Woosnam, K. (2019). Exploring tourists' perceptions of tourims impacts. *Journal of Tourism Management Perspectives, 31*, 213–235.

Jöreskog, K. G., & Sörbom, D. (1996). *LISREL 8: User's Reference Guide.* Chicago: Scientific Software International.

Kvasova, O. (2011). Socio-demographic determinants of eco-friendly tourist attitudes and behaviour. *Tourism Today, 11*, 73–95.

Kyle, G., Graefe, A., & Manning, R. (2005). Testing the Dimensionality of Place Attachment in Recreational Settings. *Environment and Behavior, 37*(2), 153–177. https://doi.org/10.1177/0013916504269654

Madrid Destino. (2019). *Anuario de Turismo Madrid.* Madrid Destino.

Martin, C. J. (2016). The sharing economy: A pathway to sustainability or a nightmarish form of neoliberal capitalism? *Ecological Economics, 121*, 149–159. https://doi.org/10.1016/j.ecolecon.2015.11.027

Meade, A. W., & Lautenschlager, G. J. (2004). A Monte-Carlo study of confirmatory factor analytic tests of measurement equivalence/invariance. *Structural Equation Modeling: A Multidisciplinary Journal, 11*(1), 60–72. https://doi.org/10.1207/S15328007SEM1101_5

Milfont, T. L., & Fischer, R. (2010). Testing measurement invariance acrosss groups: Applications in cross-cultural research. *Journal of Psychological Research, 3*(1), 111–121.

Mody, M., Suess, C., & Dogru, T. (2019). Not in my backyard? Is the anti-Airbnb discourse truly warranted? *Annals of Tourism Research, 74*, 198–203. https://doi.org/10.1016/j.annals.2018.05.004

Mody, M., Woosnam, K. M., Suess, C., & Dogru, T. (2020). Hapless victims or empowered citizens? Understanding residents' attitudes towards Airbnb using Weber's Theory of Rationality and Foucauldian concepts. *Journal of Sustainable Tourism.* https://doi.org/10.1080/09669582.2020.1834567

Nunkoo, R., Smith, S. L., & Ramkissoon, H. (2013). Residents' attitudes to tourism: A longitudinal study of 140 articles from 1984 to 2010. *Journal of Sustainable Tourism, 21*(1), 5–25. https://doi.org/10.1080/09669582.2012.673621

Öberg, C. (2018). Social and economic ties in the freelance and sharing economies. *Journal of Small Business & Entrepreneurship, 30*(1), 77–96. https://doi.org/10.1080/08276331.2017.1388954

Oskam, J., & Boswijk, A. (2016). Airbnb: The future of networked hospitality businesses. *Journal of Tourism Futures, 2*(1), 22–42. https://doi.org/10.1108/JTF-11-2015-0048

Poudel, S., Nyaupane, G. P., & Budruk, M. (2016). Stakeholders' perspectives of sustainable tourism development. *Journal of Travel Research, 55*(4), 465–480. https://doi.org/10.1177/0047287514563166

Prayag, G., Dookhony-Ramphul, K., & Maryeven, M. (2010). Hotel development and tourism impacts in Mauritius: Hoteliers' perspectives on sustainable tourism. *Development Southern Africa, 27*(5), 697–712. https://doi.org/10.1080/0376835X.2010.522832

Ribeiro, M. A., Pinto, P., Silva, J. A., & Woosnam, K. M. (2017). Residents' attitudes and the adoption of pro-tourism behaviours: The case of developing island countries. *Tourism Management, 61*, 523–537. https://doi.org/10.1016/j.tourman.2017.03.004

Ribeiro, M. A., Pinto, P., Silva, J. A., & Woosnam, K. M. (2018). Examining the predictive validity of SUS-TAS with maximum parsimony in developing island countries. *Journal of Sustainable Tourism, 26*(3), 379–398. https://doi.org/10.1080/09669582.2017.1355918

Ribeiro, M. A., Valle, P. O., & Silva, J. A. (2013). Residents' attitudes towards tourism development in Cape Verde Islands. *Tourism Geographies, 15*(4), 654–679. https://doi.org/10.1080/14616688.2013.769022

Richardson, L. (2015). Performing the sharing economy. *Geoforum, 67*, 121–129. https://doi.org/10.1016/j.geoforum.2015.11.004

Sass, D. (2011). Testing measurement invariance and comparing latent factor means within a confirmatory factor analysis framework. *Journal of Psychoeducational Assessment, 29*(4), 347–363. https://doi.org/10.1177/0734282911406661

Schmitt, N., & Kuljanin, G. (2008). Measurement invariance: Review of practice and implications. *Human Resource Management Review, 18*(4), 210–222. https://doi.org/10.1016/j.hrmr.2008.03.003

Schor, J. B. (2017). Does the sharing economy increase inequality within the eighty percent?: Findings from a qualitative study of platform providers. *Cambridge Journal of Regions, Economy and Society, 10*(2), 263–279. https://doi.org/10.1093/cjres/rsw047

Singh, R., & Mangat, N. (1996). Stratified sampling. In R. Singh & N. Mangat (Eds.), *Elements of survey sampling* (pp. 102–144). Kluwer Academic Publishers.

Sirakaya-Turk, E. (2007). Concurrent validity of the sustainable tourism attitude scale. *Annals of Tourism Research, 34*(4), 1081–1084. https://doi.org/10.1016/j.annals.2007.05.014

Sirakaya-Turk, E., Ekinci, Y., & Kaya, A. G. (2008). An examination of the validity of SUS-TAS in cross-cultures. *Journal of Travel Research*, *46*(4), 414–421. https://doi.org/10.1177/0047287507308328

Sirakaya-Turk, E., & Gursoy, D. (2013). Predictive validity of SUSTAS. *Tourism Analysis*, *18*(5), 601–605. https://doi.org/10.3727/108354213X13782245307957

Sirakaya-Turk, E., Ingram, L., & Harrill, R. (2008). Resident typologies within the integrative paradigm of sustaincentric tourism development. *Tourism Analysis*, *13*(5), 531–544. https://doi.org/10.3727/108354208788160405

Stoddard, J. E., Pollard, C. E., & Evans, E. R. (2012). The triple bottom line: A framework for sustainable tourism development. *International Journal of Hospitality & Tourism Administration*, *13*(3), 233–258. https://doi.org/10.1080/15256480.2012.698173

Su, L., & Swanson, S. (2017). The effect of destination social responsibility on tourist environmentally responsible behavior: Compared analysis of first-time and repeat tourists. *Tourism Management*, *60*, 308–321. https://doi.org/10.1016/j.tourman.2016.12.011

Tull, D. S., & Hawkins, D. I. (1993). *Marketing research: Measurement and method: A text with cases*. Macmillan.

UNWTO. (2019). *International tourism highlights*. UNWTO. https://doi.org/10.18111/9789284421152

Vandenberg, R. J., & Lance, C. E. (2000). A review and synthesis of the measurement invariance literature: Suggestions, practices, and recommendations for organizational research. *Organizational Research Methods*, *3*(1), 4–70. https://doi.org/10.1177/109442810031002

Varma, A., Jukic, N., Pestek, A., Shultz, C., & Nestorov, S. (2016). Airbnb: Exciting innovation or passing fad? *Tourism Management Perspectives*, *20*, 228–237. https://doi.org/10.1016/j.tmp.2016.09.002

Vázquez, M., Hernando, L., Gutiérrez, G., Harrington, D., & Oskam, J. (2018). *Airbnb in Europe: Amsterdam, Barcelona, Berlin, London, Madrid, Paris*. https://www.colliers.com/-/media/files/emea/emea/research/hotels/airbnb_spain_2018_v7.pdf?la=en-gb

Woosnam, K. M., Draper, J., Jiang, J. (K.), Aleshinloye, K. D., & Erul, E. (2018). Applying self-perception theory to explain residents' attitudes about tourism development through travel histories. *Tourism Management*, *64*, 357–368. https://doi.org/10.1016/j.tourman.2017.09.015

Yeager, E. P., Boley, B. B., Woosnam, K. M., & Green, G. T. (2020). Modeling residents' attitudes toward short-term vacation rentals. *Journal of Travel Research*, *59*(6), 955–974. https://doi.org/10.1177/0047287519870255

Yu, C.-P., Chancellor, H. C., & Cole, S. T. (2009). Measuring residents' attitudes toward sustainable tourism: A reexamination of the sustainable tourism attitude scale. *Journal of Travel Research*, *50*(1), 57–63. https://doi.org/10.1177/0047287509353189

Zervas, G., Proserpio, D., & Byers, J. W. (2017). The rise of the sharing economy: Estimating the impact of Airbnb on the hotel industry. *Journal of Marketing Research*, *54*(5), 687–705. https://doi.org/10.1509/jmr.15.0204

Zhang, Y., Cole, S. T., & Chancellor, C. H. (2015). Facilitation of the SUS-TAS application with parsimony, predictive validity, and global interpretation examination. *Journal of Travel Research*, *54*(6), 744–757.

Social impacts of mega-events: a systematic narrative review and research agenda

Judith Mair, P. Monica Chien, Sarah Jane Kelly and Stephanie Derrington

ABSTRACT

Whilst the economic costs and benefits of mega-events to host cities and countries are well documented, the evaluation and nature of the social impacts of such events are less clear. Although these social impacts typically include education and skills, destination branding, social cohesion, environmental sustainability and sport development, there is currently a lack of consensus on their measurement and focus upon both benefits and costs. In an evolved mega-event landscape which must align with political, social and cultural priorities, sports governing bodies, industry partners, residents and governments are increasingly concerned with qualitative impacts. Accordingly, through a systematic review of the existing literature, this paper assesses the current state of knowledge of the social costs and benefits of mega-events, including how they are identified, defined and measured, highlighting a need for further research to fully understand the social impacts of mega-events. To our knowledge, this is the first review to examine mega-event social impact evaluation, and we therefore provide a useful evidence-based guide to mega-event stakeholders in determining cases for host bids, event planning and related legacy evaluation. Our review culminates in a research agenda, providing practical guidance for future mega-event assessment.

Introduction

Mega-events have been the subject of considerable research due to their lasting effects on host communities (e.g. Gursoy et al., 2017; Zhou & Ap, 2009). Mega-events are occasional, large-scale events that exist on an international scale (Magno & Dossena, 2020). They are major/mega by virtue of size, attendance, public interest, level of financial investment, and media coverage, and have a trickle-down effect to the host community in the form of new infrastructure, economic growth and urban renewal (Lorde et al., 2011). The massive scale of these events, such as the Olympic Games, Commonwealth Games, FIFA Soccer World Cups, and World Expos, mean that they have a significant effect on a broad range of stakeholders and provide opportunities for catalysing changes (O'Brien, 2006; Tournois, 2018). These impacts start long before the event commences, extending to the bidding and planning phases, the staging of the event, and continuing well after the event (Gibson et al., 2014; Minnaert, 2012). As noted, mega-events are not all related to sport; however, as this review will demonstrate, much of the literature to date has focused on sporting mega-events.

Mega-events bring a range of benefits and development to host destinations. The existing literature has largely focused on the economic impacts of mega-events, whereas social impacts and their measures remain elusive and highly contested due to their undetermined duration and relatively qualitative nature (Pappalepore & Duignan, 2016; Ritchie et al., 2020). Social impacts have been defined as 'any impacts that potentially have an impact on the quality of life for local residents' (Fredline et al., 2003, p. 6). They can be considered as individual-level (e.g. civil liberty, participation), community-level (e.g. gentrification), host city-level (e.g. pride), and national-level (destination image/accessibility) (Ritchie et al., 2020; Smith et al., 2019). While economic benefits are clearly quantifiable (e.g. direct and indirect economic outcomes, job creation, and inbound investment), social impacts are harder to discern and measure objectively (Ritchie et al., 2009). They include intangible elements such as increased civic pride, social capital, enhanced destination image, and the development of new business networks (Mair & Duffy, 2018). Despite these benefits, there are also costs associated with the hosting of mega-events, many of which are borne by local residents. These include increased traffic, noise, and congestion (Fredline & Faulkner, 2001), as well as potential disenfranchisement from the event planning process (Chien et al., 2012). Such social costs and benefits are often referred to the Triple Bottom Line of sustainability and are closely linked with the notion of event legacy (Mair & Duffy, 2018).

Although the terms 'impacts' and 'legacies' are often used interchangeably, legacy incorporates a broader range of impacts that remain post-event to benefit the host community (Kaplanidou et al., 2013; Shipway et al., 2020). One of the most cited definitions of mega-event legacy is provided by Preuss (2007), which states that 'irrespective of the time of production and space, legacy is all planned and unplanned, positive and negative, tangible and intangible structures created for and by a sport event that remain longer than the event itself' (p. 211). Legacy remains an elusive and highly contested concept due to its undetermined duration (Pappalepore & Duignan, 2016). Some researchers have also cautioned the consideration of legacy as solely positive (e.g. Leopkey & Parent, 2017). The International Olympic Committee (IOC) further emphasises that 'delivering legacy requires strong partnerships between city leaders, the Games organisers, regional and national authorities and local communities' (International Olympic Committee (IOC), 2012, p. 58).

Despite all the expected benefits, in recent years destinations have shown a decreasing interest in bidding for and hosting mega-events (e.g. both Budapest and Hamburg withdrew bids to host the 2024 Olympic Games due to resident opposition). Their concern and reluctance have been sparked by past event hosts' failure to strategically plan and capitalise on the potential benefits (e.g. Rio 2016), and a lack of understanding of legacies (O'Brien & Gardiner, 2006; Smith et al., 2019). In response to this changing landscape, the IOC has recently released its 'New Norm' plan, with a new philosophy that invites potential hosts to present projects that best fit their sporting, economic, social, and environmental planning needs, rather than trying to fit the local context to the Games (IOC, 2018). This approach clearly emphasises the social, sporting, and environmental aspects of hosting and staging mega-events, and it is likely that other mega-events will also consider the principles of sustainability and maximal dispersal of benefits underpinning this new plan. Other mega-events, such as the Commonwealth Games, have also incorporated sustainability guidelines into their bidding documentation (Commonwealth Games Federation, n.d.). Establishing cause and effect in social and cultural changes, however, has been challenging, especially when the effects are only measured cross-sectionally and independently from each other. It would therefore be of interest for any host destinations to have a better understanding of the categorisation, scope and evaluation of social impacts associated with mega-events, so as to support their leveraging and legacy assessment.

In considering how mega-events impact society, it is important to emphasise the evolving nature of impact assessment against a backdrop of macro-environmental, political and technological shifts that are rapidly occurring and presenting new risks to mega-event assessment. Examples of such change include the ever-present risks of natural disaster, terrorism and cyber-attacks, as well as the changes wrought by the sudden appearance of the COVID-19 pandemic. A review and critique of extant social impact research is timely and relevant in this rapidly evolving context of mega-event planning and evaluation.

To this end, we provide a review of the current literature on the conceptualisation and measurement of social impacts and specifically address the following research questions:

1. What is the state of the field in terms of mega-event social impact measurement and conceptualisation?
2. What research is needed to reveal reliable social impact assessment and ensure practical measurement outcomes for stakeholders?

It is not our intention to critique the identified social impacts, but, rather, we intend to assess and map out the measurement and conceptualisation of these impacts to date.

Methodology

This paper takes a systematic approach to reviewing the literature regarding social impacts of mega-events. As Buchanan and Bryman (2011) note, systematic reviews are intended to locate existing studies, and by analysing and synthesising them and evaluating their contributions, provide clarity in relation to the state of the field. Given that many of the studies in the events field are mixed method or qualitative, a narrative synthesis approach is taken in the paper, which allows both a systematic review and a synthesis of findings that relies primarily on the use of words and text to summarise and explain the findings of the synthesis (Popay et al., 2006). The narrative synthesis approach moves beyond a simple summary of the findings of various studies in a bid to generate new insights or knowledge (Mays et al., 2005).

The PRISMA (Preferred Reporting Items of Systematic Reviews and Meta-Analysis) model developed by Moher et al. (2009) was adopted to provide a framework for deciding which studies should be included in the review and which should be excluded. This enhances the rigour of the design and the reliability of the findings and has four stages – identification, screening, eligibility and inclusion. A total of 107 papers are included in the review. Relevant studies and papers were identified by conducting online searches for a number of identified social impacts together with the words and phrases 'Olympics', 'mega-event', 'World Cup', 'Commonwealth Games', 'legacy', and 'impact'. Searches were also conducted using words and phrases related to social impacts typically associated with mega-events, including social cohesion, pride, image, diversity, diplomacy, volunteering and networks. We searched and reviewed articles that were published between 2000 and 2020 and included journals that range from tourism and events, to hospitality, sport, leisure, marketing, business, and public policy. Databases searched included Science Direct (Elsevier), EBSCO Host (Hospitality and Tourism Complete), Scopus, Web of Science, ProQuest, Sage, Business Source Complete, and Google Scholar.

Initially, we located 34 articles from tourism journals that are ranked as A* publications on the Australian Business Deans Council (ABDC) Journal Quality List, namely, *Journal of Sustainable Tourism*, *Tourism Management*, *Journal of Travel Research*, and *Annals of Tourism Research*. From the retrieved articles, a cross-citation search was undertaken to locate other relevant articles, starting with the ABDC A-ranked journals that included *Event Management*, *Journal of Destination Marketing & Management*, and *Leisure Studies*, as well as cross-disciplinary works such as *Marketing Intelligence & Planning* and *Journal of Sport Management*. Sixty-six studies were identified at this point. The next stage, screening, allowed for removal of any duplicates or non-relevant publications. Six studies were removed as a result. Following the PRISMA guidelines, the selected articles were checked for eligibility and cross-checked by looking at their reference lists to identify any studies that might have been overlooked. An additional 13 studies were added at this stage, resulting in the final list of 107 studies which were included in the review. All the articles included in this review were published in journals with an impact factor of 1.0 and above. A database was created to manage the data, which included bibliographic details of the studies, methods used, study context, and key social impacts examined (see Appendix 1 for the full list of studies). This represents the conclusion of the PRISMA model. Open coding was used to analyse the contents of the database to seek similarities and differences between and across the studies included (Miles & Huberman, 1994).

During the coding and analysis process, some impact categories emerged that were clearly recognisable and had been discussed in the literature on many occasions (e.g. volunteering; civic pride; the showcase effect on destination branding). Other impact categories identified included those that had previously been subsumed under economic impacts, but on further investigation appear to have significant impacts on local residents, their quality of life, and can act as an agent for social change (e.g. business & government networks; sports infrastructure as community connector). Finally, some social impact categories emerged that have not been explicitly considered; yet, they appear to be of importance for communities and society (e.g. disaster preparedness; and improved accessibility to infrastructure and services).

Findings

Table 1 shows the key impact categories from the extant literature that emerged from the analysis of the articles in this review, along with their definitions and measures observed. The review shows that there are clearly established social impacts and that new impacts warrant academic study as society changes (e.g. disaster preparedness and accessibility). Using these eight social impact categories as a framework, and taking a narrative approach rather than simply providing statistical details of the dates, types and sources of the articles, this review interrogates the main methods used in prior studies to identify and evaluate the social impacts of mega-events, the conceptualisation and classification of social impacts, and in some cases the theoretical underpinnings used. Given the lack of consensus on social impacts of mega-events, this study will form a comprehensive basis to inform further empirical research on sustainability of mega-events, in addition to guiding decision-making in practice, in determining viability of such events. Each of the social impact categories identified are detailed in the following sections.

Direct impacts on residents

Volunteering, education and skills

Volunteers are defined as those who offer their labour, knowledge, skills, and experience at no wage cost to an organisation. Mega-events rely heavily on volunteers, known as episodic volunteers as they generally seek only short-term or one-off volunteering positions (Lockstone & Baum, 2009). The 2012 London Olympic Games, for example, involved over 70,000 volunteers (Holmes et al., 2015) and the 2016 Rio Olympics required over 50,000 (Playthegame.org, 2016). Whilst volunteering is often associated with increased wellbeing resulting from new friendships, deeper connections with the community, the acquisition of new skills, and the protection of mental and physical health, there are no generally accepted measures of volunteering legacies (Minnaert, 2012).

Mega-events can provide a catalyst to upskill the current labour force of the host community and to retain skilled workers who originally only came to the host community for the event (Kirby et al., 2018). Upskilling through volunteer training and experiences has been linked to enhanced educational standards within the broader population, improved employability, and the

Table 1. Review key impact categories.

Impact	Definition	Key measures used	Key studies
Direct impact on residents			
Volunteering, education and skills	**Volunteering** = those who offer their labour, knowledge, skills, and experience at no wage cost to an organisation or event. **Education and skills** = learning systems or upskilling opportunities developed as part of hosting the mega-event, that provide enhanced educational achievement and improved employability or job prospect of the population.	A range of self-report surveys and interviews are used to measure volunteer motivations & satisfaction. Some post-event long-term studies have been carried out, but are difficult to operationalise. Interviews and surveys have been conducted with event volunteers to understand their perceptions of event engendered training and upskilling. Autoethnography has also been adopted to gain in-depth understanding.	Nichols and Ralston (2011): Post-event (2002 Manchester Commonwealth) survey and individual in-depth interviews revealed a social inclusion legacy from volunteering at the event. Volunteering was found result in skill development and an increase in time spent volunteering after the event. Gornostaeva & McGurck (2013): A case study of the 2012 London Olympic Games revealed that skill enhancement and other social cohesion benefits of volunteering are driven by broad and integrated social inclusion policies. Unemployed residents were found to be the least likely demographic to benefit from Olympic volunteering. Fairley et al. (2016): A post-event (2000 Sydney Olympic Games) qualitative questionnaire and in-depth interviews revealed a positive relationship between volunteering at the Olympics and continuing or increasing time spent as a volunteer after the event, and the development of transferable skills.
Social cohesion, civic pride and social capital	**Social cohesion** = the way people feel within their community, their networks and connections, and their opportunities for community participation. **Civic pride** = the positive attitude that residents have towards their local area. **Social capital** = the expectation that arises within a community of regular, honest, and cooperative behaviour, based on commonly shared norms.	All three concepts are usually measured using self-report surveys, interviews, or focus groups of residents.	Kim et al. (2006): Pre- and post-event (2002 FIFA World Cup, South Korea) surveys of residents found a significant perceptual change amongst residents in that perceptions of the expected social benefits of hosting the event were higher than the perceived actual benefits. Revealed that residents perceived the event to have generated more societal and cultural benefits as compared to economic benefits. Karadakis and Kaplanidou

(*continued*)

Table 1. Continued.

Impact	Definition	Key measures used	Key studies
			(2012): A longitudinal study of the 2010 Winter Olympic Games using surveys. Both host and non-host city residents were found to place more importance on social, rather than economic, impacts of hosting the event. Revealed that non-host city residents perceived the event as having left greater legacies than host-city residents. Minnaert (2012): Semi-longitudinal comparative case studies based on post-event (1996 Atlanta Olympic Games, 1998 Nagano Winter Olympic Games, 2000 Sydney Olympic Games, 2002 Salt Lake City Winter Olympic Games, 2004 Athens Olympic Games, 2006 Turin Winter Olympic Games, & 2008 Beijing Olympic Games) interviews of host city residents and secondary data (academic research, reports from social and charitable organisations etc.). Revealed that socially excluded groups rarely benefit from their city hosting the Olympic Games, even where the city explicitly seeks to achieve social goals through hosting the event.
Inclusion and diversity	**Inclusion** = when a variety of people feel valued and respected, have access to opportunities and resources, and can contribute their perspectives and talents to improve their community. **Diversity** = (inter alia) people of different ages, cultural backgrounds, ethnicities, genders, religions, geographical locale, SES status, and disability.	Measured using proxies, such as perceived tolerance of diversity, and more recently through tracking and analysing social media	Kaplanidou et al. (2013): Pre- and post-event (2010 FIFA World Cup, South Africa) surveys of residents of host cities. The World Cup was perceived to have increased quality of life through providing social benefits such as reducing racial segregation. Gibson et al. (2014): A pre- and post-event (2010 FIFA World Cup, South Africa) quasi-experiment on residents of various host cities revealed that residents perceived tolerance of diversity to have decreased from before to

(*continued*)

Table 1. Continued.

Impact	Definition	Key measures used	Key studies
			after the event. Kirilenko and Stepchenkova (2017): Text mining and content analysis Twitter data before, during, and after event (2014 Sochi Winter Olympic Games). National sentiment in Russia was seen to increase over the duration of the Games. It was also revealed, however, that the Olympic Games seemed to act as a 'magnifying lens' for contested issues between Russia and other countries, including in relation to inclusion and diversity within the country.
Sport participation, infrastructure & health	**Sports social impacts** = enhanced participation and community health, amenities enhancement through sporting infrastructure investment, enhanced pathways to elite sport, and the promotion of integrity and trust in sport.	Quantitative cross-sectional and longitudinal surveys post-event which capture perceived community amenity attributable to sporting infrastructure, frequency of participation, re-engagement in grassroots sports and physical activity. The direct effect of mega-events on sports participation and access to community sporting infrastructure is often also measured using resident surveys.	Veal et al. (2012): Secondary analysis of national survey data (2000 Sydney Olympic Games, 2003 Rugby Union World Cup, & 2006 Melbourne Commonwealth Games). Did not reveal any significant increase in adult sport participation following any event. Revealed a possible positive link between sport mega-events and children's sport participation, however this could not be definitively established. Chen et al. (2018): Case studies of the 2000 Sydney Olympic Games and 2008 Beijing Olympic Games revealed that hosting the Olympic Games initially has a positive effect on elite sport in the host country (i.e. following the announcement of the host city) but that without distinct government strategic planning and investment, that effect will not be sustained beyond the time of the country's home Olympics. Found that a host country's Olympic performance will likely improve in the Olympic cycles before their home Games, peak at its home Games, and

(*continued*)

Table 1. Continued.

Impact	Definition	Key measures used	Key studies
			then decline in subsequent cycles. Cleland et al. (2020): A longitudinal study of the 2014 Glasgow Commonwealth Games revealed little to no correlation between the event and increased sport participation of host city adult residents.
Impacts on the destination ecosystem			
Business & government networks	**Networks** = inter-organisational, inter-city and inter-country relationships, through which business networks and commercial investment can be accessed. A sport events network typically includes entities from tourism, sports, events, government, media and private sectors.	Qualitative and quantitative survey methods aimed at determining the extent of relationships and frequency of their interaction post-event. Strength of ties across the network has rarely been examined, and usually only in relation to ties post event.	Werner et al. (2015): an exploratory study based on the 2011 Rugby World Cup, New Zealand revealed limited collaboration between regional tourism organisations ('RTOs'), likely as a result of a shared view that increased collaboration was unnecessary and only desirable if it helped an individual RTO to attain its own objectives.
Destination branding (showcase effect)	**Destination branding** = marketing activities that are aimed at promoting a destination, through image building and reputation enhancement, to both internal and external stakeholders.	Both qualitative and quantitative methods have been used to measure visitors' as well as residents' perceptions of the host destination's brand equity dimensions (e.g. image, reputation). Examples include surveys of residents, interviews of small business, and big data analysis of visitor sentiments.	Zhou and Ap (2009): Pre- and post-event (2002 FIFA World Cup, South Korea) surveys of residents and visitors. Perceptions of the destination brand were found to depend on an individual's lifestyle and culture in addition to prior perceptions about the costs and benefits of hosting the event. Liu et al. (2014): A post-event (2008 Beijing Olympic Games) survey of non-host city (Shanghai) residents revealed that non-host city residents perceived the Olympic Games to have improved Beijing's image however host city residents did not perceive this to be a legacy of the event.
Disaster preparedness	**Disaster preparedness** = measures taken to prevent, prepare for and reduce the effects of natural and man-made disasters.	Common methods used include case study, in-depth interviews and content analysis of archival data. Studies sought to understand awareness and perceptions of event related risks from the spectator or tourist's perspective have	Toohey and Taylor (2012): A case study based on post-event (2000 Sydney Olympic Games) revealed that strict security measures were generally accepted as the price required to be paid for the international prestige of hosting an Olympic Games. Revealed legacy of increased security

(continued)

Table 1. Continued.

Impact	Definition	Key measures used	Key studies
		employed self-report surveys.	measures in host city and country following event. Walters et al. (2017): A survey conducted during event (2016 Rio de Janeiro Olympic Games) revealed that level of 'fandom' and trust in security officials at event can alleviate security concerns of spectators. Fans with a strong attachment to the event are likely to have fewer security concerns than other fans.
Accessibility and accessible tourism	**Accessibility** = the delivery of universally designed products, services and environments that allows people to function independently and with equity and dignity. **Accessible tourism** = development and provision of accessible destination experiences that enable people with access requirements, including mobility, vision, hearing and cognitive dimensions of access.	Qualitative methods such as case study and Delphi group, and quantitative methods such as survey have been used to understand aspects of accessible experiences.	Kaplanidou (2012): A post-event (1996 Atlanta Olympic Games, 2000 Sydney Olympic Games, 2004 Athens Olympic Games, & 2008 Beijing Olympic Games) survey. Accessible tourism measured by perception of availability of accessible pathways. Pappas (2017): Pre- and post-event (2012 London Olympic Games) surveys of residents, based on social exchange theory, revealed that the Olympics were perceived to have had positive social benefits including the development of information services for visitors.

development of transferable skills which can be utilised in future work, volunteering endeavours, and other areas of life (e.g. Kirby et al., 2018; Nichols & Ralston, 2011; Werner et al., 2015). The Vancouver 2010 Winter Olympic Games, for example, provided carpentry training and work experience for disadvantaged young people, indigenous people, single mothers and immigrants (Vanwynsberghe et al., 2013). Mega-events also have the potential to facilitate knowledge and skill transfers between prior and future host cities, though existing research suggests that this does not always occur for reasons of, among other things, historical issues, city rivalries, and socio-cultural differences (e.g. Beesley & Chalip, 2014).

Existing research suggests that volunteering at a mega-event results in a positive social legacy and promotes continued volunteering after the conclusion of the event (e.g. Fairley et al., 2016; Nichols & Ralston, 2011). However, further research conversely suggests that intentions to continue volunteering after a mega-event may not necessarily endure (e.g. Ralston et al., 2005; Ritchie et al., 2020). Another reason for the impermanent nature of volunteering is that efforts and intentions are stifled by a lack of adequate government funding for future events and support of organising committees to build a volunteering legacy (e.g. Nichols et al., 2017). Additionally, the link between volunteering at a mega-event and ongoing volunteer efforts is difficult to examine. Nichols and Ralston (2011) studied volunteers from the 2002 Manchester

Commonwealth Games; however, only volunteers who remained registered with Manchester Event Volunteers were surveyed, reducing the number of potential participants and adding bias in that participants' perceptions of volunteering at the Games were influenced or overshadowed by subsequent volunteering endeavours.

Volunteers at mega-events are most likely to be older people (Dickson et al., 2016), students, individuals in short-term employment contracts, and those moving between part-time or unskilled jobs (Gornostaeva & McGurk, 2013). At the same time, because they are comparatively 'hard to reach' and difficult to influence, the unemployed are the least likely to benefit from mega-event volunteering legacies (Gornostaeva & McGurk, 2013). This may be compounded by an apparent preference for skilled workers and tradespeople as mega-event volunteers (Lenskyj, 2002), presumably because this lowers training costs.

The impact of mega-events on education and upskilling has previously been measured as part of community consultation projects and volunteering studies (e.g. Fairley et al., 2016; Nichols & Ralston, 2011). Such studies have involved asking participants to self-report on their volunteer experiences and the long-term benefits of volunteering. The relevant studies do not, however, include a baseline from which to compare the education and skills of mega-event workers following the event. Moreover, those post-event studies conducted after a lengthy period of time do not appear to adequately account for the effect of other external factors on previous volunteers' skills and prospects of employment including, for example, the changing state of the overall economy (e.g. Fairley et al., 2016).

Social cohesion, civic pride and social capital

The term 'social cohesion' is commonly used by governments and policymakers to refer to an intangible concept encapsulating the way people feel within their community, their networks and connections, and their opportunities for community participation (Duffy & Mair, 2017). High levels of social cohesion are associated with good perceptions of wellbeing and quality of life in addition to a well-functioning society (Kaplanidou et al., 2013). However, social cohesion is less common in academic events research and to date, there appears to be no single accepted or validated way to measure this important social impact. In the events context, notions that underpin social cohesion are more commonly assessed using the concepts of civic pride and social capital (Gibson et al., 2014). Social cohesion, therefore, is often measured through self-report resident surveys, using Likert scales and focusing on the related notions of social capital and/or civic pride (e.g. Al-Emadi et al., 2017; Duignan et al., 2019; Li et al., 2015). The term civic pride does not represent an exclusively well-defined and understood construct; however, civic pride is linked to shared beliefs about a place which result in a shared or cohesive city image (Wood, 2006). Relevantly for the purposes of this research, civic pride relates purely to the attitudes and feelings of local residents to their local area rather than to the opinions of tourists or visitors (Wood, 2006). A number of studies have revealed a positive relationship between mega-events and residents' levels of civic pride, not only in the host city itself, but also within the host country (e.g. Chi et al., 2018; Gibson et al., 2014; Karadakis & Kaplanidou, 2012; Liu et al., 2014).

Social capital is a multi-dimensional construct, referring to the social structures and the actions and interactions within these social structures in any given society (Coleman, 1988). A full review of social capital is beyond the scope of this paper; however, it has been described as the expectation that arises within a community of regular, honest, and cooperative behaviour, based on commonly shared norms (Gibson et al., 2014). The positive outcomes from high levels of social capital can be considered as the creation of a civil society; a strong sense of community; community cooperation; and community empowerment (Ooi, Laing & Mair, 2015).

Research has found that social capital increases following a mega-event due to increased perceptions of quality of life and wellbeing (e.g. Kaplanidou, 2012; Kaplanidou et al., 2013), in addition to reinforced community spirit, heightened national identity and enhanced social unity (e.g. Kim et al., 2015; Liu et al., 2014). Importantly, studies also indicate that mega-events can have a negative effect on social capital, particularly if the event creates congestion, noise, increased crime, or other anti-social behavioural and cultural conflicts (Chiam & Cheng, 2013; Kim & Petrick, 2005). The displacement of the homeless and other marginalised groups as a result of mega-events has also been cited as a negative social capital impact (Lenskyj, 2002; Minnaert, 2012). The existing research suggests that these negative effects are felt during the event, rather than afterwards. Research into the long-term effects of mega-events on residents' social capital is still limited. As highlighted by Kaplanidou (2012), such a lack of research could be due to the difficulties of measuring the effects of a single event in light of later social, political, and economic shifts. Further research on both the positive and negative effects of mega-events on social cohesion is badly needed.

Inclusion and diversity

Social inclusion is the process of improving the terms on which individuals and groups take part in society. It occurs when a variety of people (e.g. of different ages, cultural backgrounds, genders, religions, socioeconomic status, and disability) feel valued and respected, have access to opportunities and resources, and can contribute their perspectives and talents to improve their community (Laing & Mair, 2015). Thus, whilst diversity refers to differences within a group, inclusion speaks to how those members are treated and how they feel (Nichols & Ralston, 2011; Ziakas, 2015).

Diversity is traditionally examined through surveys and interviews using surrogate indicators (e.g. Al-Emadi et al., 2017; Chen & Tian, 2015; Zhou & Ap, 2009). Proxies include, among other things, perceived tolerance of diversity, the perceptions of minority groups such as migrants, refugees, the homeless, and indigenous people in relation to collective involvement, influence in decision-making, access to resources and opportunities surrounding mega-events, and the extent of any anti-social behaviour towards particular minority groups (e.g. Chen & Tian, 2015; Gibson et al., 2014; Kim et al., 2006; Minnaert, 2012; Pappalepore & Duignan, 2016). While there is evidence from the literature of the impacts of multicultural festivals on tolerance of diversity within a community (see for example Lee et al., 2012), there is little research examining the way that mega-events affect diverse groups within the host/resident community.

There is a significant body of evidence which suggests that mega-events increase residents' perceptions of social inclusion (e.g. Kaplanidou et al., 2013; Kim & Petrick, 2005). More recently, measurement of social inclusion has also involved tracking and content-analysis of social media posts and comments (e.g. Kirilenko & Stepchenkova, 2017). The majority of this evidence, however, stems from cross-sectional surveys conducted before, during, or shortly after the conclusion of the relevant mega-event (e.g. Chiam & Cheng, 2013; Gibson et al., 2014; Kim & Petrick, 2005). Accordingly, the long-term and causal effects of mega-events on social inclusion within host-cities and host-countries remains unclear. This is also true in respect of the potential negative effects of mega-events on social inclusion which may include the widening of social and cultural divisions within communities (e.g. Lenskyj, 2002; Minnaert, 2012).

There is some evidence of a link between residents' perceptions of the effect of mega-events on social inclusion and intended or actual involvement in decision-making in relation to the event, and opportunities for participation in the event (e.g. Fairley et al.,2016; Gursoy et al., 2017; Pappas, 2017). Residents' trust in the relevant government and organising committee has also been associated with social inclusion outcomes (e.g. Chi et al., 2018; Gursoy et al., 2017; Nunkoo et al., 2018). Existing literature further indicates that favourable residents' perceptions of mega-events are linked to their support for the event and their direct experiences of the event; however, the driving factor of this relationship is unclear (e.g. Chi et al., 2018; Ouyang et al., 2017;

Prayag et al., 2013). Whilst beneficial for organisers and those lobbying for mega-events, such studies demonstrate two inherent difficulties in measuring the actual impact of mega-events on social inclusion, namely, addressing the effects of confirmation bias which exists because of the selected group of survey participants and/or the nature of questions asked (e.g. Chi et al., 2018; Duignan et al., 2019), and overcoming cultural and political factors which may influence respondents' views on the effects of an event (e.g. Li et al., 2015; Liu et al., 2014; Tournois, 2018).

Sport (participation, infrastructure and health)

Sporting social impacts associated with hosting mega-events include improved participation and community health, amenities enhancement through sporting infrastructure investment, enhanced pathways to elite sport, and the promotion of integrity and trust in sport. These outcomes are typically measured by frequency of participation, re-engagement in grassroots sports and physical activity more generally through cross-sectional and longitudinal surveys (e.g. Carmichael et al., 2013; Weed et al., 2015). The direct effect of mega-events on sports participation and access to community sporting infrastructure is also measured using resident surveys embedding multi-item Likert-style scales (e.g. Al-Emadi et al., 2017; Cleland et al., 2020; Kaplanidou, 2017). Despite the multiplicity of studies examining the effects of mega-events on sports participation, there is a clear lack of baseline data or consensus as to whether or not mega-events do in fact increase sports participation at a grassroots level (e.g. Bauman et al., 2015; Chen et al., 2018; Toohey & Taylor, 2008). Amenity and gentrification associated with event-related sporting infrastructure is commonly measured by community surveys, feedback on liveability, and studies of behaviour in using and enjoying the facilities and surroundings (e.g. Kaplanidou, 2012; Kim et al., 2015). As these measures are often examined in conjunction with quality of life assessment, there is a risk that perceptions of new infrastructure are influenced by residents' overall quality of life perceptions.

The effect of mega-events on a host country's elite sporting pathways and development is generally measured by the country's success on the world stage (e.g. Chen et al., 2018; Karadakis & Kaplanidou, 2012). Some evidence suggests that hosting mega-events can create momentum in elite sport success (e.g. Toohey & Taylor, 2008). However, there is also a lack of research as to why that momentum is generally not sustained in the years following a mega-event (Chen et al., 2018). The alleged aspirational nature of elite sport and its effect upon eliciting activity among the host population is relatively under-researched, and warrants further examination (Grix & Carmichael, 2012). The social impacts of mega sporting events therefore intersect with economic impacts in terms of positive health, social connectivity and mental resilience across the community hosting the event. Beyond the sporting infrastructure accessibility, aspirational pathways and increased participation in physical activity and sports, there is evidence that mega-events more broadly can promote positive wellbeing both physically and mentally, in addition to these quality-of-life outcomes for residents (Moon et al., 2019; Teng & Chang, 2020).

Impacts on the destination ecosystem

Business and government networks

The expansion and strengthening of business and government networks are well established as critical leveraging and legacy impacts of mega-events (Chien et al., 2018; O'Brien & Chalip, 2008). Successful network impacts depend upon inter-organisational, inter-city (including sister cities), and inter-country relationships, through which business networks and commercial investment can be accessed (O'Brien & Gardiner, 2006). A mega-event network typically includes entities from tourism, sports, events, government, media and private sectors (Parent, 2008) and these networks are usually established from existing relationships (Ziakas & Costa, 2011). The structure and nature of these relationships are under-researched in mega-event evaluation. Most previous research has focused upon the event's ability to build new networks, rather than examining the event's impact upon existing relationships and the strength of relationship ties developed (Chalip et al., 2003; Werner et al., 2015).

Impacts include monetisable and non-monetisable aspects of industry dynamics in the region, which evolve in response to rapid changes in the environment brought about by the hosting of mega-events (O'Brien & Gardiner, 2006). For example, new trade deals, business venture ecosystems, alliances, and innovation, together with strengthened international relationships and increased levels of soft power and diplomacy are positive impacts which may be caused by evolving industry dynamics (Grix et al., 2015; Preuss, 2007) facilitated by mega-events. Qualitative social impacts for business and governments in hosting regions identified by prior research include collaboration, competition, coopetition, and knowledge transfer (e.g. O'Brien, 2006; Werner et al., 2015). These non-monetisable indicators of partnerships and relationship strengths have been typically measured through surveys and interviews, drawing insights into organisational learning from external regionally-based organisations, identification of specific skills, information, knowledge transferred among organisations in the industry, and mechanisms of transfer across the network. Social network analysis has been used to analyse the complex nature of these networks, and identify how they can be enhanced through strengthening of ties, information sharing and reducing barriers (Pavlovich, 2003). Not all mega-event impacts on the business community are positive - tensions between collaboration and competition have also been assessed, in addition to whether knowledge sharing between organisations increased or decreased because of mega-events. For example, Werner and colleagues (2015) examined the improvement of relationships between Tourism Auckland and local business stakeholders, and found differing outcomes between weakly and strongly tied organisations pre and post the Rugby World Cup 2011 event. The strength of post-event ties was found to be correlated with pre-event involvement in the organisation and the strength of the interaction. Existing evidence suggests that the accrual, or lack thereof, of business benefits within host and non-host regions depends on the relationship between organising committees, governments, and key stakeholders (Fairley et al., 2016) and the sharing of common values and purposes (O'Brien & Gardiner, 2006; Ziakas, 2015). The establishment of a well-coordinated network of alliances which are integrated with leveraging initiatives and the host region's development plans has also been shown to be an essential part of ensuring business benefits extend to non-host regions (Chien et al., 2018; O'Brien, 2006).

Destination branding

Marketing activities aimed at promoting a destination through image building and reputation enhancement are generally referred to as 'destination branding' (Knott et al., 2015). This involves conveying the promise of a memorable experience unique to the particular destination (García et al., 2012). Destination branding can increase the desirability of a destination's products, services, and experiences, and therefore has important implications for development in tourism, trade, investment, education, and sport around the world (Knott et al., 2015).

Current research demonstrates that because of global media coverage, hosting a mega-event can enhance or consolidate a destination's brand image (Kaplanidou et al., 2013; Kim et al., 2015; Lee et al., 2013). Some studies suggest that evaluations of destination brand image become more positive throughout the duration of mega-events (e.g. Lepp & Gibson, 2011). However, post-event studies are less consistent, with some reporting a decline in image perceptions (e.g. Kim & Petrick, 2005; Kirilenko & Stepchenkova, 2017; Rojas-Mendez et al., 2019) whilst others suggest an increase in the strength of a destination's brand following that destination hosting a mega-event (e.g. Chen & Tian, 2015; Kim et al., 2015; Kim & Morrsion, 2005). The accuracy of some post-event studies may be called into question in light of the fact that they did not involve comparable pre-event surveys against which post-event perceptions can be objectively measured.

The strengthening or weakening of a destination's brand before, during, and after a mega-event appears to largely depend on coverage of the event and the destination itself in the mainstream media and on social media (e.g. Kirilenko & Stepchenkova, 2017; Rojas-Mendez et al., 2019). As with other social impacts, research on the effect of mega-events on destination brands is piecemeal and ad hoc. Most post-event studies are conducted within 12 months of the conclusion of the event (e.g. Kim & Petrick, 2005; Rojas-Mendez et al., 2019) and those which are more long-term do not appear to consider the impact of other external factors which may alter a destination's brand image (e.g. Kaplanidou, 2012; Liu et al., 2014).

Most elements of a destination brand, such as general awareness, brand association, attitude, reputation, perceived experience, and quality are intangible, reflecting the strength, favourability, and uniqueness of a destination's brand equity (García et al., 2012). These elements are generally assessed from residents' or visitors' perspectives, using self-reporting instruments such as surveys and interviews and, in the context of mega-events, are often examined in broader studies considering perceptions of an event's overall positive and negative impacts (e.g. Chen & Tian, 2015; Lepp & Gibson, 2011; Rojas-Mendez et al., 2019). Other elements of destination branding, such as exposure, intention to visit, and brand loyalty, are both quantifiable and monetisable. These elements can be measured by the number of times a destination brand is mentioned during event broadcasts, advertising value equivalency, media footprints, number of repeat visitors, and the amount spent by tourists in the destination (Chalip et al., 2003; Jago et al., 2010).

Disaster preparedness

This social impact refers to the measures taken to prevent, prepare for and reduce the effects of natural and man-made disasters (Miles & Shipway, 2020). Given the magnitude and international interest, mega-events are susceptible to exceptional risks that range from public health and environmental, to technology and economic risks (Miles & Shipway, 2020). While the nature and number of risks may vary with events and differ from one destination to another (Walters et al., 2017), risks are often associated with or escalated by the onset of disasters such as cyclones, earthquakes, cyber-security, and terrorist attacks, as has been evidenced in 2020 by the effects of the COVID-19 pandemic, which can have serious implications for events themselves, host communities, and a wide range of stakeholders (Toohey & Taylor, 2008; Wong & Chadwick, 2017).

To minimise the impacts of these disasters and reduce the event's vulnerability, past events have adopted processes and measures such as risk assessment, disaster simulations, security arrangements, crisis communication, disaster mitigation plans, and resource allocation (Toohey & Taylor, 2012). Notably, most of the disaster preparation and prevention measures developed for mega-events can remain with the community following the event, resulting in a long-lasting social legacy of greater community resilience against disasters, as in the case of 2000 Sydney Olympic Games (Toohey & Taylor, 2012). Many of the security measures and legislation implemented for this event remained in place after its conclusion. Such strong public and financial commitment to safety and security strengthens the community's robustness at handling future disasters, while encouraging resistance to threats physically, mentally, and managerially (Shipway, 2018; Toohey & Taylor, 2008).

Academic research considering disaster preparedness in the context of mega-events generally employs methods such as case study, in-depth interviews and content analysis of mainstream media reports (e.g. Toohey & Taylor, 2012). Findings suggest that events that do not appear to have adequate preparation can be susceptible to catastrophic risks (Wong & Chadwick, 2017) while engendering public distrust, misconceptions or fear (Konstantaki & Wickens, 2010). Some studies have also sought to understand awareness and perceptions of event related risks and disaster preparedness from the perspective of spectators and tourists, and to establish a relationship between perceived risks and event attendance (e.g. Walters et al., 2017). The collection or analysis of quantifiable data, however, remains scant, perhaps because it is only possible to evaluate the effectiveness of disaster preparedness measures if a disaster occurs. To this end, the postponement and cancellation of numerous mega-events, including the 2020 Tokyo Olympic Games due to the COVID-19 pandemic, presents an opportunity for researchers to consider the effectiveness of different disaster preparedness plans, including, for example, how and if those plans adequately protect long-term athlete wellbeing and reassure traveling sports fans (Walters et al., 2017).

Accessibility and accessible tourism

Accessibility, making destinations and experiences easier and comfortable for visitors, is a central element of any responsible and sustainable tourism policy (Darcy et al., 2010). An accessible, inclusive destination experience means that people of all abilities and ages feel welcomed as guests and are able to enjoy the experiences offered (Darcy, 2010). Specifically, development and provision of accessible destination experiences enable people with access requirements, including special needs, to function independently and with equity and dignity through the delivery of universally designed tourism products, services, and environments (Michopoulou et al., 2015).

Since mega-event organisers must consider the needs of spectators and visitors with disabilities, event hosting is said to result in accessibility legacies for host communities while facilitating the development of accessible tourism (Dickson et al., 2016). Legacies associated with disability sport events, such as enhanced venues and quality of facilities as a result of hosting the Paralympic Games, further improve quality of life for people with accessibility needs and offer a diversified tourism product to attract visitors from a growing market segment (Dickson et al., 2017). Existing research also suggests a link between increased accessibility, residential support for the event, and residents' perceived quality of life (Kaplanidou, 2012).

While an emerging stream of literature suggests that accessible tourism generates economic benefits by increasing the competitiveness of particular destinations (e.g. Domínguez et al., 2013; Vila et al., 2015), there is a paucity of research on the nexus between accessibility measures and visitors' satisfaction with particular events or destinations. Similarly, as perceived quality of life before and after mega-events is normally measured simultaneously with other social impacts, it is possible that the effect of accessible tourism on quality of life is over or understated by residents. Insights from the current literature highlight the need to consider accessibility from the whole of the visitor's journey perspective (Dickson et al., 2016). These issues warrant future research.

Discussion

While the economic impact of mega-events has been extensively examined, there has been less attention directed towards the intangible, social impacts of these events upon host destinations. Our review provides a timely overview of the state of the field in identifying, conceptualising and measuring social impacts associated with bidding and hosting mega-events. This has arisen in response to a growing need to acknowledge both positive and negative short-, medium- and long-term tangible and intangible impacts on the broader destination ecosystem (e.g. Ritchie et al., 2020). We have also identified an important distinction between those social impacts which directly affect residents, and those which have flow-on effects and unintended consequences that may not have been explicitly identified as social impacts. One key outcome from this review is the lack of a baseline measurement in many of the social impact domains identified. It is very difficult to identify and justify increases and enhancements (or conversely defend decreases and worsening) when there is no baseline measurement to work from. This is an important area for future research.

Our review also revealed the potential for positive social impacts encompassing traditional impacts of tourism accessibility, development of government-industry partnerships and networks, and emerging ideas of community pride and cohesion, regional dispersal, diversity and inclusion. The aspirational role of sport is well established, and the hosting of a mega-event elevates this role and the importance of needing to capture associated impacts in any mega-event impact assessment. Negative social impacts identified in the literature include congestion, noise, increased crime, or other anti-social behaviour and cultural conflicts and displacement of marginalised groups (Chiam & Cheng, 2013; Kim & Petrick, 2005; Minnaert, 2012). It should be emphasised that residents can simultaneously possess positive and negative evaluations of mega-events, culminating in attitudinal ambivalence (Chen et al., 2019). Note again, that the intention of this study was not to critique the social impacts themselves, but rather to provide a framework that informs which impacts are important to measure, whether they be positive, negative or neutral. The study has highlighted the importance and need for critical research that examines the impacts of mega-events on society as a whole, including the marginalised and underprivileged sections of the community.

Studies on residents' perceptions of mega-events are increasingly coming under scrutiny, revealing a need for more nuanced assessment encompassing residents' willingness to trade off temporary inconveniences for sustainable gains and examination of both personal and collective costs and benefits of hosting (Smith et al., 2019). Recent research has demonstrated the complexity of accounting for the negative and positive impacts of these events, which may even translate to resident ambivalence (Chen et al., 2019). It is clear that temporal and contextual considerations need to be reflected in evaluation, given that residents' attitudes have been demonstrated to shift and evolve over time. Social impact measurement also needs to embrace both personal and collective impacts, especially when typically considered with a political lens of government funding and, therefore, taxpayer support (Smith et al., 2019). Additionally, there is a plethora of terms and concepts that have been used (e.g. social capital, social cohesion, civic/community pride). There is significant scope for future research to determine the correlations and discrimination between these overlapping, yet distinct constructs.

The rapidly evolving landscape of mega-event management, risk and delivery has brought new impacts and risks for the sustainability of such events and their communities, as highlighted by the recent cancellation of the 2020 Tokyo Olympic Games due to the COVID-19 pandemic. Future social impact analysis will need to build consideration of how inevitable risks and preparedness for such risks, including terrorism, cyber-security, natural disasters, climate change, and pandemic will affect host stakeholders and event sustainability. Our review highlights this as an emerging social impact category. Figure 1 below summarises our novel, holistic perspective of social impacts associated with mega-events drawn from our review, across time and different target groups.

We advance the current state of knowledge of mega-events' social impacts by elucidating their complexity and interrelatedness, and between social and economic impacts, across time and contexts. For example, social impacts relating to volunteering could positively influence social cohesion and civic pride or vice versa, and volunteering and upskilling can link to economic impacts in terms of employment. Research to date has examined impacts independently and relatively statically, when in practice, impacts can compound further impacts and indirect consequences, temporally (i.e. pre, during and post-event) and spatially (i.e. at a local, regional and national level). Academic literature on these effects is still scarce, due to limited longitudinal studies that examine social impacts over an extended period. The existing longitudinal studies mainly focused on those direct impacts on residents, such as volunteering, sport participation, and social cohesion (e.g. whether people who volunteered at the mega-event continued to

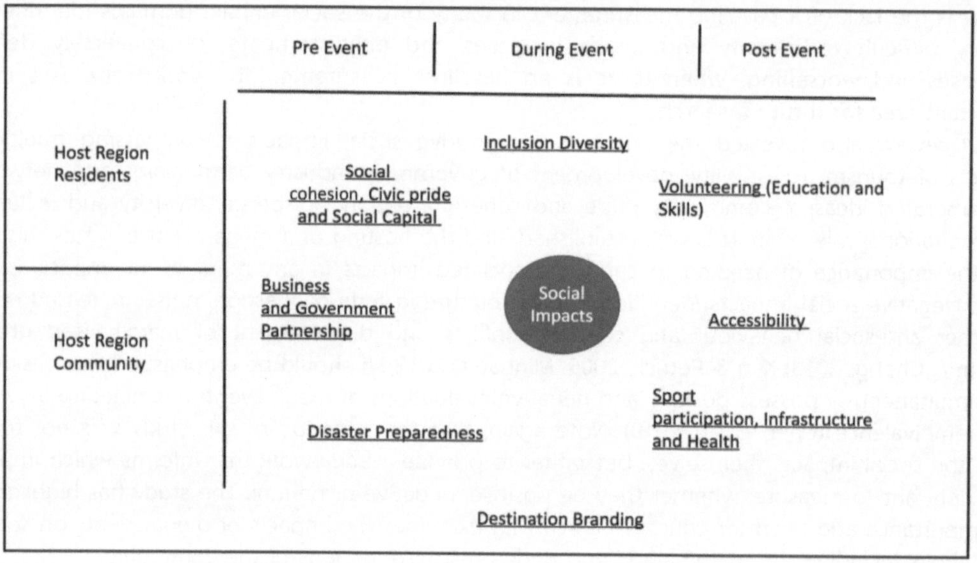

Figure 1. Social impacts of mega- events across time and contexts.

savour the benefits post-event) (e.g. Fairley et al., 2016; Shipway et al., 2020). The interrelationships between the various social impacts, as well as how they spill over to influence each other and residents indirectly, remains unexamined but vital. Hence, future research aiming to inform legacy planning should further develop the measurement of social impacts, but also the complexity of these interrelationships and how horizontal spillover occurs.

The interaction between social and economic impacts is under-researched, with our review revealing that the two categories of impacts have been traditionally examined separately and measured independently. However, the reality is that these impacts co-exist and overlap, and may even trade off, from pre, during and post analysis. To determine the long-term impact of a mega-event, future studies can be developed to investigate the inter-relationship between social and economic impacts.

Theoretically, while our review has identified the use of social exchange theory (e.g. Gursoy et al., 2017; Kaplanidou et al., 2013; Karadakis & Kaplanidou, 2012), future research could consider alternative theories (e.g. social identity theory or social dilemma theory) as overarching frameworks to understand how these social impacts can be further cultivated and sustained. For example, social identity theory can be used to understand volunteer motivation and the maintenance of volunteer network post-event. Similarly, social dilemma theory can be drawn on to understand residents' concerns and the trade-offs they are willing to make for an inclusive and diverse society. There has also been a growing body of research on sport diplomacy and development, which needs to be built into social impacts assessment (Kobierecki & Strożek, 2020). Examination of social impacts in the context of different sized mega-events is also of interest. For example, is there an optimal event size in balancing positive social impacts and sustainability of the host region? Research is needed to evaluate smaller, medium and mega-sized events, in terms of the participation and duration.

Practical implications

Our research has practical application, by providing evidence of best practice in planning bids, hosting events and evaluating legacy for relevant stakeholders, including rights holders, government, sports, industry, sponsors, residents and communities. Specifically, we advocate further research designed to capture the sophisticated and nuanced nature of social impacts, and the design and testing of novel, multi-faceted measures that align with temporal and cultural facets throughout the event life-cycle. Longitudinal and causal studies are therefore needed, rather than shorter-term, cross-sectional surveys of limited samples of stakeholders. For example, residential support is required to develop sustainable tourism in communities hosting the event, yet, to date, most social impact studies only measure a single instance of residents' social impact perceptions (Lorde et al., 2011; Ritchie et al., 2020).

Conclusion

Mega-events have enormous impacts upon the regions and cities hosting them. They are resource-intensive and often an economic and social burden for residents, who subsidise these costs as taxpayers, and experience significant disruption. However, mega-events also have potential to significantly enhance destination branding, tourism and trade, along with more indirect effects of diplomacy, community pride, and social connectivity. Mega-event evaluation throughout the process of bidding, hosting and post-event legacy, therefore, encompasses assessment and objective measurement of both tangible and intangible impacts. We provide a review of the state of the field, highlighting both positive and negative impacts and the need for further research focused upon development of more nuanced measures that capture the evolving temporal and cultural context of mega-event hosting. Importantly, there is a lack of baseline measures for most of the social impacts identified in this review.

Our research provides useful guidance to mega-event stakeholders and highlights a need to assess the event impacts through a longitudinal and systematic lens. The dearth of live sporting events and grassroots sport during the COVID-19 pandemic has illuminated the critical nature of sport for the sustainability of our communities through enhancement of social fabric, connectivity and health, and these impact categories must be integrated into any evaluation relating to hosting mega-events. While our review is largely built on mega sporting events, it has broader application to all mega-events, including cultural, social and scientific.

Disclosure statement

No potential conflict of interest was reported by the authors.

References

Al-Emadi, A., Kaplanidou, K., Diop, A., Sagas, M., Le, K. T., & Ali-Ali Mustafa, S. (2017). 2022 Qatar World Cup: Impact perceptions among Qatar residents. *Journal of Travel Research, 56*(5), 678–694. https://doi.org/10.1177/0047287516652502

Bauman, A., Bellew, B., & Craig, C. L. (2015). Did the 2000 Sydney Olympics increase physical activity among adult Australians? *British Journal of Sports Medicine, 49*(4), 243–247. https://doi.org/10.1136/bjsports-2013-093149

Beesley, L. G., & Chalip, L. (2014). Seeking (and not seeking) to leverage mega-sport events in non-host destinations: The case of Shanghai and the Beijing Olympics. *Journal of Sport & Tourism, 16*(4), 323–344.

Buchanan, D., & Bryman, A. (Eds.). (2011). *The SAGE handbook of organization research methods*. Sage.

Carmichael, F., Grix, J., & Marqués, D. P. (2013). The Olympic legacy and participation in sport: An interim assessment of Sport England's Active People Survey for sports studies research. *International Journal of Sport Policy and Politics, 5*(2), 229–244. https://doi.org/10.1080/19406940.2012.656675

Chalip, L., Green, B. C., & Hill, B. (2003). Effects of sport event media on destination image and intention to visit. *Journal of Sport Management, 17*(3), 214–234. https://doi.org/10.1123/jsm.17.3.214

Chen, F., & Tian, L. (2015). Comparative study on residents' perceptions of follow-up impacts of the 2008. *Tourism Management, 51*, 263–281. https://doi.org/10.1016/j.tourman.2015.05.029

Chen, S., Zheng, J., & Dickson, G. (2018). An organizational life cycle approach to exploring the elite sport legacy of summer Olympic host nations: The cases of China (Beijing 2008) and Australia (Sydney 2000). *The International Journal of the History of Sport, 35*(12-13), 1276–1305. https://doi.org/10.1080/09523367.2019.1583646

Chen, Z., Wang, Y., Li, X., & Lawton, L. (2019). It's not just black or white: Effects of ambivalence on residents' support for a mega-event. *Journal of Hospitality & Tourism Research, 43*(2), 283–313.

Chi, C. G. Q., Ouyang, Z., & Xu, X. (2018). Changing perceptions and reasoning process: Comparison of residents' pre- and post-event attitudes. *Annals of Tourism Research, 70*, 39–53. https://doi.org/10.1016/j.annals.2018.02.010

Chiam, M., & Cheng, E. (2013). Residents' perceptions of the inaugural Youth Olympic Games 2010: A cluster analysis. *Event Management, 17*(4), 377–389. https://doi.org/10.3727/152599513X13769392444620

Chien, P. M., Kelly, S. J., & Gill, C. (2018). Identifying objectives for mega-event leveraging: A non-host city case. *Marketing Intelligence & Planning, 36*(2), 168–184.

Chien, P. M., Ritchie, B. W., Shipway, R., & Henderson, H. (2012). I am having a dilemma: Factors affecting resident support of event development in the community. *Journal of Travel Research, 51*(4), 451–463. https://doi.org/10.1177/0047287511426336

Cleland, C. L., Ellaway, A., Clark, J., & Kearns, A. (2020). Was Glasgow 2014 inspirational? Exploring the legacy impacts of a mega-sport event via the theorized demonstration and festival effects. *Sport in Society, 23*(5), 810–831. https://doi.org/10.1080/17430437.2019.1571044

Commonwealth Games Federation. (n.d.). *Our relevance*. Retrieved July 14, 2020, from https://thecgf.com/our-relevance

Coleman, J. S. (1988). Social capital in the creation of human capital. *American Journal of Sociology, 94*, S95–S120. https://doi.org/10.1086/228943

Darcy, S. (2010). Inherent complexity: Disability, accessible tourism and accommodation information preferences. *Tourism Management, 31*(6), 816–826. https://doi.org/10.1016/j.tourman.2009.08.010

Darcy, S., Cameron, B., & Pegg, S. (2010). Accessible tourism and sustainability: A discussion and case study. *Journal of Sustainable Tourism, 18*(4), 515–537. https://doi.org/10.1080/09669581003690668

Dickson, T., Darcy, S., Johns, R., & Pentifallo, C. (2016). Inclusive by design: Transformative services and sport-event accessibility. *The Service Industries Journal, 36*(11-12), 532–555. https://doi.org/10.1080/02642069.2016.1255728

Dickson, T. J., Misener, L., & Darcy, S. (2017). Enhancing destination competitiveness through disability sport event legacies. *International Journal of Contemporary Hospitality Management, 29*(3), 924–946. https://doi.org/10.1108/IJCHM-10-2015-0530

Domínguez, T., Fraiz, J. A., & Alén, E. (2013). Economic profitability of accessible tourism for the tourism sector in Spain. *Tourism Economics, 19*(6), 1385–1399. https://doi.org/10.5367/te.2013.0246

Duignan, M. B., Pappalepore, I., & Everett, S. (2019). The 'summer of discontent': Exclusion and communal resistance at the London 2012. *Tourism Management, 70*, 355–367. https://doi.org/10.1016/j.tourman.2018.08.029

Duffy, M., & Mair, J. (2017). *Festival encounters – Theoretical perspectives on festival events*. Routledge.

Fairley, S., Gardiner, S., & Filo, K. (2016). The spirit lives on: The legacy of volunteering at the Sydney 2000. *Event Management, 20*(2), 201–215. https://doi.org/10.3727/152599516X14610017108747

Fairley, S., Lovegrove, H., & Brown, M. (2016). Leveraging events to ensure enduring benefits: The legacy strategy of the 2015. *Sport Management Review, 19*(4), 466–474. https://doi.org/10.1016/j.smr.2016.01.003

Fredline, E., & Faulkner, B. (2001). Residents' reactions to the staging of major motorsport events within their communities: A cluster analysis. *Event Management, 7*(2), 103–114. https://doi.org/10.3727/152599501108751515

Fredline, L., Jago, L., & Deery, M. (2003). The development of a generic scale to measure the social impacts of events. *Event Management, 8*(1), 23–37. https://doi.org/10.3727/152599503108751676

García, J. A., Gómez, M., & Molina, A. (2012). A destination-branding model: An empirical analysis based on stakeholders. *Tourism Management, 33*(3), 646–661. https://doi.org/10.1016/j.tourman.2011.07.006

Gibson, H. J., Walker, M., Thapa, B., Kaplanidou, K., Geldenhuys, S., & Coetzee, W. (2014). Psychic income and social capital among host nation residents: A pre-post analysis of the 2010 FIFA World Cup in South Africa. *Tourism Management, 44*, 113–122. https://doi.org/10.1016/j.tourman.2013.12.013

Gornostaeva, G., & McGurk, P. (2013). Olympic volunteering for the unemployed: Who benefits and how? *International Journal of Employment Studies, 21*(1), 5–30.

Grix, J., Brannagan, P. M., & Houlihan, B. (2015). Interrogating states' soft power strategies: a case study of sports mega-events in Brazil and the UK. *Global Society, 29*(3), 463–479. https://doi.org/10.1080/13600826.2015.1047743

Grix, J., & Carmichael, F. (2012). Why do governments invest in elite sport? A polemic. *International Journal of Sport Policy and Politics, 4*(1), 73–90. https://doi.org/10.1080/19406940.2011.627358

Gursoy, D., Yolal, M., Ribeiro, M. A., & Panosso Netto, A. (2017). Impact of trust on local residents' mega-event perceptions and their support. *Journal of Travel Research, 56*(3), 314–393. https://doi.org/10.1177/0047287516643415

Holmes, K., Hughes, M., Mair, J., & Carlsen, J. (2015). *Events and sustainability*. Routledge.

International Olympic Committee (IOC). (2012). *Olympic legacy*. IOC.

International Olympic Committee (IOC). (2018, February). *The new norm: It's a game changer*. https://www.olympic.org/news/the-new-norm-it-s-a-games-changer

Jago, L., Dwyer, L., Lipman, G., van Lill, D., & Vorster, S. (2010). Optimising the potential of mega-events: An overview. *International Journal of Event and Festival Management, 1*(3), 220–237. https://doi.org/10.1108/17852951011078023

Kaplanidou, K. (2012). The importance of legacy outcomes for Olympic Games four summer host cities residents' quality of life: 1996-2008. *European Sport Management Quarterly, 12*(4), 397–433. https://doi.org/10.1080/16184742.2012.693118

Kaplanidou, K. (2017). Health-related attitudes and behaviors among Olympic host city residents from Atlanta, Sydney, Athens, and Beijing Olympic games: Exploring potential legacies. *Event Management, 21*(2), 159–174. https://doi.org/10.3727/152599517X14878772869720

Kaplanidou, K., Karadakis, K., Gibson, H., Thapa, B., Walker, M., Geldenhuys, S., & Coetzee, W. (2013). Quality of life, event impacts, and mega-event support among South African residents before and after the 2010 FIFA World Cup. *Journal of Travel Research, 52*(5), 631–645. https://doi.org/10.1177/0047287513478501

Karadakis, K., & Kaplanidou, K. (2012). Legacy perceptions among host and non-host Olympic Games residents: A longitudinal study of the 2010 Vancouver Olympic Games. *European Sport Management Quarterly, 12*(3), 243–264. https://doi.org/10.1080/16184742.2012.680067

Kim, H. J., Gursoy, D., & Lee, S.-B. (2006). The impact of the 2002 World Cup on South Korea: Comparisons of pre- and post-games. *Tourism Management, 27*(1), 86–96. https://doi.org/10.1016/j.tourman.2004.07.010

Kim, S., & Petrick, J. F. (2005). Residents' perceptions on impacts of the FIFA 2002 World Cup: The case of Seoul as a host city. *Tourism Management, 26*(1), 25–38. https://doi.org/10.1016/j.tourman.2003.09.013

Kim, S. S., & Morrsion, A. M. (2005). Change of images of South Korea among foreign tourists after the 2002. *Tourism Management, 26*(2), 233–247. https://doi.org/10.1016/j.tourman.2003.11.003

Kim, W., Jun, H. M., Walker, M., & Drane, D. (2015). Evaluating the perceived social impacts of hosting large-scale sport tourism events: Scale development and validation. *Tourism Management, 48*, 21–32. https://doi.org/10.1016/j.tourman.2014.10.015

Kirby, S. I., Duignan, M. B., & McGillivray, D. A. V. I. D. (2018). Mega-sport events, micro and small business leveraging: Introducing the "MSE-MSB leverage model". *Event Management, 22*(6), 917–931. https://doi.org/10.3727/152599518X15346132863184

Kirilenko, A. P., & Stepchenkova, S. O. (2017). Sochi 2014 Olympics on Twitter: Perspectives of hosts and guests. *Tourism Management, 63*, 54–65. https://doi.org/10.1016/j.tourman.2017.06.007

Knott, B., Fyall, A., & Jones, I. (2015). The nation branding opportunities provided by a sport mega-event: South Africa and the 2010 FIFA World Cup. *Journal of Destination Marketing & Management, 4*(1), 46–56. https://doi.org/10.1016/j.jdmm.2014.09.001

Kobierecki, M. M., & Strożek, P. (2020). Sports mega-events and shaping the international image of states: How hosting the Olympic Games and FIFA World Cups affects interest in host nations. *International Politics*, 1–22. https://doi.org/10.1057/s41311-020-00216-w

Konstantaki, M., & Wickens, E. (2010). Residents' perceptions of environmental and security issues at the 2012 London Olympic Games. *Journal of Sport & Tourism, 15*(4), 337–357. https://doi.org/10.1080/14775085.2010.533921

Laing, J., & Mair, J. (2015). Music festivals and social inclusion – The festival organizers' perspective. *Leisure Sciences, 37*(3), 252–268. https://doi.org/10.1080/01490400.2014.991009

Lee, I. S., Arcodia, C., & Lee, T. J. (2012). Key characteristics of multicultural festivals: A critical review of the literature. *Event Management, 16*(1), 93–101. https://doi.org/10.3727/152599512X13264729827758

Lee, S.-B., Lee, C.-K., Kang, J.-S., Lee, E.-Y., & Jeon, Y. J. J. (2013). Residents' perception of the 2008 Beijing Olympics: Comparison of pre- and post-impacts. *International Journal of Tourism Research, 15*(3), 209–225. https://doi.org/10.1002/jtr.885

Lenskyj, H. J. (2002). *The best Olympics ever? The social impacts of Sydney 2000*. State University of New York Press.

Leopkey, B., & Parent, M. M. (2017). The governance of Olympic legacy: Process, actors and mechanisms. *Leisure Studies, 36*(3), 438–451.

Lepp, A., & Gibson, H. (2011). Reimagining a nation: South Africa and the 2010 FIFA World Cup. *Journal of Sport & Tourism, 16*(3), 211–230. https://doi.org/10.1080/14775085.2011.635007

Li, X., Hsu, C. H. C., & Lawton, L. J. (2015). Understanding residents' perception changes toward a mega-event through a dual-theory lens. *Journal of Travel Research, 54*(3), 396–410. https://doi.org/10.1177/0047287513517422

Liu, D., Broom, D., & Wilson, R. (2014). Legacy of the Beijing Olympic Games: A non-host city perspective. *European Sport Management Quarterly, 14*(5), 485–502. https://doi.org/10.1080/16184742.2014.947301

Lockstone, L., & Baum, T. (2009). The public face of event volunteering at the 2006 Commonwealth Games: The media perspective. *Managing Leisure, 14*(1), 38–56. https://doi.org/10.1080/13606710802551254

Lorde, T., Greenidge, D., & Devonish, D. (2011). Local residents' perceptions of the impacts of the ICC Cricket World Cup 2007 on Barbados: Comparisons of pre-and post-games. *Tourism Management, 32*(2), 349–356. https://doi.org/10.1016/j.tourman.2010.03.004

Magno, F., & Dossena, G. (2020). Pride of being part of a host community? Medium-term effects of mega-events on citizen quality of life: The case of the World Expo 2015 in Milan. *Journal of Destination Marketing & Management, 15*, 100410. https://doi.org/10.1016/j.jdmm.2020.100410

Mair, J., & Duffy, M. (2018). The role of festivals in strengthening social capital in rural communities. *Event Management, 22*(6), 875–889. https://doi.org/10.3727/152599518X15346132863229

Mays, N., Pope, C., & Popay, J. (2005). Systematically reviewing qualitative and quantitative evidence to inform management and policy making in the health field. *Journal of Health Services Research & Policy, 10*(1_suppl), 6–20. https://doi.org/10.1258/1355819054308576

Michopoulou, E., Darcy, S., Ambrose, I., & Buhalis, D. (2015). Accessible tourism futures: The world we dream to live in and the opportunities we hope to have. *Journal of Tourism Futures, 1*(3), 179–188. https://doi.org/10.1108/JTF-08-2015-0043

Miles, L., & Shipway, R. (2020). Exploring the COVID-19 pandemic as a catalyst for stimulating future research agendas for managing crises and disasters at international sport events. *Event Management, 24*(4), 537–552. https://doi.org/10.3727/152599519X15506259856688

Miles, M. B., & Huberman, A. M. (1994). *Qualitative data analysis: An expanded sourcebook*. Sage.

Minnaert, L. (2012). An Olympic Legacy for all? The non-infrastructural outcomes of the Olympic Games for socially excluded groups (Atlanta 1996 – Beijing 2008). *Tourism Management, 33*(2), 361–370. https://doi.org/10.1016/j.tourman.2011.04.005

Moher, D., Liberati, A., Tetzlaff, J., Altman, D. G., & PRISMA Group. (2009). Preferred reporting items for systematic reviews and meta-analyses: The PRISMA statement. *Annals of Internal Medicine, 151*(4), 264–269. https://doi.org/10.7326/0003-4819-151-4-200908180-00135

Moon, B. Y., Yang, S. H., Kim, K. S., & Seo, W. J. (2019). Resident-focused tourism policy: Perceived quality-of-life difference and its effect on expectations about and support for mega events. *Journal of Policy Research in Tourism, Leisure and Events, 11*(2), 220–235. https://doi.org/10.1080/19407963.2018.1516074

Nichols, G., & Ralston, R. (2011). Social inclusion through volunteering: The legacy potential of the 2012 Olympic Games. *Sociology, 45*(5), 900–914. https://doi.org/10.1177/0038038511413413

Nichols, G., Ralston, R., & Holmes, K. (2017). The 2012 Olympic ambassadors and sustainable tourism legacy. *Journal of Sustainable Tourism*, *25*(11), 1513–1528. https://doi.org/10.1080/09669582.2017.1291648

Nunkoo, R., Ribeiro, M. A., Sunnassee, V., & Gursoy, D. (2018). Public trust in mega event planning institutions: The role of knowledge, transparency and corruption. *Tourism Management*, *66*, 155–166. https://doi.org/10.1016/j.tourman.2017.11.010

O'Brien, D. (2006). Event business leveraging the Sydney 2000 Olympic games. *Annals of Tourism Research*, *33*(1), 240–261. https://doi.org/10.1016/j.annals.2005.10.011

O'Brien, D., & Chalip, L. (2008). Sport events and strategic leveraging: Pushing towards the triple bottom line. In A. G. Woodside & D. Martin (Eds.), *Tourism management*. CAB International.

O'Brien, D., & Gardiner, S. (2006). Creating sustainable mega event impacts: Networking and relationship development through pre-event training. *Sport Management Review*, *9*(1), 25–47.

Ooi, N., Laing, J., & Mair, J. (2015). Social capital as a heuristic device to explore sociocultural sustainability: A case study of mountain resort tourism in the community of Steamboat Springs, Colorado, USA. *Journal of Sustainable Tourism*, *23*(3), 417–436. https://doi.org/10.1080/09669582.2014.957211

Ouyang, Z., Gursoy, D., & Sharma, B. (2017). Role of trust, emotions and event attachment on residents' attitudes toward tourism. *Tourism Management*, *63*, 426–438. https://doi.org/10.1016/j.tourman.2017.06.026

Pappalepore, I., & Duignan, M. B. (2016). The London 2012 cultural programme: A consideration of Olympic impacts and legacies for small creative organisations in east London. *Tourism Management*, *54*, 344–355. https://doi.org/10.1016/j.tourman.2015.11.015

Pappas, N. (2017). Pre- and postevaluation of residents' participation and support of the 2012 London Olympics. *Event Management*, *21*(6), 747–770. https://doi.org/10.3727/152599517X15073047237250

Parent, M. M. (2008). Evolution and issue patterns for major-sport-event organizing committees and their stakeholders. *Journal of Sport Management*, *22*(2), 135–164. https://doi.org/10.1123/jsm.22.2.135

Pavlovich, K. (2003). The evolution and transformation of a tourism destination network: The Waitomo Caves. *Tourism Management*, *24*(2), 203–216. https://doi.org/10.1016/S0261-5177(02)00056-0

Playthegame.org. (2016). *Olympic volunteers in Rio complain about their working conditions*. Retrieved July 14, 2020. https://www.playthegame.org/news/news-articles/2016/0215_thousands-of-olympic-volunteers-in-rio-complain-about-their-working-conditions/#:~:text=More%20than%2050%2C000%20volunteers%20from,way%20to%20the%20sporting%20venues

Popay, J., Roberts, H., Sowden, A., Petticrew, M., Arai, L., Rodgers, M., Britten, N., Roen, K., & Duffy, S. (2006). Guidance on the conduct of narrative synthesis in systematic reviews. A Product from the ESRC methods programme Version, 1, b92.

Prayag, G., Hosany, S., Nunkoo, R., & Alders, T. (2013). London residents' support for the 2012 Olympic Games: The mediating effect of overall attitude. *Tourism Management*, *36*, 629–640. https://doi.org/10.1016/j.tourman.2012.08.003

Preuss, H. (2007). The conceptualisation and measurement of mega sport event legacies. *Journal of Sport & Tourism*, *12*(3-4), 207–228.

Ralston, R., Lumsdon, L., & Downward, P. (2005). The third force in events tourism: Volunteers at the XVII Commonwealth Games. *Journal of Sustainable Tourism*, *13*(5), 504–5119. https://doi.org/10.1080/09669580508668576

Ritchie, B. W., Chien, M. P., & Shipway, R. (2020). A leg(acy) to stand on? A non-host resident perspective of the London 2012 Olympic legacies. *Tourism Management*, *77*, 104031. https://doi.org/10.1016/j.tourman.2019.104031

Ritchie, B. W., Shipway, R., & Cleeve, B. (2009). Resident perceptions of mega-sporting events: A non-host city perspective of the 2012 London Olympic games. *Journal of Sport & Tourism*, *14*(2-3), 143–167. https://doi.org/10.1080/14775080902965108

Rojas-Mendez, J. I., Davies, G., Jamsawang, J., Duque, J. L. S., & Pipoli, G. M. (2019). Explaining the mixed outcomes from hosting major sporting events in promoting tourism. *Tourism Management*, *74*, 300–309.

Smith, A., Ritchie, B. W., & Chien, P. M. (2019). Citizens' attitudes towards mega-events: new framework. *Annals of Tourism Research*, *74*(C), 208–210. https://doi.org/10.1016/j.annals.2018.07.006

Shipway, R. (2018). Building resilience and managing crises and disasters in sport tourism. *Journal of Sport & Tourism*, *22*(3), 265–270.

Shipway, R., Ritchie, B. W., & Chien, P. M. (2020). Beyond the glamour: resident perceptions of Olympic legacies and volunteering intentions. *Leisure Studies*, *39*(2), 181–194.

Teng, H. Y., & Chang, S. T. (2020). Resident perceptions and support before and after the 2018 Taichung international Flora exposition. *Current Issues in Tourism*, 1–20. https://doi.org/10.1080/13683500.2020.1820455

Toohey, K., & Taylor, T. (2008). Mega Events, fear, and risk: Terrorism at the Olympic Games. *Journal of Sport Management*, *22*(4), 451–469. https://doi.org/10.1123/jsm.22.4.451

Toohey, K., & Taylor, T. (2012). Surveillance and securitization: A forgotten Sydney Olympic Legacy. *International Review for the Sociology of Sport*, *47*(3), 324–337. https://doi.org/10.1177/1012690211433456

Tournois, L. (2018). A pre-event assessment of residents' reactions to Dubai Expo 2020. *Tourism Management*, *68*, 46–51. https://doi.org/10.1016/j.tourman.2018.02.016

Vanwynsberghe, R., Surborg, B., & Wyly, E. (2013). When the games come to town: Neoliberalism, mega-events and social inclusion in the Vancouver 2010 Winter Olympic Games. *International Journal of Urban and Regional Research, 37*(6), 2074–2093. https://doi.org/10.1111/j.1468-2427.2012.01105.x

Veal, A. J., Toohey, K., & Frawley, S. (2012). The sport participation legacy of the Sydney 2000 Olympic Games and other international sporting events hosted in Australia. *Journal of Policy Research in Tourism, Leisure and Events, 4*(2), 155–184. https://doi.org/10.1080/19407963.2012.662619

Vila, T. D., Darcy, S., & González, E. A. (2015). Competing for the disability tourism market – A comparative exploration of the factors of accessible tourism competitiveness in Spain and Australia. *Tourism Management, 47*, 261–272.

Walters, G., Shipway, R., Miles, L., & Aldrigui, A. (2017). Fandom and risk perceptions of Olympic tourists. *Annals of Tourism Research, 66*, 210–215. https://doi.org/10.1016/j.annals.2017.07.012

Weed, M., Coren, E., Fiore, J., Wellard, I., Chatziefstathiou, D., Mansfield, L., & Dowse, S. (2015). The Olympic Games and raising sport participation: A systematic review of evidence and an interrogation of policy for a demonstration effect. *European Sport Management Quarterly, 15*(2), 195–226. https://doi.org/10.1080/16184742.2014.998695

Werner, K., Dickson, G., & Hyde, K. F. (2015). Learning and knowledge transfer processes in a mega-events context: The case of the 2011. *Tourism Management, 48*, 174–187. https://doi.org/10.1016/j.tourman.2014.11.003

Wong, D., & Chadwick, S. (2017). Risk and (in) security of FIFA football World Cups–outlook for Russia 2018. *Sport in Society, 20*(5-6), 583–598. https://doi.org/10.1080/17430437.2016.1159362

Wood, E. (2006). Measuring the social impacts of local authority events: A pilot study for a civic pride scale. *International Journal of Nonprofit and Voluntary Sector Marketing, 11*(3), 165–179. https://doi.org/10.1002/nvsm.21

Zhou, Y., & Ap, J. (2009). Residents' perceptions towards the impacts of the Beijing 2008. *Journal of Travel Research, 48*(1), 78–91. https://doi.org/10.1177/0047287508328792

Ziakas, V. (2015). For the benefit of all? Developing a critical perspective in mega-event leverage. *Leisure Studies, 34*(6), 689–702. https://doi.org/10.1080/02614367.2014.986507

Ziakas, V., & Costa, C. A. (2011). Event portfolio and multi-purpose development: Establishing the conceptual grounds. *Sport Management Review, 14*(4), 409–423. https://doi.org/10.1016/j.smr.2010.09.003

ỡ OPEN ACCESS

Exploring the neglected voices of children in sustainable tourism development: a comparative study in six European tourist destinations

Marko Koščak, Mladen Knežević, Daniel Binder, Antonio Pelaez-Verdet, Cem Işik, Vladimir Mićić, Katarina Borisavljević and Tina Šegota

ABSTRACT

For years, sustainable tourism has commanded the attention of academics and practitioners given its achievement through participatory planning. However, much research in the area has neglected to consider all voices of those (i.e., children) affected by tourism. This study, employing a post-test only experimental design, addresses this gap by including the voices of 498 children from six diverse European destinations concerning perceptions of tourism. Results show that children are willing to express their opinions given the opportunity. Moreover, their attitude towards tourists is negative, while they perceive locals to hold a subordinate position to tourists. In general, this study offers insight into children's attitudes to tourism from the host perspective, demonstrating that there is a unique insider's perspective that needs further research.

Introduction

Tourism is used as a development tool by governments, and it is known to impact the quality of life of individuals and the well-being of communities (Moscardo, 2012; Uysal et al., 2016). Thus, tourism needs to be seen as an important social practise taking place in a community and as an economic activity that is to be evaluated through its impacts on different community groups in order to determine its overall value. Such a view is especially needed when evaluating tourism impacts on the lives of children who live in communities where tourism represents one of the main economic activities and thus plays a significant role in their socialization process (Poria & Timothy, 2014; Yang et al., 2020). Hence, tourism can be used as a tool for development and improvement of life in a community, however, its impact on shaping children's lives cannot be neglected.

Canosa et al. (2016, p. 327) have rightfully observed that "community-focused tourism research has reached a level of maturity displayed by the theoretical and methodological eclecticism of studies in this area and the varying disciplinary influences." Nevertheless, many studies

(Byrd, 2007; Garrod et al., 2012; Šegota et al., 2017) have emphasised that not *all voices* have been heard and that more research is needed to include marginal voices of host communities, such as children and young people (Canosa et al., 2016; Yang et al., 2020). This also testifies to the concept of sustainable tourism being perceived as more than a piecemeal policy solution, which Agyeman and Evans (2004) described as 'just sustainability'. Therefore, the current generation that is undertaking the governance of tourism and sustainability (Bramwell & Lane, 2011) should be aware that at this moment there is a generation of young people whose lives have already been affected by tourism. However, their views on tourism are very often neglected (Lugosi et al., 2016; Poria & Timothy, 2014; Small, 2008; Thornton et al., 1997), making the sustainability concept incomplete.

Moreover, as Small (2008, p. 773) argues, young people need to be involved in tourism research, but their involvement should go "beyond the commercial incentive". In other words, young people's voices are usually overlooked due to methodological and ethical issues. Also, young people are very often perceived as not significantly contributing to tourism business and development. However, ending the silence of marginalized community members and increasing the interest of child-centred research will be possible with better advocacy of critical scholarship (Canosa et al., 2016).

The neglect of the role of children in sustainable tourism literature needs to be addressed. Since there are only a handful of studies that offer insight into children's attitudes towards tourism from the host perspective (for example, Canosa et al., 2018, 2019; Koščak et al., 2018; Seraphin & Green, 2019), we aim to address this gap by answering the following research questions:

RQ1. What are children's attitudes towards tourism and hospitality business in general?

RQ2. How do children perceive the impact of tourism on family life?

RQ3. Are children involved in decisions related to tourism planning and development, and to what extent?

RQ4. How do children evaluate the dynamics in their community in relation to the tourist-host relationship?

From a methodological standpoint, this study employs a single group post-test only experimental design with children between the ages of 11 and 16 in six different European tourist destinations. Children's perceptions and attitudes towards tourism were explored using a non-verbal semantic differential measurement instrument, a method that represents an innovative approach that can be applied to collecting data from children in comparative or cross-cultural studies. With our study, we aim to demonstrate that children and young people are active participants in tourism and careful observers of the environment around them. The findings may encourage both academics and tourism decision-makers to adopt a similar method used in this study to summon perceptions of tourism from the this generation of residents and to stimulate their inclusivity and engagement with the industry that affects their lives today and in the future.

The neglected voices of children in sustainable tourism development

Children's and youth's attitudes have been heavily "under-researched and undervalued in tourism studies" (Thornton et al., 1997, p. 287). Not to be mistaken, there is a considerable body of tourism research that includes children (Poria & Timothy, 2014; Yang et al., 2020). In the systematic quantitative literature review of tourism studies involving children, Yang et al. (2020, p. 238) concluded that "among the 70 papers, fewer than half (34 studies), collected empirical data". Also, we should emphasise that the research *indirectly* included children, as they were theoretically considered objects and not agents of research (Gram, 2007). For example, much recent empirical research has studied children from the perspective of how they influence family travel decision making (Khoo-Lattimore et al., 2015, 2018; Schänzel et al., 2012; Wang et al., 2016, to name a few). The role they play in holidaying decisions was found to be significant but secondary, as their perceptions were not directly surveyed. To illustrate, Khoo-Lattimore et al. (2015) summoned parents' vacation choices when they holidayed with their children, meaning that children were recognized as having an important role in the decision-making process, but were not directly asked about their choices.

By focusing our debate on the research object-agent discourse in connection to sustainability, studies that collected the empirical data *directly* from children, placing children at the centre of the research as its agents, predominately observed children in the role of tourists (Yang et al., 2020). For example, their experiences were summoned in relation to family holidays (Koo-Lattimore et al., 2015), dark tourism sites (Kerr & Price, 2018; Israfilova & Koo-Lattimore, 2019), and cruising (Radic, 2019). Moreover, among those studies that collected empirical data directly from children, but without involving any other research agents (e.g., guardians, adults, government, etc.), only 15 studies directly presented children's voices. For example, studies by Al-Frehat and Alhelalat (2015), Monterrubio et al. (2016), and Montgomery (2008) were interested in children's attitudes on child labour, while there are only 12 studies that elicited children's perception of tourism spaces (e.g. Buzinde & Manuel-Navarrete, 2013; Canosa et al., 2017, 2019; Ohashi et al., 2012; Seraphin & Green, 2019), tourists (e.g., Canosa et al., 2017; Molero et al., 2003), perception of themselves (e.g., Gamradt, 1995; Koščak et al., 2018), and impact of tourism on their health (e.g., Anderson-Fye, 2004; Dancause et al., 2011; Leatherman et al., 2010). For example, Canosa et al. (2017, 2019) used a rich collection of qualitative methods (i.e., ethnography, interviews, focus groups, etc.) to study how young people, aged 10 to 24, experience life in the Australian tourism destination Byron Bay. Drawings as a surveying technique were used by Koščak et al. (2018) and Seraphin and Green (2019). Koščak et al. (2018) surveyed children's perception of tourism development plans and whether they had an opportunity to participate in decision-making. They focused on Slovenian children, aged 11 to 12, distinguishing between three destinations with different levels of tourism development. The authors showed that children differ in their perception of tourism depending on the communities they live in. On the other hand, Seraphin and Green (2019) studied children's visions of the destinations for the future. The authors demonstrated that children are aware of the existence of cutting-edge technology and recognise its potential to shape the future of their communities.

Overall, all these studies significantly demonstrate that children and young people are not to be mistaken for 'muted' or even 'silent' social groups. Moreover, they demonstrate that living in a particular community where tourism is the main economy has implications for children's lives. Children are receptive to how tourism affects their lives, in particular, and community life, in general. This is why tourism needs to be viewed as a sustainable and responsible developmental option for such communities (Mihalič, 2016). Moreover, there are many opportunities to bring children into the spotlight of tourism research. For example, previous commentaries (e.g., Koo-Lattimore, 2015; Poria & Timothy, 2014; Yang et al., 2020) have indicated that the majority of research has explored the objective dimension of tourism impacts, but limited attention has been focused on host-children satisfaction with tourism impacts on their lives or their subjective wellbeing. However, exploring childhood experiences in tourism destinations based on children's opinions, rather than adults' or experts' assumptions, is an essential element in advancing tourism scholarship and knowledge of the industry. At the same time, such research has the capacity to contribute towards the development of a more inclusive view of tourism and its impacts (Poria & Timothy, 2014) and to help expand the researcher's capacity to conceptualize the social, economic, and cultural constructs of tourism (Carr, 2011).

Positioning children within the sustainable tourism development discourse

Resident attitudes towards tourism have received significant attention among academics and practitioners over the last fifty years. Much of this research has been based on social exchange theory that suggests that the more residents perceive that they are benefitting from tourism, the more they will have a positive perception of the industry in general and will support its development (Ap, 1992; McGehee & Andereck, 2004; Perdue et al., 1990; Sharpley, 2014). Many studies have found support for this theory (for example, Andereck et al., 2005; Ko & Stewart, 2002; Palardy et al., 2018; Su & Swanson, 2020, to name a few). However, except for a few notable examples mentioned above, resident attitude studies predominately involve the adult population and pay little attention to children and youth as residents of host communities. The reason for this limited attention appears to be due to the tendency of destination stakeholders to regard children as having no commercial value (Canosa & Graham, 2016). However, tourism is not merely an industry but also a social force with both a transformative and a disruptive power over host-communities (Higgins-Desbiolles, 2006). This is also why there are many studies that are eliciting residents' attitudes towards tourism, e.g., to enable better quality of life and minimize tourism's negative impacts on communities. However, this does not mean that children do not also form attitudes towards tourism. Therefore, it is necessary to consider the social capacity of tourism, as children will become, in due course, important business or social stakeholders in tourism destinations (Seraphin & Green, 2019). To address the gap of minimal research on children as hosts that includes directly eliciting children's attitudes towards tourism, we propose the following research question:

RQ1. What are children's attitudes towards tourism and hospitality business in general?

In light of tourism's social capacity, some studies have observed the changes tourism has on the community through the lens of social disruptive theories (McKercher et al., 2015; Perdue et al., 1999). McKercher et al. (2015) identified a U-shaped curve between place change, attitude change, and destination lifecycle, which reflects a negative-positive-negative arch. However, tourism not only disrupts life in the community, but also family life. Studies suggest that families benefit from tourism financially, but that overall family relationships are deteriorating (Chan et al., 2015). Chan et al. (2015) showed that for families where parents are employed in tourism, fatigue and incompatibility of schedules impose significant challenges to childcare and family functioning. Similarly, van Schalkwyk et al. (2006) showed that families in tourism-intensive destinations suffer from the disruption of relationships and the psycho-social development of individuals. This is because families, where parents work in the industry, do not have sufficient time to dedicate to their children and they tend to avoid social places with many tourists, as these places are where they would be connecting with tourists on a social level, which is what they wish to avoid. These insights provide a valuable resource for observing tourism and its impact on family life, however, they represent parents' views and not the views of children. Therefore, we ask the following question:

RQ2. How do children perceive the impact of tourism on family life?

Sustainable tourism development addresses the needs of visitors, the industry, the environment, and host communities (UNEP & UNWTO, 2005). and it requires "the informed participation of all relevant stakeholders" (Edgell et al., 2008, p. 195) and for all voices to be heard (Byrd, 2007; Šegota et al., 2017). In line with this approach, children's and young people's experiences should not be marginalized but included in future tourism planning and development studies (Khoo-Lattimore, 2015; Poria & Timothy, 2014; Small, 2008). However, little attention has been paid to children and to understanding if they are involved in the planning and development of tourism in their communities (Koščak et al., 2018). Therefore, the following question is posed:

RQ3. Are children involved in decisions related to tourism planning and development, and to what extent?

As the term community, a socially constructed territorial and relational structure evolving around the use of resources, implies (Prilleltensky & Prilleltensky, 2006), once a community becomes a tourism destination, the use of those resources is negotiated within the resident-tourism exchange process (Uysal et al., 2016). Therefore, studies of resident attitudes have traditionally been based on a dichotomy, such as tourists or *guests* and residents or *hosts* (Meethan, 2001). Residents as hosts accept tourism and both its positive and negative impacts as "part of the price of remaining in a tourism destination" (McKercher et al., 2015, p. 63). Such a perspective indicates structural, social, and behavioural interdependencies in the tourist-host relationships. However, little is known about how children evaluate both tourists and residents. Therefore, we ask the following question:

> *RQ4.* How do children evaluate the dynamics in their communities in relation to the tourist-host relationship?

Methods

Research design and ethical challenges of working with children

Examining the interconnectedness of tourism impacts, happiness, and satisfaction among the adult population is challenging as it is, let alone on children (Andereck & Nyaupane, 2011). Regardless, past tourism research has shown that children are a valuable source of information that is important for developing tourism communities. We decided to approach children using an experimental one-group posttest-only study. This quasi-experimental research design enables researchers to measure a dependent variable for one group of participants following a treatment without a control group (Privitera & Ahlgrim-Delzell, 2018). The one-group posttest-only design is used to gather information from participants though their recollections that are often subjective (Garson, 2013). By being interested in children's attitudes towards tourism and their participation in decision-making in their communities, which are subjective and are based on recollections, it is believed that this research design is suitable for seeking answers to our research questions.

The research took place in schools and was conducted in groups. Due to the nature of our study (i.e., children as participants), we had to address two ethical issues. The first was to inform the parents about the research and obtain their permission for the children to participate in the research. For all participating children, local research teams received all necessary agreements from local educational authorities and from parents. The second issue was the immediate motivation of children to participate in the research. Local research teams explained to the children that they were participating in a study in which they would, in some way, evaluate adults (including their parents) and were asking for their consent. They also told them that, at least in the first part of the research, they would have to clearly impart some of their views on the work of adults engaged in tourism. Thus, the basic ethical principle employed by the research teams was complete openness to children. No child refused to participate. Overall, the experiment lasted for 20 minutes.

Development of measurement instruments

In the research, the measurement instrument consisted of two parts. The first included non-verbal semantic differential (Bujas, 1967), which we used to examine children's attitudes to tourists and residents. The non-verbal semantic differential is an instrument that consists of 12 pairs of abstract drawings. Each pair of drawings belong to one of the three dominant factors that are well known in traditional verbal forms of semantic differentiation (for a detailed description see Bentler & Lavoie, 1972 and Bujas, 1967). In general, the semantic differential is a technique that largely reflects the connotative meaning of terms (Garson, 2013). In our case, they were not verbal concepts (i.e., questions or statements) but abstract drawings (i.e., the non-verbal semantic differential). These drawings reflected sets of paired opposite items, where a set of items represented important dimensions associated with children's evaluation of tourists and residents based on three factors. The first factor was Evaluation, which provided information on the grading of the object as good or bad. It indicates propensity or aversion to an object (Petrović & Kuzmanović, 1998). The second factor was Potency, which provided information on the object's strength as strong or weak. The third factor was Activity, denoting whether the object is active or passive. Each of the three factors was represented by four sets of drawings, whose meanings were standardized in the process of the test design (Bentler & Lavoie, 1972; Bujas, 1967). Once a respondent was presented with the non-verbal semantic differential, they were asked to circle the drawing that best suits their mood or feelings that connect the drawing and the concept being analyzed (Bentler & Lavoie, 1972; Garson, 2013-), which in our study were tourists and residents. The results for each factor are obtained by summing the values of the grades given to each corresponding drawing. The second part of the measurement instrument included closed-ended questions on participation in tourist activities and demographic data of children and their families.

There are many reasons why we decided to use this measurement instrument in the study. First, non-verbal semantic differentials can be understood by children who speak different languages. Its important feature is reducing social conformism because it is tough for respondents to control the answers to abstract drawings rather than words (Bujas, 1967). Moreover, when children observe the standardized drawings, they give their own meaning to them based on their own experiences, thoughts, feelings, and imaginations. They bring it all into communication with the instrument template. As much as the technique is different from a child's usual experience with various other questionnaires found in journals or online, children tend to engage with this technique, as it is usually very entertaining and attractive to them (Catterall & Ibbotson, 2000). Moreover, contrary to verbal expressions, in non-verbal sematic differentials changes in meaning over time are insignificant, making Bujas' test relevant. However, there are very few such tests. Apart from Bujas' test, the most famous is the Bentler and Lavoie test (1972), which is more extensive. It consists of an additional two factors to the classical semantic Osgood differential, while Bujas's version directly follows Osgood's differential. Therefore, we thought it would be better to use the Bujas test, because it is shorter and simpler for use in children-centered research.

The description of the study sites

As a continent, Europe is the number one tourist destination in the world, leading in international tourist arrivals (Statista, 2020). The six tourism destinations chosen for this study were selected because all of them are based in countries that have seen an increase in tourist arrivals in recent years (UNWTO, 2018) and they all represent local communities with exclusive subordination to tourism development. The following is a brief description of each destination, which was paired based on the predominant type of tourism development (i.e., sun and sea tourism, winter tourism, and health and wellness tourism).

Sun and sea tourism destination are represented by the cities of Opatija in Croatia and Malaga in Spain. The city of Opatija developed from a small fishing village to a prestigious tourism resort in 1844. Almost 50 years later, Opatija was pronounced a health resort, then saw rapid transition into an elite summer resort, known as the Nice of the Adriatic. This was due to the construction of landscape parks, luxurious villas, hotels, and similar facilities at the beginning of the 20th century and as such it was considered the cradle of Croatian tourism (Grad Opatija, 2019). The city boasted a total of 1.45 million overnight stays in 2018 (MINT, 2019), of which more than 40% were in family-owned accommodations (private rooms and apartments). When asked about their attitudes towards tourism, residents of Opatija's responses were positive (Birkić et al., 2019) and there was the belief that the application of the smart city concept in Opatija would improve their quality of life and also have a positive impact on future tourism development (Krstinić Nižić et al., 2020).

Malaga started its tourism development boom in the early 1950s. Nowadays, it is considered the second-fastest Spanish province when it comes to increase in economic growth, mostly thanks to tourism. The tourism industry activity increased by 127% in the last decade, with almost 300 cruising ships docking in Malaga's port in 2017, making it the fastest-growing port in Spain and the second fastest one in the Mediterranean (Ministerio de Fomento & Puertos del Estado, 2018). In 2018, the city recorded 27.6 million overnight stays, of which more than 70% were in hotels, followed by apartments (23%) and camps (7%) (Turismo y Planificación Costa del Sol, 2019, Instituto Nacional de Estadística, 2018, 2019). Malaga has seen rapid growth in tourist numbers in recent years, similar to other Spanish cities (e.g., Barcelona, Madrid, etc.). Hence, the general academic proposition has been that this rapid growth would result in many residents exhibiting tourism-phobia. However, a recent study by Almeida-García et al. (2019) shows that residents do not view the presence of tourists or cruise passengers as something negative, but rather they are appalled by some of their activities (e.g., binge drinking, bad behaviour by tourists, dirt and noise) that they believe could be addressed and changed by improving the city's tourism management.

Among the two winter tourism destinations, Kranjska Gora in Slovenia and Erzurum in Turkey, the former has a history of tourism development dating back to 1904. Kranjska Gora is one of the top five tourism destinations in Slovenia (MGRT, 2017). It saw the beginnings of the growth of its winter and ski tourism in 1930. Kranjska Gora quickly became a synonym for mountaineering and skiing, at present offering more than 18 different ski programs. In 2018, there was a total of 828 thousand overnight stays in Kranjska Gora in almost six thousand available beds (SURS, 2019), offered across 186 accommodation establishments, of which 85% represented privately owned apartments (Občina Kranjska Gora, 2015). According to Cigale (2009), residents of Kranjska Gora have accepted tourism with a very positive attitude, believing that tourism benefits not only individual quality of life, but also quality of life of the whole community.

Similar to Kranjska Gora, Erzurum in Turkey is another ski resort that contains two famous ski centres - Palandöken and Konaklı Ski Center. Erzurum is the highest major city in Turkey, which boasts the Palandöken ski centre that ranked 18th among the world's best ski resorts in 2012. Erzurum accounted for 1.8 million domestic and 2.3 million international overnight stays in 2018, with most of them realized in seven ski hotels (37%), followed by 59 city hotels and facilities in the city centre (Kudaka, 2018). In 2011, Erzurum hosted the Winter World University Games, the largest multi-sport event in the world apart from the Olympic Games, organized for university athletes. According to Kalkavan and Alaeddinoğlu (2017), residents of Erzurum perceived the event to be very beneficial for the city, especially its younger generations who were cheering for the city to apply to host the Winter Olympic Games.

Bad Gleichenberg in Austria and Topola in Serbia are two destinations that represent health and wellness tourism. Both destinations offer a variety of curative therapies, especially in combination with the very high mineralization of local thermal water. Bad Gleichenberg was established as a health and wellness tourism town in 1834. In that first year, 118 guests visited the booming place, which at that time was little more than a collection of villas and health clubhouses. Soon it became the favourite among the cream of European aristocracy, which by 1890 helped to swell visitations to six thousand guests (Region Bad Gleichenberg, 2020). The town saw its largest hotel built a decade ago, offering 107 rooms.

Topola is situated in the Shumadia region in Serbia and offers different tourist activities, predominately centred around health and wellness, but also around wineries, cultural events, and museums. Yearly, more than 100 thousand registered tourists visit the galleries and museums, and more than 200 thousand tourists visit a famous tourist event, the Oplenac Grape Picking (TO Topola Oplenac, 2020).

Justification and description of the sample

In this paper, the term 'young people' is frequently used interchangeably with the term 'children'. The United Nations' Convention on the Rights of the Child defines 'child' as all those under the age of 18 years (The United Nations, 2010, p. 1). The definition also acknowledges the social construction and cultural relativity of the term 'childhood', which may have a differential application in terms of the specific context, culture, or environment (Morrow, 2011). This cultural relativity is explained in the manner in which young people are placed in a conceptualised context in the studies presented in this article. Therefore, in our study, we intentionally focus on such samples within a range of tourism destinations and make a clear objective to research children's willingness to be consulted on tourism development, as well as their opportunity to actively participate in such processes. Moreover, for the purposes of this paper, we will continue using the term 'children' when addressing this social group under the age of 18 years.

Our study represents international research that involved children aged 11 to 16 who were attending school and lived in six different tourism destinations. Several factors influenced our decision to focus on children in this specific age group. First, the studies on children's attitudes towards tourism from the host perspective either sampled very young children (i.e., aged 5 to 11) (Seraphin & Green, 2019), a very wide age range (i.e., 10 to 24 years) (Canosa et al., 2018, 2019), or was very limited (i.e., 11 to 12 years) (Koščak et al., 2018). Second, children aged 11 to 16 are considered adolescents (Blakemore et al., 2010) "who can imagine things not seen or experienced" (Arrington-Sanders, 2013, p. 354). This allows them to develop the capacity to think abstractly and to form formal operational thinking characteristic i.e., meta-cognition (Arrington-Sanders, 2013). The latter was considered more suitable for the research we designed.

The sample comprised a total of 498 children, with 53.7% female and 46.3% male respondents. As seen in Table 1, most children participating in the study were 12 to 13 years of age, with an average age of the sample at 12.5 years. Nearly 80% of children came from families in which both parents were employed, with 22.3% of parents and 44% of relatives employed in the industry.

Table 1. Demographic characteristics of the sample (in %).

Demographics	Overall	Erzurum, Turkey	Bad Gleichenberg, Austria	Opatija, Croatia	Malaga, Spain	Topola, Serbia	Kranjska Gora, Slovenia
Number of participants	498 (100%)	60 (12.0%)	96 (19.3%)	84 (16.9%)	88 (17.7%)	94 (18.9%)	76 (15.3%)
Gender							
Male	46.3%	41.7%	49.5%	49.4%	40.9%	50%	44.6%
Female	53.7%	58.3%	50.5%	50.6%	59.1%	50%	55.4%
Age							
11 years	13.7%	26.7%	0%	3.6%	0%	0%	64.5%
12 years	37.8%	28.3%	43.2%	73.8%	0%	48.9%	28.9%
13 years	31.8%	35%	40%	22.6%	59.1%	24.5%	6.6%
14 and more	16.7%	10%	16.9%	0%	40.9%	25.8%	0%
Employment of parents							
Both parents	76.2%	70%	77.9%	82.1%	83%	55.3%	90.7%
Just father	16.5%	25%	15.8%	14.3%	13.6%	27.7%	2.7%
Just mother	4.6%	1.7%	3.2%	3.6%	3.4%	8.5%	6.7%
None	2.6%	3.3%	3.2%	0%	0%	8.5%	0%
Parents employed in tourism							
Both parents	22.4%	8.3%	5.3%	35.7%	20.5%	18.1%	48%
None	73%	90%	83.2%	64.3%	67%	81.9%	52%
Only one parent	4.6%	1.7%	11.%	0%	12.5%	0%	0%
Relatives employed in tourism							
Yes	44.4%	20%	43.2%	53%	46.6%	38.3%	60%
No	55.6%	80%	56.8%	47%	53.4%	61.7%	40%

Results and discussion

Children's attitudes towards tourism

Attitudes towards tourism were educed with multiple questions that were simplified so children could understand them. The first set of questions elicited their opinion on the tourism and hospitality industry in general. Children were asked if they thought tourism is a simple or difficult job. Only about 11% of children (10.6%) answered that the job is simple, 44.8% that job is demanding, and 44.6% stated they were not able to answer the question. When comparing the different destinations, the $\chi2$ test showed no statistically significant differences between samples, indicating that children have a somewhat similar perception of the weight of jobs in the tourist industry (simple job: $\chi2$ test value is 3.85, df = 5, critical value 11.070, $p = 0.05$; hard job: $\chi2$ test value is 24.20, df = 5, critical value 11.070, $p = 0.05$) with the exception of children from Turkish Erzurum who, compared to children in other destinations (see Table 2), perceive jobs in the tourism industry to be easy.

The results of the $\chi2$ test revealed exciting differences in children's perception of tourism. For example, Austrian children believe that tourism work is mostly done in offices, which is probably due to the destination type with Bad Gleichenberg's wellness and spa happening behind closed doors. On the other hand, no significant difference was found in children from Topola in Serbia, although it is a destination primarily offering wellness and spa activities, as is the case with Austria. The most significant discrepancy (according to the adjusted residual) was found in opinions on whether the work in tourism was being done outside or inside in two winter destinations. Children in Erzurum estimated that most work was done outdoors, in contrast to children in Kranjska Gora who gave a below-average rating to outside work. This difference could be

Table 2. Differences in children's attitudes towards tourism.

	Erzurum, Turkey	Bad Gleichenberg, Austria	Opatija, Croatia	Malaga, Spain	Topola, Serbia	Kranjska Gora, Slovenia	Chi-square test
In your opinion, where is most of the tourism work done?							
In the office	**4** **3.92%**	**40** **39.22%**	12 11.76%	20 19.61%	15 17,65%	**8** **7.84%**	$\chi2 = 47.88$, df = 5
With machines in factories	1 7.14%	3 21.43%	1 7.14%	1 7.14%	**6** **42.86%**	2 14.29%	$\chi^2 = 8.29$, df = 5 (n.s.)
Outside	**39** **32,77%**	**6** **5.04%**	21 17.65	26 21.65%	20 16.81%	**7** **5.88%**	$\chi^2 = 34.46$, df = 5
A little bit of everything mentioned before	**16** **6.23%**	46 17.90%	47 18,29%	40 15.56%	50 19.46%	**58** **22.57%**	$\chi^2 = 24.20$, df = 5
What kinds of characteristics must people that work in tourism have?							
They must know a lot about various things.	**33** **28.70%**	23 20.00%	**6** **5.22%**	**7** **6.09%**	**34** **29.57%**	12 10.43%	$\chi^2 = 41{,}68$, df = 5
They must know many foreign languages.	**3** **1.33%**	41 18.14%	**58** **25,66%**	**54** **23.89%**	**36** **15.93%**	34 15.04%	$\chi^2 = 50.69$, df = 5
They must have a lot of money.	**7** **46.67%**	**0** **0%**	1 6.67%	1 6.67%	4 26.67%	2 13.3%	$\chi2 = 13.4$, df = 5
They must know how to adapt to certain situations.	17 14.29%	31 26.05%	**6** **5.04%**	21 17.65%	20 16.81%	24 20.17%	$\chi2 = 17.99$, df = 5
Who do you think is more appropriate for a job in tourism?							
Women	5 7.25%	13 18.84%	10 14.49%	11 15.94%	15 21.74%	15 21.74%	$\chi2 = 6.21$, df = 5 (n.s.)
Men	**10** **28.57%**	**0** **0.00%**	2 5.71%	3 8.57%	**12** **34.29%**	8 22.86%	$\chi2$ 20.3, df = 5
Equal	38 10.98%	73 21.10%	57 16.47%	**73** **21.10%**	**57** **16.47%**	48 13.87%	$\chi2 = 16.50$, df = 5
Cannot say	7 16.28%	9 20.93%	**12** **27.91%**	**0** **0.00%**	10 23.26%	5 11.63%	$\chi2$ 20.3, df = 5

Note: Bolded responses denote statistically significant differences among destinations (at $p < 0.05$); n.s.= not significant.

attributed to ski resorts in Kranjska Gora being located in the middle of the town, while those in Erzurum are in the surrounding mountains.

Our results indicated statistically significant differences for children from two sun and sea destinations – Opatija in Croatia and Malaga in Spain. Children believe that employees' knowledge does not need to be very diverse. In contrast, children from Turkish Erzurum and Serbian Topola believe that employees' knowledge needs to be diverse. However, Turkish children do not recognize knowledge of foreign languages as necessary for the industry, which is statistically different from the other children's opinions. Turkey's domestic tourism is by far higher than its international tourism (120 Mio domestic arrivals vs 31 Mio international arrivals, OECD, 2020), hence, knowledge of foreign languages may be perceived as nonessential. Similar observations were made by Leslie and Russell (2006), where UK students expressed that foreign language skills are not essential for tourism.

When it comes to assessing whether industry employees have lots of money, children were almost unanimous in believing that those who work in tourism do not have it. This observation is similar to Canosa et al. (2019) study where children from Australia perceived tourism as attractive to migrant workers or temporary residents that are willing to work for low wages or just board. This puts the host-children in a very disadvantageous position when entering adulthood and looking for employment opportunities in the industry. Tourism is, unfortunately, a labour-intensive industry with meagre wages (Dwyer, 2018; Løseth, 2018) and our results show that this has been noticed by children. Additionally, children evaluated if employees need to adapt to certain situations. Except for Croatian children, the others believe that adaptation is necessary. Opatija, on the other hand, is a well-established destination with a very long history of tourism. With its typical customers knowing precisely what they can expect, but also with businesses seeing success in their *modus operandi*, such as the "*business as usual* paradigm" (Dwyer, 2018, p. 3, *original emphasis*), very few situations are created where employees need to act quickly and adapt to newness. On the topic of gender equality, children in all destinations think that both sexes are equally suitable for work in tourism, while only children in Erzurum and Topola think in statistically higher proportions that tourism is a job for men. In line with this, tourism research shows that gender inequalities in the industry still exist (Xu, 2018) regardless of some studies finding support for the initial hypothesis that tourism can help in improving women's statuses by providing job opportunities and increasing their income (Gibson, 2001).

Children's synthesis of tourism and family life

The opinion on the impact of tourism on family life was elicited from children with several different questions (see Table 3). Concerning one of the biggest problems for family life associated with tourism, we pointed out working hours. Hence, we asked children about their opinion on whether people mostly engage in tourism activities in the daytime. Children from all destinations do not differ in perceiving little activity occurring only in the morning or only in the afternoon. Most of them perceive tourism activity taking place both in the mornings and afternoons and even nights. More specifically, 75% of Turkish children saw that tourism requires almost all day and night commitment, with similar observations reported by Kusluvan and Kusluvan (2000) for Turkish undergraduate tourism and hospitality students.

When asked whether the rhythm of work affects family life, Turkish, Croatian, Serbian, and Spanish children statistically significantly differed in affirmative responses from children in Austria and Slovenia. Many children believe that tourism negatively affects family life. This corresponds to similar findings in the studies of Chan et al. (2015) and van Schalkwyk et al. (2006), which report tourism impacts families in a way that it intensifies and transforms their schedules. In combination with children's observations that tourism demands significant commitment, the results point to the industry's transformative power over family life.

Table 3. Differences in children's perception of tourism in connection to family life.

	Erzurum, Turkey	Bad Gleichenberg, Austria	Opatija, Croatia	Malaga, Spain	Topola, Serbia	Kranjska Gora, Slovenia	Chi-square test
What kind of working hours do people who work in tourism have?							
In the morning	2	1	2	5	2	5	$\chi2 = 5.24$, $df = 5$ (n.s.)
	11.76%	5.88%	11.76%	29.41%	11.76%	29.41%	
In the afternoon	5	3	2	1	8	3	$\chi^2 = 8.55$, $df = 5$ (n.s)
	22.73%	13.64%	9.09%	4.55%	36.36%	13.64%	
In the morning and afternoon	**8**	48	41	**26**	48	**44**	$\chi^2 = 35.19$, $df = 5$
	3.72%	22.33%	19.07%	**12.09%**	22.33%	**20.47%**	
Mornings, afternoons and even nights	**45**	41	34	**57**	**34**	**24**	$\chi^2 = 16.31$, $df = 5$
	19.15%	17.45%	14.47%	**24.25%**	**14.47%**	**10.21%**	
Do you think that working hours like that affect family life negatively?							
Yes	**42**	37	**19**	**54**	**14**	27	$\chi^2 = 35.03$, $df = 5$
	21.76%	19.17%	**9.84%**	**27.98%**	**7.25%**	13.99%	
No	**4**	**27**	**25**	12	18	13	$\chi2 = 22.63$, $df = 5$
	4.04%	**27.27%**	**25.25%**	12.12%	18.18%	13.13%	
I do not know	**14**	**29**	36	**22**	**60**	36	$\chi2 = 37.91$, $df = 5$
	7.11%	**14.72%**	18.27%	**11.17%**	**30.46%**	18.27%	
Would you like to work in tourism someday?							
Yes	38	8	17	12	20	15	$\chi2 = 29.96$, $df = 5$
	34.55%	7.27%	15.45%	10.91%	18.18%	13.64%	
No	11	55	37	44	31	32	$\chi2 = 31.03$, $df = 5$
	5.24%	26.19%	17.62%	20.95%	14.76%	15.24%	
I do not know	11	31	26	32	42	29	$\chi2 = 18.02$, $df = 5$
	6.43%	18.13%	15.20%	18.71%	24.56%	16.96%	

Note: Bolded responses denote statistically significant differences among destinations (at $p < 0.05$); n.s.= not significant.

Most children who would like to work in tourism are from Erzurum, while those with the least ambition to work in the industry are from Bad Gleichenberg. We can speculate that these differences have to do with the economic development of the regions and countries, as well as employment opportunities. Therefore, tourism in Turkey creates three times more jobs than in Austria, with international visitors expected to spend more than in Austria (WTTC, 2019a, 2019b). In general, similar to the findings of Canosa et al. (2019), our study also found that children recognize that the tourism and hospitality industry creates employment opportunities, but at the same time these are not viewed through 'rose colored glasses'. That is, the industry is perceived to demand all-day commitment and to have a negative impact on family life, the two reasons why children are either predominately indecisive or opposed to employment in the industry.

Involvement of children in tourism development

Our results showed that 64% of children were relatively unfamiliar with the tourism industry, with 18% declaring they know virtually nothing about tourism and 18% stating that they are familiar with tourism on a relatively large scale. Also, we asked children to express their involvement in tourism planning and development of communities in which they live.

As evident in Table 4, children differ in how much they know about the plans for further tourism development. Children from Opatija, Bad Gleichenberg, and Kranjska Gora were shown to know more about tourism development plans for the towns in which they live than children in other destinations. All three former destinations have a common origin and culture, and tourism development has occurred at almost the same time. Similarly, in Opatija and Kranjska Gora, almost 25% of children have been invited by local authorities to state their opinions on future tourism development, while to others the invitation was extended to only 14% or 20%. However, the general conclusion is that children are not generally informed about future tourism

Table 4. Children's involvement in tourism development.

	Erzurum, Turkey	Bad Gleichenberg, Austria	Opatija, Croatia	Malaga, Spain	Topola, Serbia	Kranjska Gora, Slovenia	Chi-square test
Do you know what tourism development plans are in the town you live in?							
Yes	7 8.14%	21 24.42%	**23** **26.74%**	**9** **10.47%**	**8** **9.30%**	18 20.93%	$\chi2 = 17.81$, df = 5
A little	**13** **6.34%**	41 20.00%	40 19.51%	43 20.98%	35 17.07%	33 16.10%	$\chi2 = 17.82$, df = 5
No	**40** **19.90%**	32 15.92%	**20** **9.95%**	36 17.91%	**49** **24.38%**	24 11.94%	$\chi2 = 16.82$, df = 5
Have you been invited to state your opinion about tourism development plans?							
Yes	10 10.00%	13 13.00%	25 25.00%	17 17.00%	13 13.00%	22 22.00%	$\chi2 = 10.16$ df = 5 (n.s.)
No	48 12.47%	81 21.04%	**59** **15.32%**	69 17.92%	75 19.48%	**53** **13.77%**	$\chi2 = 13.04$, df = 5
Would you like to be a part of the discussion about future tourism development?							
Yes, very much	**11** **7.64%**	**8** **5.56%**	**36** **25.00%**	21 14.58%	**40** **27.78%**	28 19.44%	$\chi2 = 35.42$, df = 5
No, I am not interested	21 15.56%	28 20.74%	**16** **11.85%**	27 20.00%	22 16.30%	21 15.56%	$\chi2 = 4.33$, df = 5 (n.s.)
I do not know what to say	28 13.80%	**58** **27.10%**	32 14.95%	40 18.69%	**30** **14.02%**	26 12.15%	$\chi2 = 20.06$, df = 5

Note: Bolded responses denote statistically significant differences among destinations (at $p < 0.05$); n.s.= not significant.

development (Koščak et al., 2018). However, similar to Canosa et al. (2018) study in Australia, children wish to be part of the tourism discussion. Almost one third expressed their desire to participate in such discussions (29.21%). Moreover, children from Opatija and Topola are statistically significantly different from others in their wish to participate in tourism development. We speculate that these differences are rooted in society and culture, as both destinations were historically endowed with socialism in times of Yugoslavia. At that time, many public interest issues led to wide-ranging public debates, especially in smaller local communities (Tollefson, 2002), a praxis that might have remained important to this day.

Concerning giving their opinions about community planning and development, the children in this study reported that they were not invited to share their thoughts even though they would like to do so. This could be problematic as being discouraged from sharing their thoughts about the future of the community could lead to apathy when they enter adulthood. This creates fertile ground for a community in which the majority of residents are passive observers of the future development of their community and few individuals are orchestrators of the same (Šegota et al., 2017). To avoid this, the literature calls for greater empowerment of residents (Boley & McGehee, 2014; Joo et al., 2020; Mihalič et al., 2016; Palardy et al., 2018), which could be applied to young residents at an early age so that they are listened to and included in the planning of the future of their community.

Eliciting children's feelings about tourism

The development of children in environments where the tourism industry plays an important role has been conditioned by many socio-cultural, political, economic, and ideological forces at play (Knežević et al., 2019). Development in such an environment provokes a series of emotional elements in connection to tourists and locals. The host-tourist relationship inevitably leaves a mark on the development of children, as well as their decisions regarding their future professions. To study these emotional elements, we used non-verbal semantic differential scales and asked the children how they felt about tourists and residents. Here, we must remind the reader that, when interpreting data, one should always keep in mind that this experimental technique

Table 5. Children's feelings towards tourists and residents.

Factor and object	Erzurum, Turkey M (SD) R	Bad Gleichenberg, Austria M (SD) R	Opatija, Croatia M (SD) R	Malaga, Spain M (SD) R	Topola, Serbia M (SD) R	Kranjska Gora, Slovenia M (SD) R
Potency	**6.90 (3.59)**^{bcdf}	2.58 (4.57)^{a}	1.99 (5.21)^{ae}	3.20 (4.94)^{a}	4.44 (3.95)^{cf}	1.77 (5.79)^{ae}
of tourists	1	1	1	2	2	1
Activity	3.10 (3.10)^{e}	2.16 (4.82)^{e}	1.32 (5.76)^{e}	3.48 (5.08)^{f}	**5.37 (4.37)**^{abcf}	0.50 (6.40)^{e}
of tourists	2	2	2	1	**1**	5
Evaluation	−0.75 (3.26)^{e}	0.93 (4.22)^{e}	0.53 (4.38)^{e}	1.40 (4.69)^{e}	**3.77 (3.99)**^{abcdf}	0.67 (5.35)^{e}
of tourists	3	3	3	3	**3**	4
Activity	−3.13 (3.12)^{cdef}	−1.25 (3.76)	−0.13 (3.97)^{a}	**0.18 (4.18)**^{a}	0.37 (2.91)^{a}	0.53 (3.62)^{a}
of residents	6	9	5	**4**	5	5
Potency	**−2.45 (3.97)**	−0.52 (3.81)^{f}	−0.47 (4.42)^{f}	−1.09 (3.74)^{f}	−1.97 (3.81)	−3.30 (4.85)^{bcd}
of residents	**4**	6	6	8	9	9
Evaluation	−4.92 (3.50)^{bcdef}	**−0.84 (4.13)**^{a}	−1.10 (4.65)^{a}	−0.84 (6.07)^{a}	−0.47 (4.33)^{a}	−0.57 (4.07)^{a}
of residents	9	**7**	9	7	7	7

Note: M = Mean; SD = Standard deviation, R = overall rating. The subscripts represent statistically significant differences (at p = 0.000) between destinations, where subscript a = Erzurum, b = Bad Gleichenberg, c = Opatija, d = Malaga, e = Topola, f = Kranjska Gora. The highest results for factor and object are indicated with bold for each row.

is a projective type. The drawings are standardized; however, children give their own meaning to the drawings based on their own experiences, thoughts, feelings, and imaginations.

Our results reveal that, in general, children gave tourists much better ratings than residents on all three factors. Residents are perceived as being more passive, evaluated worse than tourists, and seen as weak, whereas tourists are evaluated more positively, but this evaluation is lower than their potency and activity. Moreover, the results reveal that potency is perceived as the most crucial tourist trait, followed by activity and evaluation. This suggests that children recognize the change or disruption tourists bring to communities (Graburn, 1983), which is also consistent with the results from studies by Canosa et al. (2019) and Koščak et al. (2018).

The detailed results of data analysis are presented in Table 5. The data were analyzed using One Way ANOVA, coupled with the Scheffe Post Hock Multiple Comparisons test.

In evaluating tourists, children's perceptions were very consistent among destinations. The latter is evident from the results of the overall ratings, where tourists were rated in the top three places, with relatively few cross-cutting differences in ratings. Estimates differed in the amount, but it should be emphasized that in the non-verbal semantic differential scales, the ranking of the object is the most important indicator. However, statistically significant differences in grades given by Erzurum children for tourist potency are two to three times higher than in other destinations. This indicates that the strength of the change tourist activity imposes on the local community is more significant. Also, among former Yugoslav countries, there are statistically significant differences in children grading the potency of tourists, with children from Serbia grading potency two times higher than their peers in Slovenia and Croatia. Such results could be attributed to both Opatija and Kranjska Gora being established tourism destinations with a high number of visitors (Grad Opatija, 2019; MGRT, 2017). Hence, children from these two local communities might perceive tourist activity not to be intense as their communities are established or matured destinations. On the other hand, Topola receives few visitors and its tourism is only developing compared to the other two destinations. This is similar to many studies on resident attitudes towards tourism in which it was reported that residents accept tourist activity as part of the everyday in established and mature destinations, whereas in new destinations residents are much more attentive to the activity impacting daily life (McKercher et al., 2015). Moreover, evaluation of the tourists is low, with the exception of children in Topola, whose evaluation score is statistically different from other destinations, i.e., more positive. What is important to note is that it was evident that children do not have some extremely positive emotions associated with tourists. Apart from the rating 6.9 for potency given to tourists by Erzurum children and the rating 5.75 for the activity given by Topola children, all other ratings are below 50% of the possible estimate (12). The fact that children consider tourists to be strong probably suggests that they perceive tourists to have the power to manage their lives. The result of the correlation coefficient supports these conclusions, as Spearman's rank correlation coefficient between potency and activity factors is $r_s = 0.511$ (statistically significant at $p = 0.001$ level).

In evaluating residents, children's most pronounced experience of the local population is the experience of the activity, with an average rank of 5.67. The latter ranked significantly lower than the tourist activity. We would speculate that this is related to the role of the local community in a tourism destination. In our case, as in the case of Canosa et al. (2019) and Koščak et al. (2018), children perceive residents as marginal community groups as opposed to tourists, indicating towards their subordinate role in the resident-tourist exchange process. Moreover, children in most surveyed destinations do not perceive tourism as a desirable life orientation. This has been suggested by the position of the potency and evaluation factors of residents and the Spearman's rank correlation coefficient between evaluation and activity factors ($r_s = 0.138$, $p = 0.001$). Interestingly, we found a negative correlation between the evaluation of residents and the potency of tourists ($r_s = -0.127$, $p = 0.001$). This suggests that children give a lower rating in the evaluation of residents whenever they rate tourist potency higher.

Such findings suggest that children have internalized the widespread view that tourists disrupt residents' lives with their visits to the community and, moreover, that residents subordinate their daily lives to tourist experiences. Canosa et al. (2018) suggest that for the creation of healthy and socially sustainable host communities, children need to be involved in community planning and development. Not only are they quickly reaching their adulthood, which will officially grant them more power to influence tourism decision-making, but they are also future consumers of community resources and providers of tourist experiences, which puts them in a position of powerful negotiators when it comes to how these resources are used across different groups, including visitors (Seraphin & Green, 2019). Therefore, to make them advocates of tourism in the future, the community needs to see them as equal contributors to decision-making to make their experience of the guest-host relationship as positive as possible.

Conclusion

In ten years from now, we will be able to evaluate tourism's contribution to developmental goals set in the Sustainable Tourism Programme and the 2030 Agenda for Sustainable Development and the Sustainable Development Goals. More specifically, many goals set in these agendas wish to see young people live in inclusive and developed societies, where sustainable tourism serves as the tool to improve quality of life, promote inclusivity, and advance knowledge (WTO & UNDP, 2017). However, one of the crucial steps towards a more inclusive society is to listen to all voices. This study represents one step toward the latter: we listened to the opinions of children in six diverse European destinations for whom tourism represents an immanent future as an employment opportunity or/and a force that shapes their quality of life once they reach adulthood. However, it is important to understand that children will enter adult community life with attitudes and feelings they carry from childhood (van Schalkwyk et al., 2006), and the latter are something that we know little about. Moreover, this study is very important as it sets the basis for understanding children's attitudes towards tourism in destinations with three predominant types of tourism development (i.e., sun and sea tourism, winter tourism, and health and wellness tourism).

We designed the study as experimental, one-group posttest-only research with the aim to better understand children's attitudes towards tourism and its impacts on family life, inclusiveness in decision-making, and feelings towards the tourist-host relationship that have been playing out in tourism communities. The results of our research bring to light both positive and negative effects of tourism, which the industry will have to address if it wishes to be considered as contributing to, instead of impeding sustainable development goals. More specifically, our results show that children notice very early on that being employed in the tourism and hospitality industry means one is doing a challenging job, and the higher the economic development of a tourist area, the more tourism is perceived as a difficult job. Moreover, children see tourism as a disruptive power for family life and for the community. The results show that children feel negatively towards tourists. In our case, children perceive residents as a marginal community group as opposed to tourists, indicating their subordinate role in the resident-tourist exchange process (Canosa et al., 2019; Koščak et al., 2018). Moreover, children in most surveyed destinations do not perceive tourism as a desirable life orientation and they feel the strength of the change tourist activity imposes on their local communities. There are two probable reasons for this. One is that children feel part of the local population, which includes their family members, who they also perceive to be subordinated to tourists. Such perception certainly does little to contribute to changing the perception of tourism becoming a desired life orientation and a force for social change. Another reason may arise from the fact that children are not involved in tourism development and planning. In fact, we can conclude that their involvement in the process has been largely ignored. This approach to children, which excludes them from tourism planning and developmental processes, in no way contributes to sustainability in general or/and to tourism being a tool for an inclusive society in particular (Canosa et al., 2019).

Our study is not without limitations. We see these as being related to our research design, in which we used a non-verbal semantic differential measurement instrument. In the latter, children are exposed to the standardized drawings, to which they give their own meaning based on their experiences. However, we have not explored what the standardized drawings mean to children and what associations they have in relation to their experiences and lived realities. This could be overcome in future studies if qualitative methods would complement the use of non-verbal semantic differential measurement instrument to give children the opportunity to explain why they chose certain images instead of others. Further, some questions posed to children might be viewed as too simplistic. Other researchers are likely to encounter different attitudes of tourism if the study was designed differently (i.e., using standard Likert-type scale answers and measurement scales commonly used in resident attitudes research). However, they should be encouraged to explore a different approach to research that includes children, such as ours. Moreover, our results showed that some children do not know how to answer some questions we asked or were indecisive. We see this as a limitation that could be explored in the future. We did not delve further into understanding children's indecisiveness or where their 'I do not know' answers come from by asking additional questions. However, future work, if it uses similar approach to ours by providing respondents with the indecisive options as answers, may be interested in uncovering the origins of children's indecisiveness. The latter could provide valuable insights into what kind of information children need to make informed decisions and form attitudes.

As for future tourism research, our study sets the groundwork for understanding children's attitudes towards tourism in three different types of tourism destinations (sun and sea, winter, and health and spa destinations). New research could take the same approach and examine differences across destinations in different stages of the life cycle. It would also be interesting to see whether children's attitudes change if they are involved in tourism planning and how this inclusion varies across different stages of planning and implementation, as well as different ages. Further, it would be interesting to examine how attitudes towards tourism are affected by parent-child relationships and if attitudes are passed on intergenerationally.

In general, the above findings add to the literature on sustainable tourism development, particularly by demonstrating that children have their views and are willing to express themselves given the opportunity. Our study demonstrates that children have a unique insider's perspective, which tourism managers and tourism planning officials should not neglect but instead include in their planning process in order to make tourism development sustainable, empowering, and inclusive.

Acknowledgement

We would like to thank the teachers in primary schools in the cities of Opatija, Malaga, Kranjska Gora, Erzurum, Bad Gleichenberg and Topola for motivating their students to participate in our research and for preparing adequate conditions for the accomplishment of research. We would also like to thank all the students in these cities who accepted our invitation and participated with the utmost seriousness in the realization of the research. The way these children have contributed tells us that they are ready to participate in the development of the destinations they live in.

Disclosure statement

No potential conflict of interest was reported by the authors.

ORCID

Marko Koščak http://orcid.org/0000-0002-2332-5639
Mladen Knežević http://orcid.org/0000-0001-6737-2114
Daniel Binder http://orcid.org/0000-0001-5604-2566
Antonio Pelaez-Verdet http://orcid.org/0000-0003-2545-5776
Cem Işik http://orcid.org/0000-0001-5125-7648
Vladimir Mićić http://orcid.org/0000-0002-0719-8481
Katarina Borisavljević http://orcid.org/0000-0002-1320-8156
Tina Šegota http://orcid.org/0000-0001-7994-1396

References

Agyeman, J., & Evans, B. (2004). 'Just sustainability': The emerging discourse of environmental justice in Britain? *The Geographical Journal*, *170*(2), 155–164. https://doi.org/10.1111/j.0016-7398.2004.00117.x

Al-Frehat, M. M., & Alhelalat, J. A. (2015). A study on the educational status of tourism child labourers in Petra, Jordan. *Mediterranean Journal of Social Sciences*, *6*(4), 20–29.

Almeida-García, F., Cortés-Macías, R., & Balbuena-Vázquez, A. (2019). Tourismphobia in historic centres: The case of Malaga. *Boletín de la Asociación de Geógrafos Españoles*, *83*(83), 1–32. https://doi.org/10.21138/bage.2823

Andereck, K. L., & Nyaupane, G. P. (2011). Exploring the nature of tourism and quality of life perceptions among residents. *Journal of Travel Research*, *50*(3), 248–260. https://doi.org/10.1177/0047287510362918

Andereck, K. L., Valentine, K. M., Knopf, R. C., & Vogt, C. A. (2005). Residents' perceptions of community tourism impacts. *Annals of Tourism Research*, *32*(4), 1056–1076. https://doi.org/10.1016/j.annals.2005.03.001

Anderson-Fye, E. P. (2004). A "coca-cola" shape: Cultural change, body image, and eating disorders in San Andrés, Belize. *Culture, Medicine and Psychiatry*, *28*(4), 561–595. https://doi.org/10.1007/s11013-004-1068-4

Ap, J. (1992). Residents' perceptions on tourism impacts. *Annals of Tourism Research*, *19*(4), 665–690. https://doi.org/10.1016/0160-7383(92)90060-3

Arrington-Sanders, R. (2013). Adolescent psychosocial, social, and cognitive development. *Pediatrics in Review*, *34*(8), 354–359. https://doi.org/10.1542/pir.34-8-354

Bentler, P. M., & Lavoie, A. L. (1972). A nonverbal semantic differential. *Journal of Verbal Learning and Verbal Behavior*, *11*(4), 491–496. https://doi.org/10.1016/S0022-5371(72)80032-X

Birkić, D., Primužak, A., & Erdeljac, N. (2019). Sustainable tourism development of coastal destination – The role and the significance of local residents. *TOSEE - Tourism in South East Europe*, *5*, 101–119. https://doi.org/10.20867/tosee.05.21

Blakemore, S.-J., Burnett, S., & Dahl, R. E. (2010). The role of puberty in the developing adolescent brain. *Human Brain Mapping*, *31*(6), 926–933. https://doi.org/10.1002/hbm.21052

Boley, B. B., & McGehee, N. G. (2014). Measuring empowerment: Developing and validating the Resident Empowerment through Tourism Scale (RETS). *Tourism Management*, *45*, 85–94. https://doi.org/10.1016/j.tourman.2014.04.003

Bramwell, B., & Lane, B. (2011). Editorial: Critical research on the governance of tourism and sustainability. *Journal of Sustainable Tourism*, *19*(4–5), 411–421. https://doi.org/10.1080/09669582.2011.580586

Bujas, Z. (1967). Graphic form of Osgood's semantic differential. *Acta Instituti Psyhologici Universiti Zagrabiensis*, *49–63*, 5–12.

Buzinde, C. N., & Manuel-Navarrete, D. (2013). The social production of space in tourism enclaves: Mayan children's perceptions of tourism boundaries. *Annals of Tourism Research*, *43*, 482–505. https://doi.org/10.1016/j.annals.2013.06.003

Byrd, E. T. (2007). Stakeholders in sustainable tourism development and their roles: Applying stakeholder theory to sustainable tourism development. *Tourism Review*, *62*(2), 6–13. https://doi.org/10.1108/16605370780000309

Canosa, A., & Graham, A. (2016). Ethical tourism research involving children. *Annals of Tourism Research*, *61*, 219–221. https://doi.org/10.1016/j.annals.2016.07.006

Canosa, A., Graham, A., & Wilson, E. (2018). Growing up in a tourist destination: Negotiating space, identity and belonging. *Children's Geographies*, *16*(2), 156–168. https://doi.org/10.1080/14733285.2017.1334115

Canosa, A., Graham, A., & Wilson, E. (2019). My overloved town: The challenges of growing up in a small coastal tourist destination (Byron Bay, Australia). In C. Milano, J. M. Cheer & M. Novelli (Eds.), *Overtourism: Excesses, discontents and measures in travel and tourism* (pp. 190–204). Routledge.

Canosa, A., Moyle, B., & Wray, M. (2016). Can anybody hear me? A critical analysis of young residents' voices in tourism studies. *Tourism Analysis*, *21*(2), 325–337. https://doi.org/10.3727/108354216X14559233985097

Canosa, A., Wilson, E., & Graham, A. (2017). Empowering young people through participatory film: A postmethodological approach. *Current Issues in Tourism*, *20*(8), 894–907. https://doi.org/10.1080/13683500.2016.1179270

Carr, N. (2011). *Children's and families' holiday experience*. Routledge.

Catterall, M., & Ibbotson, P. (2000). Using projective techniques in education research. *British Educational Research Journal*, *26*(2), 245–256. https://doi.org/10.1080/01411920050000971

Chan, S. T. M., Kwok, C. L. Y., & Siu, J. Y. N. (2015). The Macau family-in-transition: The perceived impact of casino employment on family relationships among dealer families. *The Open Family Studies Journal*, *7*(1), 86–95. https://doi.org/10.2174/1874922401507010086

Cigale, D. (2009). Attitude of the population of Slovenian Alpine tourist resorts toward tourism and its impacts. *Razgledi*, *31*, 93–113.

Dancause, K. N., Dehuff, C., Soloway, L. E., Vilar, M., Chan, C., Wilson, M., Tarivonda, L., Regenvanu, R., Kaneko, A., Garruto, R. M., & Lum, J. K. (2011). Behavioral changes associated with economic development in the South Pacific: Health transition in Vanuatu. *American Journal of Human Biology: The Official Journal of the Human Biology Council*, *23*(3), 366–376. doi:10.1002/ajhb.21146 https://doi.org/10.1002/ajhb.21146

Dwyer, L. (2018). Saluting while the ship sinks: The necessity for tourism paradigm change. *Journal of Sustainable Tourism*, *26*(1), 29–48. https://doi.org/10.1080/09669582.2017.1308372

Edgell, D. L., Sr., Jr. DelMastro, A. M., Smith, G., & Swanson, J. R. (2008). *Tourism policy and planning, yesterday today and tomorrow*. Butterworth–Heinemann.

Gamradt, J. (1995). Jamaican children's representations of tourism. *Annals of Tourism Research*, *22*(4), 735–762. https://doi.org/10.1016/0160-7383(95)00024-7

Garrod, B., Fyall, A., Leask, A., & Reid, E. (2012). Engaging residents as stakeholders of the visitor attraction. *Tourism Management*, *33*(5), 1159–1173. https://doi.org/10.1016/j.tourman.2011.11.014

Garson, D. G. (2013). Research design. In G. David Garson (ed.), *Blue book series* (pp. 1–53). Statistical Associates Publishing.

Gibson, H. J. (2001). Gender in tourism: Theoretical perspectives. In Y. Apostolopoulos, S. Sönmez, & D. J. Timothy (Eds.), *Women as producers and consumers of tourism in developing regions* (pp. 19–46). Praeger.

Graburn, N. H. H. (1983). Editorial comments. *Annals of Tourism Research*, *10*(1), 1–3. https://doi.org/10.1016/0160-7383(83)90045-2

Grad Opatija. (2019). *Turizam* [online]. http://www.opatija.hr/hr/turizam/

Gram, M. (2007). Children as co-decision makers in the family? *Young Consumers*, *8*(1), 19–28. https://doi.org/10.1108/17473610710733749

Higgins-Desbiolles, F. (2006). More than an "industry": The forgotten power of tourism as a social force. *Tourism Management*, *27*(6), 1192–1208. 2005.05.020 https://doi.org/10.1016/j.tourman

Instituto Nacional de Estadística. (2018). *Encuesta de ocupación hotelera*. [online]. http://www.ine.es/jaxiT3/Tabla.htm?t=2076

Instituto Nacional de Estadística. (2019). *Indicadores urban audit para municipios mayores de 20.000 habitantes*. [online]. https://www.ine.es/jaxiT3/Datos.htm?t=30143

Israfilova, F., & Khoo-Lattimore, C. (2019). Sad and violent but I enjoy it: Children's engagement with dark tourism as an educational tool. *Tourism and Hospitality Research*, *19*(4), 478–487. https://doi.org/10.1177/1467358418782736

Joo, D., Woosnam, K. M., Strzelecka, M., & Boley, B. B. (2020). Knowledge, empowerment, and action: Testing the empowerment theory in a tourism context. *Journal of Sustainable Tourism*, *28*(1), 69–85. https://doi.org/10.1080/09669582.2019.1675673

Kalkavan, A., & Alaeddinoglu, V. (2017). The research of winter Olympics: Example of Erzurum. *Journal of Current Research on Social Sciences*, *7*(2), 271–286. https://doi.org/10.26579/jocress-7.2.20

Kerr, M. M., & Price, R. H. (2018). "I know the plane crashed": Children's perspectives in dark tourism. In P. R. Stone, R. Hartmann, T. Seaton, R. Sharpley and L. White (Eds.), *The Palgrave handbook of dark tourism studies* (pp. 553–583). Palgrave Macmillan.

Khoo-Lattimore, C. (2015). Kids on board: Methodological challenges, concerns and clarifications when including young children's voices in tourism research. *Current Issues in Tourism*, *18*(9), 845–858. https://doi.org/10.1080/13683500.2015.1049129

Khoo-Lattimore, C., delChiappa, G., & Yang, M. J. (2018). A family for the holidays: Delineating the hospitality needs of European parents with young children. *Young Consumers*, *19*(2), 159–171. https://doi.org/10.1108/YC-08-2017-00730

Khoo-Lattimore, C., Prayag, P., & Cheah, B. L. (2015). Kids on board: Exploring the choice process and vacation needs of Asian parents with young children in resort hotels. *Journal of Hospitality Marketing & Management*, *24*(5), 511–531. https://doi.org/10.1080/19368623.2014.914862

Knežević, M., O'Rourke, T., Šegota, T., & Koščak, M. (2019). Inter-generational concepts of sustainability and the role of children in local tourism destinations. In M. Koščak & T. O'Rourke (Eds.), *Ethical and responsible tourism: Managing sustainability in local tourism destinations* (pp. 129–141). Routledge.

Ko, D.-W., & Stewart, W. P. (2002). A structural equation model of residents' attitudes for tourism development. *Tourism Management*, *23*(5), 521–530. https://doi.org/10.1016/S0261-5177(02)00006-7

Koščak, M., Colarič-Jakše, L., Fabjan, D., Kukulj, S., Založnik, S., Knežević, M., O'Rourke, T., & Prevolšek, B. (2018). No one asks the children, right? *Tourism: An International Interdisciplinary Journal, 66*(4), 396–410.

Krstinić Nižić, M., Vodeb, K., & Šverko Grdić, Z. (2020). The smart city concept for sustainable development of a tourist destination. *Pomorski Zbornik, 58*(1), 111–123. https://doi.org/10.18048/2020.58.07

Kudaka. (2018). *Erzurum Kış Turizmi Değerlenedirme Raporu* [online]. https://www.kudaka.org.tr/ekler/e4520-Erzurum-Kis-Turizmi-Degerlendirme-Raporu.pdf

Kusluvan, S., & Kusluvan, Z. (2000). Perceptions and attitudes of undergraduate tourism students towards working in the tourism industry in Turkey. *Tourism Management, 21*(3), 251–269. https://doi.org/10.1016/S0261-5177(99)00057-6

Leatherman, T. L., Goodman, A. H., & Stillman, T. (2010). Changes in stature, weight, and nutritional status with tourism-based economic development in the Yucatan. *Economics and Human Biology, 8*(2), 153–158. doi:10.1016/jehb.2010.05.008

Leslie, D., & Russell, H. (2006). The importance of foreign language skills in the tourism sector: A comparative study of student perceptions in the UK and continental Europe. *Tourism Management, 27*(6), 1397–1407. https://doi.org/10.1016/j.tourman.2005.12.016

Løseth, K. (2018). Knowledge development in adventure tourism businesses – the influence of serious leisure. *Annals of Leisure Research, 21*(5), 575–591. https://doi.org/10.1080/11745398.2017.1406808

Lugosi, P., Robinson, R. N. S., Golubovskaya, M., & Foley, L. (2016). The hospitality consumption experiences of parents and carers with children: A qualitative study of foodservice settings. *International Journal of Hospitality Management, 54*, 84–94. https://doi.org/10.1016/j.ijhm.2016.01.012

McGehee, N. G., & Andereck, K. L. (2004). Factors predicting rural residents' support of tourism. *Journal of Travel Research, 43*(2), 131–140. https://doi.org/10.1177/0047287504268234

McKercher, B., Wang, D., & Park, W. (2015). Social impacts as a function of place change. *Annals of Tourism Research, 50*, 52–66. https://doi.org/10.1016/j.annals.2014.11.002

Meethan, K. (2001). *Tourism in global society: Place, culture, and consumption*. Palgrave Macmillan.

MGRT. (2017). *Strategija trajnostne rasti slovenskega turizma 2017–2021* [online]. http://www.mgrt.gov.si/si/delovna_podrocja/turizem/strategija_trajnostne_rasti_slovenskega_turizma/

Mihalič, T. (2016). Sustainable-responsible tourism discourse – towards 'responsustable' tourism. *Journal of Cleaner Production, 111*(Part B), 461–470.

Mihalič, T., Šegota, T., Knežević Cvelbar, L., & Kuščer, K. (2016). The influence of the political environment and destination governance on sustainable tourism development: A study of Bled, Slovenia. *Journal of Sustainable Tourism, 24*(11), 1489–1505. https://doi.org/10.1080/09669582.2015.1134557

Ministerio de Fomento, & Puertos del Estado. (2018). *Memorias anuales 2017*. [online]. http://www.puertos.es/Memorias_Anuales/2017/index.html

MINT. (2019). *S novogodišnje zdravice u Opatiji: U Hrvatskoj lani 6,5 posto više turista i 4 posto više noćenja* [online]. https://mint.gov.hr/vijesti/s-novogodisnje-zdravice-u-opatiji-u-hrvatskoj-lani-6-5-posto-vise-turista-i-4-posto-vise-nocenja/17776

Molero, F., Navas, M'a S., GonzáLez, J. ' L., AlemáN, P., & Cuadrado, I. (2003). Paupers or riches: The perception of immigrants, tourists and ingroup members in a sample of Spanish children. *Journal of Ethnic and Migration Studies, 29*(3), 501–518. https://doi.org/10.1080/13691830305617

Monterrubio, C., Rodríguez-Muñoz, G., & Durán-Barrios, J. M. (2016). Social dimensions of child tourism labour: Listening to children's voices in two Mexican rural communities. *International Journal of Tourism Policy, 6*(2), 147–165. https://doi.org/10.1504/IJTP.2016.077969

Montgomery, H. (2008). Buying innocence: Child-sex tourists in Thailand. *Third World Quarterly, 29*(5), 903–917. https://doi.org/10.1080/01436590802106023

Morrow, V. (2011). *Understanding children and childhood* (Centre for Children and Young People Background Briefing Series, No. 1, 2nd ed.). Southern Cross University, Centre for Children and Young People.

Moscardo, G. (2012). Building social capital to enhance the quality-of-life of destination residents. In M. Uysal, R. R. Perdue, & M. J. Sirgy (Eds.), *Handbook of tourism and quality-of-life research: Enhancing the lives of tourists and residents of host communities* (pp. 403–422). Springer.

Občina Kranjska Gora. (2015). *Strategija razvoja turizma občine Kranjska* Gora 2015-*2025* [online]. https://www.kranjska-gora.si/media/dokumenti/TURIZEM_STRATEGIJA_K_4%2012%202015.pdf

OECD. (2020). OECD. Stats [online]. https://stats.oecd.org/

Ohashi, Y., Ohashi, K., Meskanen, P., Hummelin, N., Kato, F., & Kynäslahti, H. (2012). What children and youth told about their home city in digital stories in 'C my city! *Digital Creativity, 23*(2), 126–135. https://doi.org/10.1080/14626268.2012.709942

Palardy, N. P., Boley, B. B., & Gaither, C. J. (2018). Residents and urban greenways: Modeling support for the Atlanta BeltLine. *Landscape and Urban Planning, 169*(May), 250–259. https://doi.org/10.1016/j.landurbplan.2017.09.006

Perdue, R. R., Long, P. T., & Allen, L. (1990). Resident support for tourism development. *Annals of Tourism Research, 17*(4), 586–599. https://doi.org/10.1016/0160-7383(90)90029-Q

Perdue, R. R., Long, P. T., & Kang, Y. S. (1999). Boomtown tourism and resident quality of life. *Journal of Business Research*, *44*(3), 165–177. https://doi.org/10.1016/S0148-2963(97)00198-7

Petrović, N. B., & Kuzmanović, B. R. (1998). Konotativno značenje politički relevantnih pojmova kod učenika srednjih škola. *Psihologija*, *31*(4), 365–384.

Poria, Y., & Timothy, D. J. (2014). Where are the children in tourism research? *Annals of Tourism Research*, *47*, 93–95. https://doi.org/10.1016/j.annals.2014.03.002

Prilleltensky, I., & Prilleltensky, O. (2006). *Promoting well-being: Linking personal, organizational, and community change*. John Wiley and Sons.

Privitera, G. J., & Ahlgrim-Delzell, L. (2018). Quasi-experimental and single-case experimental designs. In Gregory J. Privitera and Lynn Ahlgrim-Delzell (Eds.), *Research methods for education* (pp. 333–370). SAGE Publications.

Radic, A. (2019). Towards an understanding of a child's cruise experience. *Current Issues in Tourism*, *22*(2), 237–252. doi: 10.1080/13683500.2017.1368463

Region Bad Gleichenberg. (2020). *Thermenland Styria Region* [online]. https://www.bad-gleichenberg.at/en/region/badgleichenberg/

Schänzel, H., Yeoman, I., & Backer, E. (2012). *Family tourism: Multidisciplinary perspectives*. Channel View.

Šegota, T., Mihalič, T., & Kuščer, K. (2017). The impact of residents' informedness and involvement on perceptions of tourism impacts: The case of the destination Bled. *Journal of Destination Marketing & Management*, *6*(3), 196–206. https://doi.org/10.1016/j.jdmm.2016.03.007

Seraphin, H., & Green, S. (2019). The significance of the contribution of children to conceptualising the destination of the future. *International Journal of Tourism Cities*, *5*(4), 544–555. https://doi.org/10.1108/IJTC-12-2018-0097

Sharpley, R. (2014). Host perceptions of tourism: A review of the research. *Tourism Management*, *42*(1), 37–49. https://doi.org/10.1016/j.tourman.2013.10.007

Small, J. (2008). The absence of childhood in tourism studies. *Annals of Tourism Research*, *35*(3), 772–789. https://doi.org/10.1016/j.annals.2008.06.002

Statista.com. (2020). *Travel and tourism in Europe - Statistics & facts* [online]. https://www.statista.com/topics/3848/travel-and-tourism-in-europe/#dossierSummary__chapter4

Su, L., & Swanson, S. R. (2020). The effect of personal benefits from, and support of, tourism development: The role of relational quality and quality-of-life. *Journal of Sustainable Tourism*, *28*(3), 433–454. https://doi.org/10.1080/09669582.2019.1680681

SURS. (2019). *Prihodi in prenočitve domačih in tujih turistov, občine, Slovenija, mesečno* [online]. https://pxweb.stat.si/pxweb/Dialog/statfile2.asp

The United Nations. (2010). *The United Nations Convention on the rights of the child. Ireland: Children's rights alliance [online]*. https://www.childrensrights.ie/sites/default/files/submissions_reports/files/UNCRCEnglish_0.pdf

Thornton, P. R., Shaw, G., & Williams, A. M. (1997). Tourist group holiday decision-making and behaviour: The influence of children. *Tourism Management*, *18*(5), 287–297. https://doi.org/10.1016/S0261-5177(97)00017-4

TO Topoa Oplenac. (2020). *Istorija* [online]. https://topolaoplenac.org.rs/srb/istorija/

Tollefson, J. W. (2002). Language rights and the destruction of Yugoslavia. In J. W. Tollefson (ed.), *Language policies in education: Critical issues* (pp. 179–201). Lawrence Erlbaum Associates.

Turismo y Planificación Costa del Sol. (2019). *Observatorio Turístico de Málaga - Costa del Sol*, 2018 [online]. https://www.costadelsolmalaga.org/5773/estudios-estadisticas

UNEP & UNWTO. (2005). *Making tourism more sustainable, a guide for policy makers*. [online]. http://www.unep.fr/scp/publications/details.asp?id=DTI/0592/PA

UNWTO. (2018). *European Union tourism trends*. UNWTO. https://doi.org/10.18111/9789284419470

Uysal, M., Sirgy, M. J., Woo, E., & Kim, H. L. (2016). Quality of life (QOL) and well-being research in tourism. *Tourism Management*, *53*, 244–261. https://doi.org/10.1016/j.tourman.2015.07.013

van Schalkwyk, G. J., Tran, E., & Chang, K. (2006). The impact of Macao's gaming industry on family life. An exploratory study. *China Perspectives*, *64*(March-April), 1–17. https://doi.org/10.4000/chinaperspectives.603

Wang, G., Xu, L., & Wang, C. (2016). Modern residential outdoor space for children with inter-generational parents: A case study in Beijing. *International Journal of Tourism Cities*, *2*(3), 206–220. https://doi.org/10.1108/IJTC-03-2016-0004

WTO & UNDP. (2017). *Tourism and the Sustainable Development Goals – Journey to 2030* [online]. https://www.e-unwto.org/doi/book/10.18111/9789284419401

WTTC. (2019a). *Austria - 2019 annual research: Key highlights* [online]. https://www.wttc.org/economic-impact/country-analysis/country-data/

WTTC. (2019b). *Turkey - 2019 annual research: Key highlights* [online]. https://www.wttc.org/economic-impact/country-analysis/country-data/

Xu, H. (2018). Moving toward gender and tourism geographies studies. *Tourism Geographies*, *20*(4), 721–727. https://doi.org/10.1080/14616688.2018.1486878

Yang, M. J. H., Yang, E. C. L., & Khoo-Lattimore, C. (2020). Host-children of tourism destinations: Systematic quantitative literature review. *Tourism Recreation Research*, *45*(2), 231–246. https://doi.org/10.1080/02508281.2019.1662213

Index

Page numbers in **bold** refer to tables and those in *italic* refer to figures.